REFORMING PHILOSOPHY

REFORMING PHILOSOPHY

A Victorian Debate
on Science and Society

LAURA J. SNYDER

THE UNIVERSITY OF CHICAGO PRESS
CHICAGO AND LONDON

The University of Chicago Press, Chicago 60637
The University of Chicago Press, Ltd., London
© 2006 by The Univeristy of Chicago
All rights reserved. Published 2006.
Paperback edition 2014
Printed in the United States of America

23 22 21 20 19 18 17 16 15 14   2 3 4 5 6

ISBN-13: 978-0-226-76733-8   (cloth)
ISBN-13: 978-0-226-21432-0   (paper)
ISBN-13: 978-0-226-76735-2   (e-book)
DOI: 10.7208/chicago/9780226767352.001.0001

Library of Congress Cataloging-in-Publication Data

Snyder, Laura J.
    Reforming philosophy : a Victorian debate on science and society / Laura J. Snyder.
        p. cm.
    Includes bibliographical references and index.
    ISBN 0-226-76733-7 (hardcover : alk. paper)
    1. Mill, John Stuart, 1806–1873. 2. Whewell, William, 1794–1866. 3. Philosophy,
English—19th century. 4. Science—Philosophy—History—19th century. 5. Political
science—Philosophy—History—19th century. I. Title
BI607.S69 2006
192—dc22

                                                              2005035508

FOR LEO

CONTENTS

ACKNOWLEDGMENTS

While working on this project I have accumulated many debts to generous people and institutions. I must first thank my teachers: Dennis Schmidt, Sam Schweber, Mark Hulliung, Peter Achinstein, Stephen Barker, Kenneth Schaffner. I am especially grateful to Jim Kloppenberg, whose example as a teacher, scholar, and mentor continues to serve as a model for me. I have also been fortunate to find a community of scholars and friends who have provided invaluable support and encouragement. Ernan McMullin, Bob Richards, Michael Ruse, David Hull, Dan Garber, Giuliano Pancaldi, and Jim Lennox have helped the project along in many tangible and intangible ways.

I am grateful to the following institutions for grants and fellowships while I was researching and writing this book: the U.S. Fulbright Commission, the Andrew Mellon Foundation, University of Pennsylvania and its Philosophy Department, University of Chicago and its Fishbein Center for the History of Science, Cambridge University, Clare Hall College, Cambridge, the National Endowment for the Humanities, and Istituto Studi Avanzati at the University of Bologna. St. John's University has been especially generous in granting me leaves and financial support. In particular, I would like to thank Arthur Gianelli, Willard Gingerich, Jeffrey Fagen, and Julia Upton. I am happy to acknowledge the important work of staff members of libraries at St. John's University, University of Chicago, Columbia University, University of Pennsylvania, Trinity College, Cambridge, the Royal Society of London, the British Library, and University of Bologna, especially Jonathan Smith and Kate Harcourt.

The manuscript was read, in whole and in part, at various stages, by Giovanni Boniolo, Pietro Corsi, Giovanni Giorgini, Jim Lennox, Bob Richards, Michael Ruse, and an anonymous referee for the University of Chicago Press.

Their comments and suggestions greatly improved the book. I can only hope, of course, that the final result finds favor with them.

I thank the Master and Fellows of Trinity College, Cambridge, for permission to quote from the Whewell collection and reprint two images of Whewell. The Royal Society of London kindly allowed me to quote from the correspondence of John Herschel. The National Portrait Gallery gave permission to reprint one image of Mill, as did Thoemmes Press. The cover image was provided by the Science and Society Picture Library.

I am grateful to my editors at University of Chicago Press, Christie Henry and Catherine Rice, for their excitement about this project and their patience in waiting for its completion. Lucille Hartman provided expert help in preparing versions of the manuscript. Some last-minute, long-distance research aid was provided by Jonathan Tsou, Susan Morris, and Magdalena Mayo.

My son, Leo, arrived during the final stages of work on the manuscript. Because he reformed my life so completely and wonderfully, this book is dedicated to him.

Bologna
January 2005

PROLOGUE

In the middle of the nineteenth century, John Stuart Mill and William Whewell engaged in an important debate that was played out within the pages of successive editions of their major works. This debate was wide-ranging, encompassing philosophy of science (for which the debate is chiefly known today), moral philosophy, economics, and politics. It was a debate between two fascinating men, who were both seen as being among the brilliant thinkers of the time. William Wordsworth believed that Mill was one of the most remarkable men of the age.[1] The astronomer and philosopher John Herschel described his friend Whewell by claiming that "a more wonderful variety and amount of knowledge in almost every department of human inquiry was perhaps never in the same interval of time accumulated by any man."[2] Alfred, Lord Tennyson more succinctly called him a "lion-like man."[3] Both men produced bodies of work that are incredible in terms of breadth as well as quantity. Mill most famously wrote on scientific method, political economy, moral philosophy, and politics, but also penned essays on literature, history, poetry, the classics, and religion. Whewell's oeuvre is perhaps even more impressive. He is best known today for his multivolume works on the history of science and the philosophy of science. But he also wrote on (in alphabetical order!) architecture, crystallography, educational reform, geology, international law, mechanics, mineralogy, morality, natural theology, political economy, and the science of the tides, and translated Plato's dialogues as well as German novels and poetry.

1. Reported in a letter from Julius Charles Hare to William Whewell, October 25, 1838, Whewell Papers (hereafter WP), Add.Ms.a206 f. 173.
2. John Herschel, "The Reverend William Whewell, D.D.," p. liii.
3. Alfred, Lord Tennyson, quoted in Hallam, Lord Tennyson, *Alfred, Lord Tennyson*, p. 38.

Mill's and Whewell's lives were extraordinary as well, in a way that was perhaps more possible in the nineteenth century than now. Whewell, long-time Master of Trinity College and tutor to a generation of Cambridge-educated mathematicians and men of science, was himself involved in scientific research and was an influential early member of the British Association for the Advancement of Science. His breadth of expertise is suggested (though not exhausted) by the fact that he held, at different times, the Professorship of Mineralogy and the Chair of Moral Philosophy at Cambridge. One of his more striking accomplishments was the invention—upon the poet S. T. Coleridge's prodding—of the English word *scientist*.[4] Whewell's fame as a terminological innovator caused Michael Faraday to seek his advice, which resulted in the new words *cathode, anode,* and *ion.* Mill never held a university appointment, spending his working career at the British East India Office, yet through his writings he exerted an influence upon university students that was unparalleled. It was said of him that "Mr. Mill is the author who has most powerfully influenced nearly all the young men of the greatest promise."[5] Even the dry and ostensibly apolitical *System of Logic* became a "sacred book for students who claimed to be genuine liberals," according to Leslie Stephen.[6] Later in life Mill served a term in Parliament, where he impressed William Gladstone as being a "saint of rationalism."[7]

Given their remarkable lives and the scope of their work, it is not surprising that the debate between Whewell and Mill would range over the most important issues of the day. Mill himself described Britain in those days as being in a state of "intellectual anarchy," in which old systems of thought were being abandoned and new ones were yet to be fully accepted. This was especially so in the areas of science, morality, and politics. As Richard Yeo has shown, it was a time in which the very meaning and status of science were matters of contention.[8] Both Mill and Whewell took on the task of "defining" science, and much of their debate over science involved the characterization of its method and aims. It was also a time in which the hegemony of utilitarian moral theories (which had been dominant in England since the eighteenth century) was being questioned. Whewell, through his

---

4. Whewell described his introduction of the term, which occurred at the 1833 meeting of the British Association of the Advancement of Science, in his "Mrs Somerville on the Connexion of the Sciences." For more details on the term and its reception, see Sidney Ross, "Scientist."

5. Henry Fawcett, "His Influence at the Universities," in E. H. Fox Bourne et al., *John Stuart Mill,* p. 74.

6. Leslie Stephen, *The English Utilitarians,* vol. 3, *John Stuart Mill,* pp. 75–76.

7. Gladstone made this remark to W. L. Courtney when asked for his thoughts on Mill. Courtney, *Life of Mill,* p. 141.

8. See Richard Yeo, *Defining Science.*

writings and by his teaching moral philosophy to a new generation of students at Cambridge, was partially responsible for this movement away from the ethical systems of William Paley and Jeremy Bentham. In response to his own recognition of the flaws of Benthamite morality, Mill developed a variant type of utilitarianism that incorporated some of Whewell's insights. Another area of strong contention in nineteenth-century Britain was political reform, especially as concerned the obligations of the state to the poorest members of society. This controversy led to a pervasive popular interest in political economy, as seen by the success of Harriet Martineau's *Illustrations of Political Economy*, which taught a version of the science through simple tales of common experience and which reached monthly sales of over 10,000 (in a time when Dickens's works were considered successful if they sold 3,000 copies). So, again, it is no surprise to find Mill and Whewell in the midst of a debate over political economy.

The dispute between Mill and Whewell, then, captured the "spirit of the age," as Mill might have said; it concerned fundamental intellectual conflicts that were raging throughout the nineteenth century in Britain. Indeed, there is a very real sense in which one can understand much about the intellectual spirit of Victorian Britain through study of the Mill-Whewell controversy. Yet, to this point, no one has considered the debate between Mill and Whewell in its entirety. Their conflict over the methods of science is examined in only a handful of journal articles. There are even fewer articles that address their debates in moral philosophy and in political economy. In the last fifteen years there have been several important monographs on Whewell (by Menachem Fisch, James Henderson, and Richard Yeo), but none of these consists in a sustained discussion of all these aspects of Whewell's thought, and none attempts to understand him in the context of his debate with Mill. In this book I show that we can gain a new and richer understanding of the ideas of Mill and Whewell by looking at the argumentative context in which each man developed his views. Moreover, we can use this debate as a means of illuminating significant currents of thought in the early Victorian age.

In order to do so, and yet write a book of reasonable length, it has been necessary to concentrate on Whewell and Mill while diminishing the role of other prominent thinkers who were equally enmeshed in the contemporary debate over science and society. Richard Whately and the Oxford school, John Herschel, and others who play only "supporting roles" in my narrative obviously could take center stage in another. But what makes the particular debate between Whewell and Mill so worthy of this concentration is that both men were philosophical authorities engaged in a self-conscious attempt to reform the philosophy and cultural conditions of the age. Understanding their debate can therefore shed light on both the philosophical and the cultural climate of early Victorian Britain. As John Burrow has described it, writing intellectual history is "eavesdropping on the conversations of the past."[9] In order to write part of the intellectual history of this time, I focus on one particularly iconic conversation between two fascinating thinkers.

Writing about such prolific individuals is difficult, and frankly I began this project intending to limit my examination to their debate over science. But it emerged in the course of my research that this was an untenable way to draw the battle line between them. The differences between Mill and Whewell over induction in science cannot be understood without setting them in the broader context of their conflicts over the proper way to reform society. Both men believed that proper scientific method could aid in renovating not only science but also morality and politics. Mill pointed to this when, in his *Autobiography*, he admitted that his intention in writing *System of Logic* was to attack the "intuitionist" philosophy in its stronghold, namely in mathematics and natural science, in order to banish it from moral and political realms. Mill regarded the intuitionist philosophy as a looming danger. He believed that "the notion that truths external to the mind may be known by intuition or consciousness, independently of observations and experience, [was] . . . the great intellectual support of false doctrines and bad institutions." He disdained intuitionism as "an instrument devised for consecrating all deep seated prejudices."[10] The purely empirical philosophy of science Mill presented in the *Logic* was developed for expressly political ends: to promote an epistemology more suited to the reforms of society he sought. At the same time, Whewell believed that his "antithetical epistemology," which combined empirical and idealist elements, could be useful in the battle against utilitarianism and the Ricardian system of political economy. To understand fully the dispute between Mill and Whewell over science, then, it is necessary to look to the larger political and moral context of the debate.

This understanding of the debate explains an anecdote reported by Mill's friend and biographer Alexander Bain: when presented with the opportunity finally to meet his famous opponent, Mill refused. So, strangely enough, the Mill-Whewell debate was one in which the debaters never met. For Mill the conflict was personal as well as philosophical. He saw their differences as

---

9. John Burrow, "The Languages of the Past and the Language of the Historians: The History of Ideas in Theory and Practice," John Coffin Memorial Lecture, 1987; quoted in Donald Winch, *Riches and Poverty*, p. 28.

10. John Stuart Mill, *Autobiography, Collected Works* (hereafter *CW*), 1:233.

concerning not only logic and scientific method but also the ideas that would have the most power in shaping society. Mill (rather unfairly, as we will see) considered Whewell, a member of the educational and scientific "establishment," a purveyor of opinions that served to justify the status quo, including cruelty to animals, forced marriages, and slavery. During the course of their controversy over moral philosophy, Whewell was unable to persuade Mill that he did not, in fact, endorse slavery (nor, of course, did he support the other two). He was understandably offended that Mill claimed otherwise. Whewell may therefore have had some satisfaction toward the end of his life, when he received the only letter Mill ever sent to him. Responding to a note in which Whewell had praised some elements of Mill's *Examination of Sir William Hamilton's Philosophy*, Mill wrote,

> But a still greater cause of satisfaction to me from receiving your note is that it gives me an opportunity on which, without impertinent intrusion, I may express to you how strongly I have felt drawn to you by what I have heard of your sentiments respecting the American struggle, now drawing to a close, between liberty and slavery, and between legal government and a rebellion without justification or excuse. No question of our time has been such a touchstone of men—has so tested their sterling qualities of mind and heart, as this one: and I shall all my life feel united by a sort of special tie with those, whether personally known to me or not, who have been faithful when so many were faithless.[11]

Although he was thus convinced that Whewell favored the antislavery forces of the Northern states, Mill never recorded, so far as I know, any change in his opinion of Whewell's moral philosophy. Neither did Mill decide to go to Cambridge and meet his famous antagonist. The divide was too wide; the debate had gone on too long.

11. John Stuart Mill to William Whewell, May 24, 1865, WP Add.Ms.a.209 f. 48(1).

"Reforming the Philosophy of the Age"

We are no longer young men. I hope your turn [for a church preferment] will come before long, for I have yet to make out my case by reforming the philosophy of the age, which I am going to set about in reality. I dare say you laugh at my conceit, but you and I are friends too old and intimate I hope for me to mind that.
—*William Whewell to Hugh James Rose*, 1836

From the winter of 1821, when I first read Bentham, and especially after the commencement of the Westminster Review, I had what might truly be called an object in life; to be a reformer of the world.
—*John Stuart Mill*, Autobiography

It was an age of reform. Britain in the early and mid-nineteenth century saw numerous reforms of government and society, reforms that extended the political franchise, lessened legally sanctioned discrimination of the Catholic minority, altered the economic policies of the nation, and expanded educational opportunities. Therefore it is no surprise that two of the age's most eminent individuals described themselves as reformers. Both William Whewell and John Stuart Mill intended to reform philosophy as well as the world; indeed, by reforming the "philosophy of the age," they thought they could effect social and political change. Thus Whewell, in explaining his desire to construct a new philosophical system, wrote to Julius Charles Hare, "I believe we want such systems more than anything else, because at the root of all improved national life must be a steady conviction of the reason, and the reason cannot acquiesce in what is not coherent, that is, systematic."[1] At the same time Mill believed he could "reform the world" by reforming philosophy. In his essay on Alexis de Tocqueville he claimed that "economical and social changes, though among the greatest, are not the only forces which shape the course of our species; ideas are not always the mere signs and effects of social circumstances, they are themselves a power in history."[2] And the moment for exerting such power was at hand. "There never was a time," he wrote, "when ideas went for more in human affairs than they do now."[3]

Both Whewell and Mill, then, saw themselves as engaged in a project to reform philosophy and, by doing so, to reform the larger society. But they each conceived of this project, and the desired result, in different ways. Thus there arose a debate between them, one that covered the main areas they both targeted for transformation: morality, politics, and science. It may initially seem odd to find science on this list. However, Mill and Whewell each considered the improvement of science instrumental in the reformation of morality and politics. Mill's desire to reform philosophy began with the intention to reform society by reestablishing morality and politics on different grounds; he then extended this project to include science as well. As we saw in the prologue, Mill came to believe that he had to defeat intuitionism in its "stronghold"—that is, natural science and mathematics—in order for his reformist philosophy to gain victory in morality and politics. For Whewell, on the other hand, the reform of science was always part of his project, if not its starting point. But both men saw the reform of science as central to the reform of society.

---

1. William Whewell to Julius Charles Hare, March 13, 1842, WP Add.Ms.a.215 f. 266. In a letter to Hare four years earlier, while Whewell was working on his moral philosophy, he wrote, "I sometimes think she is rather a hard task-mistress, this Reforming Philosophy of mine; for she carries me into regions where I can hardly expect any contemporary friends to follow me—at least for a long time"(January 7, 1838; in Isaac Todhunter, *William Whewell, D.D.*, 2:267).

2. John Stuart Mill, "De Tocqueville on Democracy in America (II)," *CW* 18:197–98.

3. John Stuart Mill to Robert Barclay Fox, September 9, 1842, *CW* 13:544. Here Mill was sounding a theme common at the time. To take another example of this connection between knowledge and reforming society, Samuel Bailey wrote in 1829 that "the acquisition of knowledge has become an object of immense interest and importance . . . the welfare of society in a thousand ways is deeply implicated in the rectification of error and the discovery of truth" (*Discourse on Various Subjects*, p. 1). Auguste Comte had expressed this theme as well in his first lecture of the *Cours de Philosophie Positive*. For a recent discussion of this concept of knowledge in early and mid-Victorian Britain, see Alan Rauch, *Useful Knowledge*.

Mill's Project of Reform

Mill started life as an experiment to confirm his father's ideas about educational reform.[4] The philosopher James Mill had his son reading Greek at age three and Latin at eight (at which age Mill also began teaching his younger sister Greek). At the age of twelve, Mill read six of Plato's dialogues (including the *Theaetetus*, which, Mill later noted, could have been omitted, because "it was totally impossible that I should understand it"—apparently this was not the case for the other five).[5] When he was thirteen, he studied logic, including Hobbes's work, and went through a "complete course of study" in political economy.[6] His first published work appeared in 1822, when he was sixteen: a criticism of Robert Torrens's theory of value in political economy. Mill never had any formal schooling. James Mill had been urged to send his son to Trinity College, Cambridge (indeed, John's godfather had left £500 in his will for this purpose),[7] but he refused, viewing the Oxbridge system as socially and politically corrupt. Had he sent John, it is possible that Whewell would have been his tutor.

In 1821, John Mill tells us in his *Autobiography*, he became caught up in the desire to "reform the world" on Benthamite principles. Mill believed that such a project entailed spreading the doctrine of Jeremy Bentham's moral philosophy and its associated political consequence: namely, that social and political institutions be subjected to the test of utility, that is, the test of whether they provided the greatest happiness for the greatest number. To further this project, Mill and his father wrote for the new Benthamite periodical, the *Westminster Review*, which Bentham had founded in 1824.[8] John Mill also formed a society for those who shared his zeal for the new philosophy, baptizing Bentham's doctrine with a word he had found in John Galt's novel *Annals of the Parish*: "utilitarianism."[9] The Utilitarian Society never

---

4. See Jack Stillinger, "John Mill's Education."
5. John Stuart Mill, *Autobiography*, CW 1:9. See also Myles Burnyeat, "What Was the 'Common Arrangement'?"
6. John Stuart Mill, *Autobiography*, CW 1:31.
7. See Susan F. Cannon, *Science in Culture*, p. 31.
8. By the time Bentham decided to start the journal, James Mill was working as an examiner in the India House, and could not be connected in any official capacity with a journal seen as being on the side of political agitation. Thus, although Bentham wanted the senior Mill as editor, he chose instead John Bowring, whom the Mills did not like (see G. L. Nesbitt, *Benthamite Reviewing*, pp. 28–29).
9. John Stuart Mill, *Autobiography*, CW 1:81. For an alternative etymology of this name, see G. L. Nesbitt, *Benthamite Reviewing*, p. 24.

FIGURE I. Portrait of John Stuart Mill. By permission of
Thoemmes Press.

consisted of more than ten members and broke up within a few years, but
Mill nonetheless benefited by being brought into contact with other clever
young men. Moreover, he bestowed a name to a movement, much as Whewell
provided terminology for new sciences (and for the man of science himself).

In 1826, when he was twenty, Mill suffered a mental breakdown that
lasted for several years. This crisis, and Mill's recovery from it, had a profound
effect on his reforming project. He described the crisis as beginning from a
kind of depression, which caused him to ask the following question:

> "Suppose that all your objects in life were realized; that all the changes in
> institutions and opinions which you are looking forward to, could be com-
> pletely effected at this very instant: would this be a great joy and happiness
> to you?" And an irrepressible self-consciousness distinctly answered,
> "No!" At this my heart sank within me: the whole foundation on which
> my life was constructed fell down. All my happiness was to have been
> found in the continual pursuit of this end. The end had ceased to charm,

and how could there ever again be any interest in the means? I seemed to have nothing left to live for.[10]

He seemed no longer to care about reforming the world. The education he had received, and indeed the philosophic outlook of Bentham and his father, tended, he now saw, "to wear away the feelings."[11] Mill knew rationally that the greatest good was the happiness of the greatest number, but he felt no feeling of sympathy or benevolence toward his fellow beings. Indeed, Mill realized that he fit the stereotype of the Benthamite "reasoning machine."[12] The gloom did not dissipate until he read Jean-François Marmontel's *Memoirs;* coming to the passage relating the death of Marmontel's father, Mill was moved to tears. "From that moment," Mill relates, "My burthen grew lighter. The oppression of the thought that all feeling was dead within me, was gone. I was no longer hopeless; I was not a stock or a stone. I had still, it seemed, some of the material out of which all worth of character, and all capacity for happiness, are made."[13] Mill was also comforted by reading William Wordsworth for the first time, which he did in 1828. He later explained that "what made Wordsworth's poems a medicine for my state of mind, was that they expressed . . . states of feeling, and of thought coloured by feeling. . . . They seemed to be the very culture of the feelings, which I was in quest of. . . . From them I seemed to learn what would be the perennial sources of happiness, when all the greater evils of life shall have been removed."[14] From this time on, Mill tells us, "the cultivation of the feelings became one of the cardinal points in my ethical and philosophical creed." He announced this new way of thinking in a speech in 1829, in which he claimed that poetry is an important part of education, for its beneficial effects on both the intellect and the feelings. He told Thomas Carlyle that "poetry is higher than Logic, and . . . the union of the two is Philosophy,"[15] thereby signaling a lifelong attempt to seek a synthesis of intellect and feeling, what he termed the "maintenance of due balance among the faculties." During this period Mill also became enchanted with Goethe's notion of *Vielseitigkeit,* or "many-sidedness." For a time, he admitted, this notion "possessed" him.[16] Mill took "many-

10. John Stuart Mill, *Autobiography, CW* 1:139.

11. Ibid., 1:141.

12. Referring to the utilitarians, Thomas Carlyle complained in his "Signs of the Times" (1829) that men had "grown mechanical in head and heart, as well as in hand" (p. 444).

13. John Stuart Mill, *Autobiography, CW* 1:145.

14. Ibid., 1:151.

15. John Stuart Mill to Thomas Carlyle, July 5, 1833, *CW,* 12:163.

16. John Stuart Mill to Thomas Carlyle, January 12, 1834, *CW* 12:205.

sidedness" to mean seeing all points of view in order to find the portions of truth residing in different, even contrary, systems of thought.[17] Many of his old friends and acquaintances could not understand this new direction; John Bowring said of Mill that he "was most emphatically a philosopher, but then he read Wordsworth and that muddled him, and he has been in a strange confusion ever since."[18]

Around this time, Mill was introduced to the ideas of S. T. Coleridge (and to the man himself)[19] by his friends John Sterling and F. D. Maurice, both of whom he met in 1828.[20] His exposure to the views of Coleridge influenced Mill's conception of his reformist project in two ways. One concerned Coleridge's notion of an "intellectual clerisy." Coleridge believed that civilization is grounded in the "cultivation of the people," which he described as the "harmonious development of those qualities and faculties that characterize our humanity." In order to bring about and maintain this cultivation, it was necessary to have a "permanent class or order"[21] that would lead the way. At one time, priests exercised the required moral and intellectual supremacy. Now, Coleridge believed, this role would fall to philosophers, once they became worthy to possess it.[22] Under the influence of Coleridge's thought, Mill came to believe that in normal times, the people would follow philosophers as their leaders and guides. In his series of essays entitled "The Spirit of the Age," he noted that the "natural" state of society is one in which "the opinions and feelings of the people are, with their voluntary acquiescence, formed *for* them, by the most cultivated minds which the intelligence and morality

17. S. R. Letwin notes that Mill's notion of many-sidedness "was in fact almost the contrary of Goethe's 'Vielseitigkeit,' which he believed himself merely to have translated" (*The Pursuit of Certainty*, p. 237).

18. Reported by Caroline Fox, *Memories of Old Friends*, p. 113; journal entry for August 7, 1840. James Fitzjames Stephen similarly wrote, "I am falling foul . . . of John Mill in his modern and more human mood . . . which always makes me feel that he is a deserter from the proper principles of rigidity and ferocity in which he was brought up" (Leslie Stephen, *The Life of James Fitzjames Stephen*, p. 308).

19. Mill claimed to have met Coleridge several times. See Christopher Turk, *Coleridge and Mill*, p. 51.

20. John Sterling (1806–44), a writer and poet, was president of the Cambridge Union Society and member of the Apostles (an elite secret society founded at Cambridge in 1820, supposedly composed of the twelve brightest students) while at Cambridge. He was a frequent associate of Coleridge. F. D. Maurice (1805–72) was also an Apostle, and a friend of Tennyson's. After gaining renewed religious faith he had an ecclesiastical career, during which he was a proponent of Christian socialism. In 1866 Maurice was elected to the Knightsbridge professorship of casuistry, moral theology, and moral philosophy at Cambridge, which had been held decades earlier by Whewell.

21. S. T. Coleridge, *Collected Works*, 10:42–43.

22. On this relation between the views of Mill and Coleridge, see Christopher Turk, *Coleridge and Mill*.

of the times calls into existence."[23] Only in a time of transition, such as Britain was then experiencing, did the people refuse the guidance of intellectuals. Because of this they were often led to support morally and intellectually bankrupt social policies, including those allowing slavery. It was up to the intellectual class to become worthy of their role as leaders, and to convince the people of their worthiness. Once they had done this, society at large would support the reforms deemed necessary by the philosophers; indeed, this was the sole way evils like slavery could be eradicated. Later, in his 1861 *Considerations on Representative Government,* Mill claimed that it was not economic influences that had destroyed slavery, but rather "the spread of moral convictions."[24] The philosopher's role, he believed, is to develop and spread the ideas and convictions required to reform society. Mill argued for the intellectual and moral education of all members of society, in order that everyone would come to recognize the value of these ideas and agree with the need for reforms based on them.

Mill's reading of Coleridge influenced his work in another important way as well. He came to believe that the philosophy of Coleridge, though wrong in some of its central positions, could provide a corrective to the excesses of the Benthamite position. His attempt to unite the two resulted in a pair of essays he published in 1838 and 1840, in which he compared Bentham and Coleridge. Although Mill had claimed to have given up on the ideal of many-sidedness some years earlier, it is clear that this model provided the motivation for his desire to find the half-truths expressed in these two opposed systems.[25] In the essay on Bentham, Mill praised his former mentor for being a "great reformer of philosophy" who had, however, grasped only half the truth.[26] He described Bentham's error as "one-sidedness."[27] In the Coleridge essay, Mill

---

23. John Stuart Mill, "The Spirit of the Age, V(I)," *CW* 22:307. In a letter to the Saint-Simonian Gustave d'Eichthal from this period, Mill explained that "the people, i.e., the uninstructed, shall entertain the same feelings of deference and submission to the authority of the instructed, in morals and politics, as they at present do in the physical sciences" (November 7, 1829, *CW* 12:40).

24. John Stuart Mill, *Considerations on Representative Government, CW* 19:382.

25. In a letter to Carlyle, Mill explained that when he was possessed by the ideal of many-sidedness, "I saw, or seemed to see, so much of good and of truth in the most positive parts of the most opposite opinions and practices, could they but be divested of their exclusive pretensions, that I scarcely found myself called upon to *deny* anything but Denial itself" (see John Stuart Mill to Thomas Carlyle, January 12, 1834, *CW* 12:205). However, Mill never abandoned this ideal. Years later, in his *On Liberty,* Mill noted that two conflicting views may "share the truth between them"; indeed, this belief led Mill to his claims about the importance of allowing free debate on all issues. In his *Auguste Comte and Positivism,* published in 1865, Mill claimed that "M. Comte has got hold of half the truth" (*CW* 10:313).

26. John Stuart Mill, "Bentham," *CW* 10:83.

27. Ibid., 10:112.

explained that "the two men are each other's 'completing counterpart': the strong points of each correspond to the weak points of the other. Whoever could master the premises and combine the methods of both, would possess the entire English philosophy of their age."[28]

Understanding what Mill meant by mastering the premises and combining the methods of both men is crucial to understanding his project of reform in science, morality, and politics, as well as comprehending its ultimate outcome. In the essay on Coleridge, Mill detailed three main differences between the Benthamite and Coleridgian schools: the "purely abstract" or "philosophical" difference, the "concrete" or "practical" difference, and the political difference. On the level of the "philosophical," the school of Bentham followed the Lockean epistemological position that sensations are the sole materials of knowledge. The opposing view is that of Coleridge's school, which maintained that the human mind has the capacity to perceive the nature of "things in themselves," that it can recognize truths not known by our senses, truths which are a priori. Here Mill characterized himself as being firmly in the camp of the Benthamites, claiming that "we see no grounds for believing that anything can be the object of knowledge except our experience, and what can be inferred from our experience by the analogies of experience itself; nor is there any idea, feeling, or power in the human mind, which . . . requires that its origin should be referred to any other source."[29] Thus it was not at the level of the epistemological that Mill desired to combine the insights of Coleridge and Bentham. (He did, however, point out that even here Coleridge and his followers performed a crucial service to philosophy by forcing the Benthamites to reject the "shallowest" doctrines of Étienne de Condillac for the more sophisticated empiricism of David Hartley.)[30]

It was in the realms of the practical and the political that Mill believed it possible to unite the two approaches. In the essay on Coleridge, Mill explained that the Benthamites followed the philosophy of the eighteenth century in advocating a rationalist, deductive approach to the sciences of society. He claimed that "they committed the common error, of mistaking

28. John Stuart Mill, "Coleridge," *CW* 10:121. Mill may have been influenced by reading Carlyle's review of Croker's edition of Boswell's *Life of Johnson*. Carlyle claimed that Johnson was the ancestor of the Tories and Hume of the Whigs, and that whoever could synthesize the ideas of the two would become the "whole man of a new time." See John Stuart Mill, *Early Draft of the Autobiography*, *CW* 1:182 (interestingly, Mill omitted reference to having read this article in the final version of the *Autobiography*); Thomas Carlyle, "Boswell's *Life of Johnson*"; and Nicholas Capaldi, *John Stuart Mill, A Biography*, pp. 139–40.

29. John Stuart Mill, "Coleridge," *CW* 10:128–29.

30. Ibid., 10:130.

the state of things with which they had always been familiar, for the universal and natural condition of mankind."[31] Mill's father was guilty of this error;
this is why Mill claimed that James Mill was the "last [representative] of the
eighteenth century."[32] In the essay on Bentham, Mill elaborated on this blun-
der. Moral and political philosophy, as well as the sciences of society, require
as their materials knowledge of the properties of man and his position in the
world.[33] However, Mill claimed, Bentham neglected to gaze beyond his own
mind to see how men really were. He had knowledge only of his own mind—
a poor ground for generalization, since his own human deficiencies were rife.
"Knowing so little of human feelings, he knew still less of the influences by
which these feelings were formed," Mill wrote. "All the more subtle workings
both of the mind upon itself, and of external things upon the mind, escaped
him; no one, probably, who, in a highly instructed age, ever attempted to give
a rule to all human conduct, set out with a more limited conception either of
the agencies by which human conduct *is* or of those by which it *should* be,
influenced."[34] Bentham developed a "philosophy of universal human nature,"
based on his experience of his own limited nature, which he applied as a first
principle in moral philosophy, politics, and the social sciences. Mill suggested that it was because of a certain lack of confidence that Bentham took
this deductive approach: his self-confidence required "systematic unity."[35]
The deductive approach, Mill argued, could bear no fruit; it could accomplish
nothing in political and social philosophy. For instance, by not taking note of
differences in national characters, and differences of these characters over
time, Bentham erred in claiming that there was one type of government best
for all peoples and at all times in history: namely, democratic rule. Mill
thought this historically simplistic. "Is it," he asked, "the proper condition
of man, in all ages and nations, to be under the despotism of Public Opinion?"[36] Rather, in Mill's view, particular nations in different times required
distinct modes of government.[37] In the *Logic*, Mill characterized Bentham's
flawed approach as the "Geometrical Method."

31. Ibid., 10:132.
32. See John Stuart Mill, *Autobiography*, CW 1:213.
33. John Stuart Mill, "Bentham," CW 10:89.
34. Ibid., 10:93.
35. Ibid., 10:111. Although Mill seems here to disdain the need for "systematic unity," in the
Autobiography he emphasized his own concern with consistency: "I found the fabric of my old and
taught opinions giving way in many fresh places, but was incessantly occupied with weaving it
anew" (CW 1:163).
36. John Stuart Mill, "Bentham," CW 10:107.
37. In an essay from 1834, Mill had criticized the Benthamites in similar terms. See his
"Review of Miss Martineau's Summary of Political Economy," CW 4:225. Besides Coleridge, other

This eighteenth-century deductive construction of the sciences of man
and society was countered by the school of Coleridge, which promoted a philosophy more in keeping with the contemporary age. This was, Mill noted,
Coleridge's greatest contribution to philosophy. The nineteenth-century
approach to the social sciences was "inductive and Baconian," and the members of the "Germano-Coleridgian school" were the first to examine in detail
the inductive laws governing the existence and growth of human society.[38]
They were historically sensitive, examining human behavior in the varying
contexts of time and place. Moreover, they examined the forces that actually
do bind particular societies together, such as education, culture, and love of
country. The Coleridgians did not assume that the conditions currently
obtaining in society were those that must universally obtain. Instead, they
developed both a philosophy of history and a "philosophy of human culture"
far richer than what the eighteenth-century school could provide. As his
evaluations of Coleridge and Bentham indicate, Mill himself concluded that
even in the realms of moral and political philosophy a more inductive
approach was required.[39] Yet a purely inductive method was not adequate: in
the *Logic*, Mill called this view the "Chemical, or Experimental Method,"
and criticized its assumption that by observation and experiment alone we
can discover the complex laws of human behavior in society. A method combining both Benthamite deductive reasoning and Coleridgian inductivism
was required.

In the same way, Mill came to believe it was necessary to bring together the
political insights of Bentham and Coleridge. Mill viewed them as "two sorts of
men—the one demanding the extinction of the institutions and creeds which
had hitherto existed; the other that they be made a reality: the one pressing
new doctrines to their utmost consequences; the other reasserting the best
meaning and purposes of the old."[40] By the time he wrote this pair of essays,
Mill had seen that the Benthamite agenda of the Philosophical Radicals
ignored the importance of bringing into being a society whose citizens were
cultivated. It was not enough, Mill eventually realized, to reform institutions
without reforming the people themselves. Thus he aimed at a "complete ren-

influences on Mill's view of historical diversity include the French historians Jules Michelet and
François-Pierre-Guillaume Guizot, as well as the Saint-Simonians and Comte.
    38. John Stuart Mill, "Coleridge," *CW* 10:139.
    39. And indeed, in book 6 of the *Logic*, published only a few years after the essay on Coleridge,
Mill argued that all the "moral sciences" should proceed according to the methods of the physical
sciences, that is, the inductive methods he had outlined in book 3.
    40. John Stuart Mill, "Coleridge," *CW* 10:145–46.

ovation of the human mind,"[41] one that could be brought about by a kind of moral and intellectual education and nurtured by a certain type of liberty. Only when this had been effected could the goals of the Benthamites be fully realized. Mill desired to unify what was most useful of the Benthamite program for political reform with the Coleridgian desire to form a cultivated society.[42]

Mill's project, then, was to bridge both the practical (methodological) and the political differences between Bentham and Coleridge. Recognizing this is crucial for understanding Mill's work in the natural sciences, moral philosophy, economics, and politics. In particular, putting Mill's body of work in the context of this project resolves many of the "strange confusions" with which he has been charged by critics, from Bowring to those of the present day.

Whewell's Project

Whewell was born in Lancaster in 1794, the eldest child of a master carpenter. Like Mill, Whewell was a prodigy. While at the "Blue School" in Lancaster, the master of the grammar school noticed Whewell's evident intelligence and offered to teach him for free at his school. According to Richard Owen, the comparative anatomist, his lifelong friendship with Whewell began at the school when the embarrassed Whewell was sent to the much younger Owen to learn "the meaning of the mysterious word viz."[43] Soon, Whewell was examined by a tutor from Trinity College, Cambridge, who predicted that the young boy would one day place among the top six wranglers (the holders of first-class degrees). Whewell was then sent to Heversham Grammar School in Westmorland, some twelve miles to the north, where he would be able to qualify for a closed exhibition to Trinity. (In the nineteenth century and earlier, these "closed exhibitions," or scholarships, were set aside for the children of working-class parents.) Whewell studied at Heversham for two years, and received private coaching in mathematics from John Gough, the blind mathematician to whom reference is made in Wordsworth's "Excursion." Although Whewell did win the exhibition, it did not provide full resources for a boy of his family's means to attend Cambridge; money had to be raised in a

41. John Stuart Mill, "Grote's *History of Greece* (V)," *CW* 25:1162.

42. As Joseph Hamburger has noted in *John Stuart Mill on Liberty and Control*, the dominant view of Mill as holding the same views as the modern-day individualist liberal relies on only *half* his outlook: the Benthamite half, which recognized the need for individuals to participate in politics, but not the Coleridgian part, which stressed the need for moral and intellectual authority in order to bring about the "moral regeneration" Mill sought (see p. 228). We will explore this theme further in chapter 4.

43. Leslie Stephen, who relays this account, questions its truth (see "Whewell, William").

public subscription to supplement the scholarship money. He thus came up to Trinity in 1812 as a "sub-sizar" (scholarship student).[44]

Almost as soon as he arrived, Whewell became involved with the Analytical Society, which had been formed by Charles Babbage, John Herschel, and others the year before. The society was founded with the purpose of reforming the teaching and practice of mathematics at Cambridge. Its members wished to introduce Continental methods of mathematics to a university which had excluded progress since the time of Newton. (Essentially, they wished to introduce Lagrange's algebraic and formalistic version of the calculus, which included replacing Newton's fluxion dot notation with Leibniz's "d" notation.)[45] In 1814 Whewell won the Chancellor's prize for his epic poem "Boadicea," following in the footsteps of his mother, Elizabeth, who had published poems in the local papers. Yet, contrary to the worries of his old schoolmaster, this attention to the arts did not cause Whewell to neglect the mathematical side of his training;[46] in 1816 he proved his mathematical prowess by placing as second wrangler and second Smith's prizeman.[47]

While at Cambridge, he became close friends with Richard Jones (the future political economist) and Herschel, each of whom had great effect on his future thinking about science. He also formed friendships with Hugh James Rose, an admirer of Wordsworth, and Julius Charles Hare and Connop Thirlwall, men greatly interested in and influenced by the German Romantic movement.[48] After a short postgraduate period as a private tutor, Whewell

44. In letters to his father during these student days, Whewell often expressed concern about money, assuring his father that his expenses were as low as possible. (See, for instance, the letter of January 18, 1814, in which Whewell justifies the high expense for Newton's *Principia*, "a book that I should unavoidably have to get sooner or later" [in Janet Stair Douglas, *The Life and Selections from the Correspondence of William Whewell, D.D.*, p. 11].) Whewell's status as a sub-sizar meant that he would have had to serve the other boys at dinner. See Martha Garland, *Cambridge before Darwin*, p. 11, and Sheldon Rothblatt, *The Revolution of the Dons.*

45. For more on this see Menachem Fisch, "'The Emergency Which Has Arrived,'" and Harvey Becher, "Woodhouse, Babbage, Peacock and Modern Algebra" and "William Whewell and Cambridge Mathematics."

46. See Janet Stair Douglas, *The Life and Selections from the Correspondence of William Whewell, D.D.*, pp. 4–5.

47. The top wranglers competed for the first and second Smith's prizes, awarded after an examination designed to select the "best proficients in Mathematics and Natural Philosophy." See Andrew Warwick, *Masters of Theory*, pp. 55–56. Later second wranglers included J. C. Maxwell, W. K. Clifford, and Lord Kelvin (see Margaret Schabas, *A World Ruled by Number*, p. 119).

48. Hugh James Rose (1795–1838) became known as a Christian apologist and member of the High-Church Hackney Phalanx. As select preacher at Cambridge from the mid-20s to mid-30s, he delivered a number of popular and influential sermons. Julius Charles Hare (1795–1855) became classical lecturer at Cambridge in 1822; two of his students (and friends) were John Sterling and F. D. Maurice. With his brother Augustus he wrote the popular *Guesses at Truth*. Connop Thirlwall (1797–1875) was forced to resign his assistant tutorship for his liberal views. Hare and Thirlwall pro-

FIGURE 2. Portrait of William Whewell by G. F. Joseph, 1836. By permission of the Master and Fellows of Trinity College, Cambridge.

won a college fellowship in 1817. Unlike Herschel, Jones, and others of his friends, he did not choose to leave Cambridge to become a lawyer or curate, the other careers typically pursued by Cambridge graduates needing to earn a

---

duced a translation of the first two volumes of Barthold Niebuhr's *History of Rome*. Thirlwall himself wrote an eight-volume *History of Greece*.

living. Herschel originally chose the former of these (before giving it up to return to science); this option was probably not available to Whewell, for securing a position as a lawyer required a fairly large investment of money that his family did not have.[49] But he could have followed Jones in attempting to attain a parish living; this would have left him some time to pursue scientific interests. Instead, Whewell chose to remain at Cambridge, suggesting both that he did not feel any great calling to serve the church, and that he wanted to do more with science than merely dabble in it during his free time.[50]

After winning his fellowship, Whewell was assistant tutor from 1818 until 1823, and tutor from 1823 until 1838. He was elected to the Royal Society in 1820, and ordained a priest—as required for Trinity Fellows to maintain their position after an initial seven-year period—in 1825. He took up John Henslow's Chair in Mineralogy in 1828 (when Henslow vacated it for the Botany Professorship), and resigned it in 1832. During the early 1830s Whewell was instrumental in the formation of the British Association for the Advancement of Science (BAAS), the Statistical Section of the BAAS, and the Statistical Society of London. In 1838 he became Professor of Moral Philosophy. Shortly after his marriage to Cordelia Marshall on October 12, 1841, he was named Master of Trinity College, having been recommended to the queen by Prime Minister Robert Peel. Whewell was Vice-Chancellor of the University in 1842 and again in 1855. In 1848 he played a large role in establishing the Natural and Moral Sciences Triposes at the university. Whewell engaged in scientific research, winning a medal from the Royal Society for his work on the science of the tides. He corresponded with the most eminent scientists of his day, published his own translations of German novellas and poetry, and with his popular translations of Plato's dialogues was partly responsible for a "Platonic revival" in Britain in the 1850s.[51] Toward the end of his life Whewell was asked to tutor the young Prince of Wales (the future Edward VII) in political economy.

As this brief summary of his life indicates, Whewell's rise from what we would probably call the middle (or lower middle) class to the intellectual and social elite was spectacular. If this success tended to make him a bit arrogant, and even imperious, in his later life, it also, I think, imbued him with a sense of mission. Whewell very much saw himself as having a "calling" to reform

49. For details, see David Valone, "The Dark and Tangled Recesses of Knowledge."

50. In later years Whewell considered taking a parish living rather than ending his days as an unmarried Trinity fellow, but worried about losing his freedom to write his philosophical works. See letter to Julius Charles Hare, December 15, 1840, WP Add.Ms.a.215 f. 53.

51. See Frank M. Turner, *The Greek Heritage in Victorian Britain*, pp. 371–72.

philosophy; he referred to this as "my real task in life,"[52] and indeed as his "vocation."[53] So while Bentham and James Mill quite deliberately made a reformer out of John Mill, Whewell came more naturally to the role.

Like Mill, Whewell wished to reform morality and politics as well as science. He objected to utilitarianism, agreeing with Carlyle's depiction of the doctrine as "Pig Philosophy." He accepted the Professorship in Moral Philosophy in 1838 in part to counter the prevailing utilitarian view at Cambridge, and wrote his major works in ethics in order that there be alternative texts to replace William Paley's *Moral and Political Philosophy* as required reading for undergraduates. Moreover, Whewell objected to the political consequences derived from utilitarianism by the Philosophical Radicals. (As I argue later, he was not as conservative politically as Mill believed him to be, but he was certainly no radical.) His rejection of Ricardian economics was based, in part, on his rejection of the political agenda to which it was put to use by the Benthamites. But he also rejected it because of its deductive methodology, which was in sharp contrast with the "induction project" he, Jones, and Herschel had begun while they were all at Cambridge.

From his undergraduate days, Whewell shared with Jones and Herschel the goal of defining and promulgating the proper method of physical science: a reformed Baconian inductivism. He later recalled that the *Novum Organum* was one of the favorite topics of discussion between the three of them during their time together at Cambridge.[54] As early as 1818 Whewell had decided that his role in life had to do, at least in part, with the more metaphysical aspects of studying science; thus he told Herschel that he recognized he was not destined to be a great scientist, but rather one of those "lookers on, who, not making a single experiment to further the progress of

52. See letter to Julius Charles Hare, December 15, 1840. In this letter Whewell asked for Hare's advice on whether he should leave the university and take a parish living in Masham. He was feeling lonely and unhappy in the college, and preferred not to remain a bachelor (as required for Trinity Fellows) all his life. Hare responded by agreeing that Whewell's calling "seems to me to be that of a doctor, rather than a pastor," and assured Whewell that giving up his work in inductive philosophy "would be an evil thing for England." Hare rather presciently observed that "what I should wish for would be a post where you might fulfill that ministry—the Mastership of Trinity, a deanery, or something of that sort" (December 17, 1840, in Janet Stair Douglas, *The Life and Selections from the Correspondence of William Whewell, D.D.*, pp. 210–11). Whewell remained in college; less than a year later, he took up the Mastership of Trinity.

53. In his 1831 review of Herschel's *Preliminary Discourse,* Whewell referred to "the history and philosophy of physical science" as his "vocation" ("Modern Science—Inductive Philosophy," p. 374). Later, he wrote to Herschel that "the reform of our philosophy is the work to which I have the strongest vocation, and which I cannot give up if I were to try" (letter to Herschel, April 9, 1836, in Isaac Todhunter, *William Whewell, D.D.*, 2:234).

54. William Whewell, Prefatory Notice to *Literary Remains,* by Richard Jones, p. xix.

science, employ ourselves with twisting the results of other people into all possible speculations mathematical, physical and metaphysical."[55] Around the same time he wrote to Jones that he was dreaming of "undertakings metaphysical, philological, mathematical, and others which I would execute if I had time."[56] While writing and soon after publishing his 1819 textbook, *Elementary Treatise on Mechanics,* Whewell was thinking about (and arguing with Jones about) philosophical questions concerning mechanics, such as whether the laws of motion were necessary truths.[57] In October of 1825 he wryly cautioned Jones against publicly accusing him of being too metaphysical, because "if you do so you may easily give people an impression which you will not be able to remove when I have convinced you, as I certainly shall at the first opportunity, that everything which I believe is most true, philosophical, and inductive."[58] By this time, Whewell was already concerned with his *reputation* as a methodologist of science. Throughout his career he continued to consider his work in reforming science paramount; thus he later complained to Jones about his lectures on moral philosophy taking time away from his "prescribed course of writing about induction."[59]

I would argue, then, that Whewell's interest in reforming scientific method arose early, and was not secondary to his interests in morality, politics, and theology. In taking this position I differ from other commentators, who have claimed that his interest in reforming science derived from his desire to counter utilitarianism in morality and politics, or his wish to promote a particular theological position.[60] In the case of morality, it is noteworthy that in 1826 Rose chided Whewell for excessively celebrating mathematics and the physical sciences, because such an attitude contributed, in Rose's view, to the appeal of utilitarianism. Whewell was chastised, that is, for *not* making the attack on utilitarianism his primary object. Indeed, at least part of his criticism of utilitarian moral philosophy was that it had a negative effect on the

<hr/>

55. William Whewell to John Herschel, November 1, 1818, in Isaac Todhunter, *William Whewell, D.D.,* 2:28–29.

56. William Whewell to Richard Jones, August 21, 1818, in ibid., 2:27.

57. See letters between Whewell and Jones in 1819–22, especially those in the Whewell Papers.

58. William Whewell to Richard Jones, October 17, 1825, WP Add.Ms.c.51 f. 23.

59. See, for example, William Whewell to Richard Jones, May 2, 1835, WP Add.Ms.c.51 f. 181.

60. For example, Donald Winch claims that Whewell's criticisms of Ricardian political economics were based solely on his rejection of the political position of the Philosophical Radicals, and that his criticisms of its deductive methodology were merely a "guise" (*Riches and Poverty,* p. 377). Perry Williams similarly argues that Whewell's fight against utilitarianism and political radicalism, and his defense of Christianity, were the main motivations for his interest in scientific method, and that this interest did not even arise until the late 1820s and '30s ("Passing on the Torch"). David Valone argues that Whewell's interest in political economy led him to his interest in scientific method in the early 1820s ("The Dark and Tangled Recesses of Knowledge"; see p. 129).

study of science: Whewell argued that "the love of knowledge ought not to be degraded so far as to be weighed ounce for ounce against the pleasures of sense" and blamed England's lack of dominance in physical science on its tradition of utilitarianism.[61] The utilitarians' preference for practical application over pure knowledge—or for Art over Science—had slowed the pace of England's scientific progress relative to other countries, such as France and Germany, which were not beset by this philosophy.

Similarly, Whewell's early writings on theological topics were motivated by the desire to defend science. He was inspired to write about theology by Rose's criticisms of science in a sermon preached on Commencement Sunday in July of 1826. Rose argued that a link existed between the cultivation of science and the emergence of religious doubts. Science, because of its concern with the material world, was a natural ally of utilitarianism, which was concerned with what is materially useful. On these grounds Rose challenged the tradition of science education at Trinity; instead of concentrating on science and mathematics, the Trinity student, he thought, should study divinity, literature, and history.[62] Whewell countered that in fact science *supported* religion. As he wrote to Rose some months after the sermon, "To tell the truth I am persuaded that there is not in the nature of science anything unfavourable to religious feelings, and if I were not so persuaded I should be much puzzled to account for our being invested, as we so amply are, with the faculties that lead us to the discovery of scientific truth—It would be strange if our Creator should be found to be urging us on in a career which headed to a forgetfulness of him."[63] In a series of sermons Whewell preached at St. Mary's in February 1827, and in his 1833 Bridgewater Treatise, he gave fuller versions of this argument.[64] In the Bridgewater Treatise, he argued that induction is not only the methodology most likely to lead to truths, it is also most likely to lead to Truth—to belief in a Supreme Deity. He succinctly detailed the relevant difference between inductive and deductive thinkers in a letter to Jones:

61. See William Whewell, "Modern Science—Inductive Philosophy," p. 404.

62. See David Valone, "The Dark and Tangled Recesses of Knowledge," pp. 185–98. As Valone notes, in the printed version Rose retreated somewhat from his stronger position. There he allowed that there might be a place for mathematics in the curriculum, but maintained his objection to the tendency to "exalt it above other pursuits, to make it the exclusive object, to force it on the attention of all, and to devote to it minds capable of far better and higher things" (quoted in ibid., p. 198).

63. William Whewell to Hugh James Rose, November 19, 1826, WP R.2.99 f. 26.

64. The Earl of Bridgewater bequeathed money for the publication of a series of books illustrating the "Power, Wisdom, and Goodness of God, as manifested in the Creation." Whewell's was the first published, and one of the most successful.

The deductive people go on following, illustrating, expanding, a given notion which of the nature of it must be defined and limited and so restricted to the range of our primitive knowledge—But the minds that feel a conviction of principles of unity as yet undetected, that believe in the existence of truths wider than they can limit by phrases habitually current, and that assent to the possibility of a connection among laws that seem far asunder, while they acknowledge their ignorance what the connection is; these are minds which have the best chance of discovering new principles and new generalizations and such habits of thought lead naturally to the persuasion of a supreme principle of unity and connexion.[65]

Science need not be abandoned for being irreligious, Whewell argued, because the proper, inductive, view of science leads to belief in God.[66]

These early theological writings, then, can be read not as Whewell's attempt to "save" religion from science, but rather to save science from the criticism of religious men such as Rose. Prior to Rose's 1826 sermon, Whewell's letters, notebooks, and published writings contain little on theological topics. It is noteworthy that just prior to this time, Whewell had needed to decide whether he would take orders in the church, as required to continue his fellowship past the initial seven-year period. That he waited until the deadline was upon him, and that it was not an obvious choice for him, speaks to his relative lack of interest in theological matters up until the mid-1820s. However, this is not to suggest that Whewell remained uninterested in theological questions. Indeed, as we will see, when he fully developed his view of science Whewell gave it a theological foundation. My point here is merely that theological concerns did not motivate his interest in developing and promulgating a view of scientific method.

In the case of politics, apart from the occasional worried reference to events in France, there are no mentions of national or international politics

---

65. William Whewell to Richard Jones [dated October 31, 1832, by archivist], WP Add.Ms.c.51 f. 143. See also his *Astronomy and General Physics, Considered with Reference to Natural Theology* (Bridgewater Treatise), pp. 230–57, and William Whewell to Richard Jones, March 24, 1833, WP Add.Ms.c.51 f. 154.

66. Whewell maintained this position throughout his career. In the third edition of his *History of the Inductive Sciences*, published in 1857, Whewell referred to the Bridgewater Treatise and expressed his current agreement with its claims: "In another work I have endeavoured to show that those who have been discoverers in science have generally had minds, the disposition of which was to believe in an intelligent Maker of the universe; and that the scientific speculations which produced an opposite tendency, were generally those which, while they might deal familiarly with known physical truths, and conjecture boldly with regard to the unknown, did not add to the number of solid generalizations" (3:388).

at all in Whewell's early letters or notebooks. (He did discuss university politics, especially at the time he was trying to obtain a professorship.) He took notes on reviews of economic treatises, on T. R. Malthus's *Essay on the Principle of Population*, and on Jane Marcet's *Conversations in Political Economy* in 1817,[67] but already by this time he was criticizing the deductive nature of Ricardian economics and the non-Baconian "anticipatory" nature of Malthus's population principle. So again, it does not seem that this early interest in political economy sparked his interest in scientific method; rather, it seems to have stemmed from this interest.

Unlike Mill, then, Whewell's desire to reform science was not purely *motivated* by moral, theological, or political concerns. Yet neither was his interest in reforming science unconnected to these concerns, as we will see in the coming chapters. Like his philosophical hero, Francis Bacon, Whewell sought a kind of "logic" that could "embrace everything"; that is, a philosophy that would apply to all areas of thought. Even as an undergraduate Whewell had set for himself the goal of "universal knowledge."[68] In 1817 he was president of the Cambridge Union Debating Society, which had been founded two years earlier, where debates ranged over political, philosophical, literary, and religious topics. He was involved in the early days of the Cambridge Philosophical Society, founded by Adam Sedgwick and Henslow in 1819, which embodied this ideal of universal knowledge; papers were presented on mathematical and scientific subjects, as well as on political economy (by Jones), German architecture (by Whewell), and literature.[69] After his student days, Whewell spent the rest of his life trying to attain his lofty goal; his bibliography and biography speak to his relative success in this endeavor. Further, he wished to combine all the areas of his "universal knowledge" into one grand system, tied together by a single philosophy. Bacon too had wished to create such a grand system; he noted in the *Novum Organum*, "It may be asked . . . whether I speak of natural philosophy only, or whether I mean that the other sciences, logic, ethics, and politics, should be carried by this method. Now I certainly mean what I have said to be understood of them all; and as the common logic . . . extends to all science; so does mine also, which proceeds

67. See David Valone, "The Dark and Tangled Recesses of Knowledge," p. 88.
68. In a letter to his old headmaster, Whewell referred to his last letter, which "conveyed to you certain yearnings after the whole circle of the sciences, certain ecstatic aspirations after universal knowledge, certain indefinite desires to approximate something like omniscience" (William Whewell to George Morland, December 15, 1815, in Isaac Todhunter, *William Whewell, D.D.*, 2:10).
69. Whewell remained involved in the Cambridge Philosophical Society for the remainder of his life. He was secretary of the society several times until 1836, and president between 1843 and 1845. For more on the history of the society, see A. Rupert Hall, *The Cambridge Philosophical Society*.

by induction, embrace everything."[70] In discussing Bacon in *On the Philosophy of Discovery*, Whewell remarked, "it cannot be denied that the commanding position which Bacon occupies in men's estimation arises from his proclaiming a reform in philosophy of so comprehensive a nature; a reform, which was to infuse a new spirit into every part of knowledge."[71] This was Whewell's project as well.[72]

Plan of the Book

In chapter 1 I show that Whewell and Jones were engaged from their early undergraduate days in what they believed to be their "mission," namely, to "bring the people to the right way of viewing induction."[73] As they noted, everyone appealed to "inductive method" as the proper way to do science, but very few understood correctly in what this method consisted. Whewell, with his friend's help, developed a view of induction that incorporates both empirical and rational elements of knowledge. Because of the rationalist aspect of Whewell's view, it has often been argued that his epistemology is essentially a British version of Kantianism. I argue that this is not the case; there are important differences between Whewell's epistemology and Kant's. I address as well the claim that Whewell was strongly influenced by German romanticism as it was expressed in England in the work of Coleridge. I show that he was, on the contrary, extremely critical of Coleridge. I explain how Whewell's inductive methodology followed in the tradition of Bacon's while at the same time "renovating" it. Finally, I discuss his striking claim that necessary truths can be discovered by empirical science.

After this detailed discussion of Whewell's philosophy of science, we can better understand Mill's view, which was, as he admitted in the *Autobiography*, deliberately presented in opposition to Whewell's work. In chapter 2 I discuss the reforms Mill proposed in science to counter the view of Whewell. I show that his overriding desire was to expel the intuitionist philosophy from

---

70. Francis Bacon, *Works*, 4:112.

71. William Whewell, *On the Philosophy of Discovery*, p. 126.

72. Bacon had begun with the desire to reform law, and through this, politics, but then saw that this project could be extended to include the reform of natural philosophy as well. An important part of his proposed reform of the natural philosophy was the shift from a model of *scientia* as an esoteric pursuit, conducted by isolated individuals guarding their secret knowledge (here, Bacon had in mind the alchemists), to a model of a public pursuit, conducted by groups of individuals sharing their results, and the useful inventions to be drawn from them, with the state. As part of this proposed shift to *scientia* as a public enterprise, Bacon was concerned with the reform of the natural philosophers themselves, and thus with moral philosophy. See Stephen Gaukroger, *Francis Bacon and the Transformation of Early-Modern Philosophy*, pp. 37ff.

73. William Whewell to Richard Jones, February 25, 1831, WP Add.Ms.c.51 f. 99.

its "stronghold" in physical science and mathematics, because he saw this as being the crucial precondition for reforming moral and political philosophy. The intuitionist epistemology led to political and social conservatism, Mill believed, by reassuring people that what they believed deeply must be true and necessary. Indeed, intuitionism allowed "every inveterate belief and every intense feeling" to be "its own all-sufficient voucher and justification."[74] If he could demonstrate that knowledge of physical science and even mathematics did not require any a priori axioms, Mill hoped, then he would have proved the superfluity of a priori elements in morality and political philosophy. It is because of this, I argue, that Mill developed an ultra-empiricist, phenomenalist epistemology, and rejected necessity in mathematics and causal relations. He thus "radicalized" induction. I also discuss Mill's "final and most elaborate protest against the Intuitionist school,"[75] his *Examination of Sir William Hamilton's Philosophy* (1865). In this work, Mill departed from the "commonsense" form of realism he endorsed in the *Logic*, and supported instead a position close to Berkeley's immaterialism. I show how this metaphysical stance is related to Mill's epistemological phenomenalism. I end the chapter by discussing Whewell's response to the work of "young Mill." Many of Whewell's criticisms of Mill arise from their differing positions on the relation between the history of science and writing about scientific method.

In chapter 3 I discuss the controversy between Mill and Whewell over the confirmation of scientific theories. This debate is related to their differing opinions over the nature of inductive inference and the type of knowledge we can reach inductively. As I describe in chapter 1, Whewell believed we could make reliable probable inferences to empirical truths about unobserved causes and entities. Mill, on the other hand, as I show in chapter 2, denied us even probable knowledge of unobserved causes and entities. Thus Whewell accepted, and Mill rejected, the claim that through science, we can gain knowledge of the underlying causal structure of nature. This difference is reflected in their respective views about classification. Whewell maintained that scientists discover groupings of things that reflect the causal structure of the physical world; that is, scientists discover natural kinds. Moreover, he believed theories that were confirmed show that they have uncovered the causal structure of nature. His most important confirmation criterion, "consilience," signifies that a theory has causally unified different natural kinds.

74. John Stuart Mill, *Autobiography*, CW 1:233.
75. See Leslie Stephen, *The English Utilitarians*, vol. 3, *John Stuart Mill*, and John Stuart Mill, *System of Logic*, CW 7:60.

Conversely, Mill believed that scientists group things based only on conjunctions of observed properties that reflect no underlying causal structure. (Although Mill claimed to accept "the reality of kinds in nature," he in fact rejected natural kinds.) Since Mill denied that we can have knowledge of the causal structure of nature, it is not astonishing that he opposed Whewell's "tests of hypotheses." Charles Darwin, however, seems to have embraced Whewell's criterion of consilience. I discuss the way in which Darwin made use of this criterion in his *Origin of Species*, and how Whewell and Mill reacted to Darwin's theory of evolution by natural selection.

In chapter 4 I examine more closely the moral and political context in which the debate between Mill and Whewell over science developed. I discuss the evolution of Mill's political views as he moved from his early admiration of Benthamism to the realization of its shortcomings. Under the influence of Coleridge, Auguste Comte, and the Saint-Simonians, Mill began to understand that the Benthamite approach to politics was flawed.[76] In order truly to reform society, Mill came to believe, there must be a "change of character . . . in the uncultivated herd."[77] Individuals, not merely political institutions, needed to be reformed. Mill, I argue, believed that this transformation of the mind could be brought about by a certain kind of education and nurtured by a particular type of liberty.

I next discuss Whewell's politics. As we have seen, Mill believed that Whewell's philosophy was the paradigmatic example of an antireformist view justifying the current status quo. But Whewell's actual political stance was not as conservative as Mill supposed. Whewell's view of political change was similar to his view of scientific change: there are no scientific revolutions—nor should there be political revolutions—completely discontinuous and unrelated to what came before. For instance, he supported the extension of the franchise that resulted from the passage of the Reform Bill of 1832, but thought that further reforms ought to come slowly. Moreover, Whewell's view of morality was not so different from Mill's, and in fact may have influenced the way in which Mill reconfigured Benthamism. Both men eschewed Bentham's claim that pleasure was the sole determinant of virtuous action, erecting instead moral philosophies that stressed the importance of creating morally excellent characters which would naturally find happiness in acting

76. Mill met Gustave d'Eichthal, a disciple of Claude-Henri Saint-Simon, in 1828. D'Eichthal had heard Mill speak at the Debating Society, and later sent him literature of the Saint-Simonians in an attempt to enlist him in their cause. Mill was most struck by their view that societal forms and institutions must be seen in their historical context. See John Robson, *The Improvement of Mankind*, pp. 76–77.

77. John Stuart Mill, *Early Draft of Autobiography, CW* 1:112, 238.

virtuously. Both believed that such characters could be fashioned by a proper education—one aimed toward the cultivation of minds. And both had hopes that delivering this kind of education to a greater number of people would lead to an improved society. Mill's mischaracterization of both Whewell's politics and his morality arose from his misconception of Whewell's epistemology, specifically its notion of necessary truth.

We have seen that both Mill and Whewell believed that reforming society required the renovation of science—a connection that is most apparent in their writings on political economy, which I examine in chapter 5. Whewell, along with his friend Jones, endorsed an inductive methodology in political economy, which fell in with their more general "induction project." But this methodology also conformed to their moral and political views: they saw the inductive methodology in political economy as a means to fight against the utilitarian moral philosophy and the political consequences the utilitarian radicals drew from it. Moreover, unlike the utilitarians, they believed, following Adam Smith and Malthus, that political economy is a branch of morality and politics, and so must concern man not only as an economic abstraction but also as a moral, social, and intellectual being. Turning to Mill, I show that his views of political economy changed over time as he moved from his early Benthamite days to the more nuanced position of his later years. Ironically, by accepting the Coleridgian view of the complexity of human nature, Mill was led to support the deductive, geometrical method in political economy, the very same process he had criticized Bentham for endorsing. But Mill saw this method as a temporary measure, until we could learn more about the complicated laws of man's behavior. His conception of a cultivated society eventually led him to focus, in his final work on political economy, on the Art rather than the Science of economics. He retained a Ricardian methodology in the science of political economy while utilizing a more inductive one in discussing the practical results of economics for society. Like Whewell, Jones, Malthus, and Smith, but unlike David Ricardo, Mill eventually accepted the importance of admitting moral, social, and political considerations into his discussion of political economy. Indeed, he had come to agree with Whewell and Jones that even the political economist must be concerned with reforming society at large.

I conclude the book by showing that understanding the Mill-Whewell debate, and the particular positions expressed by each within it, has a value that goes beyond the historical. Those interested in reforming science and society today can learn much from their controversy. In particular, Mill's conception of liberty, far from being an exemplar of the "negative liberty" position, indicates many of the problems with this stance. And Whewell's notion of inductive reasoning, which is richer than that presented by Mill and his modern followers, can provide a starting point for developing a view of the kind of reasoning used in discovering scientific theories.

Throughout the book I demonstrate that common ways of interpreting the ideas of Mill and Whewell are flawed, partly because their works have been read out of their contexts. By situating these works within the argumentative context of the debate between the two men, we learn that Mill's views arose, and took the shape they did, in great part because of his worries about the kind of philosophy Whewell's represented for him. Whewell, then, was an important influence on Mill, in ways that have not previously been noted. Mill's philosophy of science was designed for the purpose of defeating Whewell's "intuitionism"—thus Whewell's view was a stimulus, albeit a negative one, for Mill's position. But, as I show in this book, Whewell's philosophy was also a crucial positive influence on the development of Mill's ideas. In both moral philosophy and political economy, Mill's stance came to resemble Whewell's over the course of their dialectical engagement. Scholars working on Mill have, I believed, ignored an important source for understanding certain shifts in Mill's positions. At the same time, conflicts with Richard Whately, the Ricardian political economists, and the utilitarians influenced Whewell's positions and the ways in which he expressed them. So looking to these argumentative contexts is necessary as well.

Much is also learned by examining the debate between Mill and Whewell in the types of contexts Quentin Skinner and others have pointed to as crucial for studying the history of ideas, particularly the social, political, and "individual" contexts.[78] Without attending to the social and political context of the controversy, it is impossible to understand the motivations Mill and Whewell each had for developing their positions. For instance, the increase in the number of the indigent poor, concerns over the amounts of money being paid out in relief, and worries about working-class revolutions throughout Europe were all factors in stimulating the interest in political economy felt by Whewell, his friend Jones, and Mill, as well as inspiring the particular positions they each came to hold. Mill's evolving engagement with the Philosophical Radicals is also key to understanding the evolution of his own philosophical views. Further, as Skinner and his followers have also stressed, the historian of philosophy must understand the individual context of particular

---

78. See, for instance, Quentin Skinner, "Meaning and Understanding in the History of Ideas" and the introduction to *The Foundations of Modern Political Thought*, and J. G. A. Pocock, *Politics, Language and Time*. For evaluations of Skinner's approach, see the essays in James Tully, ed., *Meaning and Context*.

arguments and claims within the whole body of work of a writer. It is not possible to have proper insight into Mill's philosophy of science by reading only *System of Logic* and ignoring *On Liberty* and "Utilitarianism," not to mention his review of Whewell's moral philosophy and the *Autobiography*. Only in this way can his arguments (some of them, on the face of it, rather weak) against Whewell's epistemology and scientific methodology be understood. At the same time, *On Liberty*, read alone, is often taken to be endorsing a political stance similar to some forms of modern-day individualistic liberalism, but in the context of Mill's other writings it becomes clear that this is not the case. Similarly, a comprehensive understanding of Whewell's philosophy of science requires familiarity with his other works, including his writings on political economy and moral philosophy. Finally (although this is underemphasized by Skinner) the historian of philosophy must also be concerned with the private history of the philosopher under study—that is, with his or her interests and background—in order to recover his or her intentions in writing the work. Whewell's vast correspondence, especially with Jones, is a treasure trove of information about the "induction project" they had devised as undergraduates, and about the development of this project over the course of Whewell's long career. Whewell's experience in scientific research, and his mathematical work, is also relevant to the philosophical view he proposed. For example, his initial encounter with the mineralogical work of Friedrich Mohs in Germany provided the impetus for his attempts to bring about a union between empiricism and a priorism. Understanding this enables the historian of ideas to fend off idealist interpretations of Whewell's view. Likewise, it is necessary to understand Mill's political interests in order to comprehend fully his philosophy of science. As we will see, his view of science was developed more for political than for epistemological motives. For example, his concept of causation in the physical and human sciences was shaped not only by his general rejection of intuitionism, but also by his desire to allow humans the freedom to reform themselves in ways necessary for the renovation of society. Knowing about his desire to build a bridge between Bentham and Coleridge is also required to comprehend the development of his philosophy, and the rather peculiar forms it sometimes takes.

Thus, this book can be seen as an extended illustration of—and argument for—a particular methodological approach to the history of philosophy. Only by using this methodology can the historian of ideas uncover the real ideas and intended arguments of her subjects. And only after this is done can the modern-day philosopher truly engage with these ideas and arguments, in order to see whether they can offer us any wisdom today. This is why, for example, a philosopher of science interested in mining the debate between

Mill and Whewell for insights on discovery and confirmation should—indeed, must—be concerned as well with their broader debate over morality and politics, and the context in which this debate occurred. Barring this, the philosopher is looking not at the debate between Mill and Whewell on science but rather at an artificial construction, a conflict between two positions that are mere set-pieces devised by modern writers. And those interested primarily in the Victorian debate over reforming society must likewise be concerned with the debate over reforming science, which was conceived as an integral part of the wider reformist project.

CHAPTER ONE

Whewell and the Reform
of Inductive Philosophy

Those who theorize rightly are in the end the lords of the earth.
—*William Whewell to Richard Jones, 1836*

From his days as an undergraduate at Trinity College to his final years spent as Master of that institution, Whewell regarded his project as the reform of the inductive philosophy, a reform meant to apply to all areas of knowledge. Whewell intended that his reformed inductive philosophy would provide the groundwork for the reshaping of more than natural science; morality, politics, and economics would also be transformed. Armed with the proper inductive method, then, philosophers could be the "lords of the earth."

Whewell's desire to reform all of philosophy by the renovation of induction mirrored—and was influenced by—Francis Bacon's project. This influence began early. As undergraduates, Whewell, Richard Jones, and John Herschel were drawn together by their shared interest in the works of Bacon. The three of them agreed with Bacon that proper inductive method must be established for every area of thought. Even after receiving their degrees, when they were no longer gathered together at Cambridge, they all continued to express this as their shared goal. In 1831, Jones claimed to be seeking an "outline of reasoning . . . inductively on almost all subjects."[1] While he was engaged in his astronomical observations in the mid-1830s at the Cape of Good Hope, Herschel told Whewell that he was working on "an enquiry into the moral nature of men," which he hoped would culminate in "the inductive con-

---

1. Richard Jones to William Whewell, February 25, 1831, WP Add.Ms.c.52 f. 21.

struction of a system of Ethicks."[2] Whewell informed Jones that the "hyper-physical branches of knowledge . . . art, language, political economy, morals and the like," as well as the natural branches, could be improved by reforming induction.[3] All three set for themselves the task of carrying out this reform. Whewell and Jones were especially serious about this endeavor, and saw it as consisting in two parts. The first was to *define* a "true idea of induction." In an early notebook entry, Whewell lamented "that the true idea of induction has not been generally fixed and agreed upon must I think be very obvious."[4] Bacon's own view, though a useful starting point, was itself flawed; it needed to be renovated. (Thus, in later years, Whewell gave the title *Novum Organon Renovatum* to one volume of the third edition of his *Philosophy of the Inductive Sciences, Founded upon Their History.*) The second task was "to get *the people* into a right way of thinking about induction,"[5] that is, to publicize the nature and value of induction in all areas of thought. In one of his many letters to Jones, Whewell called induction the "true faith," and wondered how it could "best be propagated."[6] Both the task of defining induction, and that of popularizing it, were seen as part of the project of reforming all of philosophy, and thus as necessary for the reform of society. Jones told Whewell that "systems palpably mischievous and immoral" were attracting adherents, due to an inadequate comprehension of scientific induction and its relevance to morality, politics, and economics.[7] Whewell agreed. Some years later, referring to his *Philosophy of the Inductive Sciences*, he informed Julius Charles Hare that he was planning to write "a philosophy such as shall really give a right and wholesome turn to men's minds."[8]

Whewell and Jones felt that they were engaged in a "battle" against the "downwards mad," those who preferred a deductive approach in the natural and moral sciences.[9] One of these "deductive savages" was Richard Whately, Fellow of Oriel College, Oxford, who became archbishop of Dublin in 1831.[10]

2. John Herschel to William Whewell, May 9, 1835, WP Add.Ms.a.207 f. 26. In his *Preliminary Discourse on the Study of Natural Philosophy,* Herschel noted that the methods of the physical sciences could be applied fruitfully to problems in moral and social philosophy (see pp. 72–74).

3. William Whewell to Richard Jones, August 5, 1834, WP Add.Ms.c.51 f. 174.

4. See notebook dated June 28, 1830, WP R.18.17 f. 12, pp. v–ix.

5. William Whewell to Richard Jones, February 25, 1831, WP Add.Ms.c.51 f. 99.

6. Ibid.

7. See Richard Jones to William Whewell, September 27, 1827, Add.Ms.c.52 f. 15, and Richard Yeo, *Defining Science*, p. 62.

8. William Whewell to Julius Charles Hare, December 25, 1833, in Isaac Todhunter, *William Whewell, D.D.*, 2:175.

9. See William Whewell to Richard Jones, January 20, 1833, WP Add.Ms.c.51 f. 149; and July 22, 1831, Add.Ms.c.51 f. 110.

10. William Whewell to Richard Jones, February 19, 1832, WP Add.Ms.c.51 f. 129.

Early that year Jones wrote to Whewell after seeing the third edition of Whately's *Elements of Logic*, objecting to Whately's "strange notion" that induction was a type of deductive reasoning.[11] Indeed, Whately had claimed that *all* forms of reasoning could be assimilated to the syllogism.[12] Induction, for example, was said to be reasoning in which the major premise, which is generally suppressed, can be expressed as "what belongs to the individual or individuals we have examined, belongs to the whole class under which they come."[13] Thus, to take Whately's example, if we find, from an examination of the history of several tyrannies, that each of them lasted a short time, we conclude that "all tyrannies are likely to be of short duration." In coming to this conclusion we make use of a suppressed major premise, namely, "what belongs to the tyrannies in question is likely to belong to all."[14] Whately admitted that some would complain that his notion of induction was too narrow in that it did not account for how the *minor* premises are obtained, that is, for how it was ascertained that each of the examined tyrannies were "short-lived." But he distinguished between "Logical Discoveries," which occur when syllogistic reasoning alone is used to deduce a conclusion from known premises, and "Physical Discoveries," which involve more than syllogistic reasoning because various methods are used for ascertaining the premises—including observation, experiment, the selection and combination of facts, abstraction of principles, and others.[15] On Whately's view logic was only concerned with reasoning from premises, not with ascertaining the premises. Thus inductive logic was also only concerned with reasoning from premises, or with Logical—not Physical—discovery.[16]

Whewell and Jones saw Whately's characterization of induction as more than just a point of logic. Rather, it presented a potential obstacle for the reform of philosophy they sought. Many people associated induction with sci-

---

11. Richard Jones to William Whewell, February 24, 1831, WP Add.Ms.c.52 f. 20. In this letter Jones also criticized the appendix of the work, written by Nassau Senior, which claimed that political economy was a deductive science starting from assumed definitions. The debate over this issue will be examined in chapter 5.

12. Richard Whately, *Elements of Logic*, p. 207.

13. Ibid., p. 209.

14. Ibid.

15. Ibid., pp. 234–36.

16. As Pietro Corsi notes, Whately thus distinguished between two senses of induction: one having to do with the collection of facts, the other being a form of inference. The first was not of concern to Whately. Rather, his aim was to demonstrate that the second was "under the dominion of logic" (*Science and Religion*, p. 151). This aim arose in the context of attacks on the Oxford curriculum from those, such as John Playfair, who argued that progress in the mathematical sciences had been impeded by the reliance on Aristotelian logic. Whately and his teacher, Edward Copleston, defended the curriculum and the prominence of Aristotelian logic within it (see also Raymie McKerrow, "Richard Whately and the Revival of Logic in Nineteenth-Century England").

entific method, even though they were unclear about the precise meaning of the term.[17] If people accepted Whately's definition of induction, they might be led to the erroneous conclusion that science is essentially deductive, concerned only with deducing conclusions from assumed "first principles." Whewell and Jones linked this deductive view of science with that of the Scholastic Aristotelians;[18] indeed, Whately had explicitly framed his work as a defense of Aristotelian logic against the "confused" views of induction that resulted from its connection with Bacon.[19] Because of their emphasis on the syllogism, Whewell complained, the Scholastics "talked of experiment" but "showed little disposition to discover the truths of nature by observation of facts."[20] In a notebook from 1830, Whewell wrote of the Aristotelian method in terms reminiscent of Bacon's criticisms: such a method "could lead to no such truths, and in the development of physical science especially was entirely barren. . . . The business of speculative men became, not *discovery*, but *argumentation*."[21] Later, in his *History of the Inductive Sciences*, he would refer to the Middle Ages as a "stationary period" in science. Whewell and Jones believed that the correct view of induction needed to be brought before the public in order to prevail against sterile deductive approaches to scientific knowledge. Whewell expressed his "confidence" to Jones that "by and by the whole world will think [the deductive definition of science] as nonsensical as we do." But before this could happen, he and Jones would need to spread the "true faith."[22]

17. Whewell complained that people praised Bacon's inductive philosophy, and considered it the proper method for science, but without having a clear understanding of what this philosophy entails (notebook, WP R.18.17 f. 12, pp. v–viii; see also p. xi).

18. In his earlier works, such as the *Topics*, Aristotle elaborated methods for the "discovery of knowledge," while in later works such as the *Prior* and *Posterior Analytics* his concern was with methods of demonstration, or presentation of results. In Scholastic philosophy, Aristotle's method of presentation—which was mainly syllogistic—came to dominate discussions of methods for discovering natural knowledge.

19. See Richard Whately, *Elements of Logic*, pp. ix and 9, and Raymie McKerrow, "Richard Whately and the Revival of Logic in Nineteenth-Century England," p. 172. In a letter to Jones, Whewell claimed that Whately and his followers were even worse than Aristotle, "far more immersed in verbal trifling and useless subtlety" (April 7, 1843, WP Add.Ms.c.51 f. 227). Pietro Corsi has noted that Whately and the "Oxford Noetics" used Dugald Stewart's writings in defense of Oxford's educational system, with its emphasis on Aristotelian logic, a position opposed to Stewart's own. See Corsi's "The Heritage of Dugald Stewart."

20. William Whewell, *On the Philosophy of Discovery*, p. 48.

21. William Whewell, notebook, WP R.18.17 f. 12, p. 94. See also a notebook dating from 1831 to 1832 (WP R.18.17 f. 15, p. 24). Thirty years later, in the final edition of *The Philosophy of the Inductive Sciences*, Whewell criticized Descartes on similar grounds. See *On the Philosophy of Discovery*, p. 163.

22. William Whewell to Richard Jones, February 25, 1831, WP Add.Ms.c.51 f. 99.

Whewell's Antithetical Epistemology

We have seen that the early influence of Bacon strongly inclined Whewell to inductive, empirical views of epistemology. At the same time, however, he had a deep appreciation for a priori, deductive forms of reasoning, due to his interest in mathematical sciences. His work in mechanics exposed him to a physical science which seemed to incorporate both empirical and a priori elements.[23] But it was his experience studying mineralogy abroad that struck Whewell with the need to combine empirical and a priori elements in an epistemology and scientific methodology. In 1825, when John Henslow vacated the chair of Mineralogy at Cambridge to take up the Botany Professorship, Whewell announced himself a candidate for the position. Although he had published mathematical papers on crystallography, Whewell did not have much empirical knowledge of mineralogy; thus he went abroad to study with experts such as Friedrich Mohs.[24] He was strongly impressed with the "German school" in mineralogy, especially with the elegant mathematical treatment given by Mohs of a science that Whewell had previously considered purely empirical. He wrote to his friend Hugh James Rose that "I am afraid . . . that I may not bring back my faith as untainted as you have done: for I find my mineralogical supernaturalismus giving way in some respects. It may be possible to bring about a union between the two creeds [that is, the a priori and the empirical], which I hope will not be such a thing in science as you hold it to be in faith."[25]

Whewell eventually reconciled the empirical and a priori elements of science in an epistemology that is "antithetical" in that it expresses what Whewell called the Fundamental Antithesis, or dual nature, of knowl-

---

23. Whewell's first book was a text on dynamics, published in 1819. He went on to produce four more textbooks on mechanics, a book on analytical statics, and a text on the differential calculus: *An Elementary Treatise on Mechanics* (1819), *The First Principles of Mechanics with Historical and Practical Illustrations* (1832), *On the Free Motion of Points and on Universal Gravitation, including the Principal Propositions of Books I and III of the "Principia": The First Part of a Treatise on Dynamics* (1832), *An Introduction to Dynamics Covering the Laws of Motion and the First Three Sections of the "Principia"* (1832), *On the Motion of Points Constrained and Resisted, and the Motion of Rigid Bodies* (1834), *Mechanical Euclid* (1837), *Analytical Statics* (1833), and *The Doctrine of Limits with Its Applications Namely Conic Sections, the First Three Sections of Newton, the Differential Calculus* (1838).

24. This was not unusual at the time. When Adam Sedgwick was elected to the Woodwardian professorship of geology in 1818, he knew little of the discipline and needed to learn it quickly. See Martha Garland, *Cambridge before Darwin*, p. 96.

25. William Whewell to Hugh James Rose, August 15, 1825, in Isaac Todhunter, *William Whewell, D.D.*, 2:60. See also Menachem Fisch, *William Whewell, Philosopher of Science*, p. 67.

edge.[26] According to Whewell's mature epistemology, all knowledge involves an ideal, or subjective, element as well as an empirical, or objective, element.[27] Although his experience with mineralogy had sparked Whewell's desire to "bring about a union between the two creeds" of empiricism and a priorism, his notebook writings on induction prior to 1831 present induction as a purely empirical process, consisting in enumerative induction of observed instances.[28] He had not, up to this point, found a way to synthesize the a priori elements of scientific knowledge with the empirical epistemology he wished to follow Bacon in endorsing. By February of 1831, while working on his review of Herschel's *Preliminary Discourse on the Study of Natural Philosophy*, Whewell for the first time used a metaphor that indicates his initial attempt to combine these empirical and a priori or purely rational elements of science. He explained,

> Induction agrees with mere Observation in accumulating facts, and with Pure Reason in stating general propositions; but she does *more* than Observation, inasmuch as she not only collects facts, but catches some connexion or relation among them; and *less* than pure Reason . . . because she only declares that there *are* connecting properties, without asserting that they *must* exist of necessity and in all cases. If we consider the facts of external nature to lie before us like a heap of pearls of various forms and sizes, mere Observation takes up an indiscriminate handful of them; Induction seizes some thread on which a portion of the heap are strung, and binds such threads together.[29]

He asked Jones to look over his draft.[30] Jones's assessment proved to be valuable for Whewell:

> I go along with you in your use of the word induction only I fear to a certain extent—I do not myself like to oppose it to or contrast it with either

26. Menachem Fisch refers to Whewell's epistemology as "antithetical" in his "Necessary and Contingent Truth in William Whewell's Antithetical Theory of Knowledge" and his *William Whewell, Philosopher of Science.*

27. However, it took Whewell some years after his return from Germany—and much help from his friend Jones—before he hit on this solution for incorporating both elements of science. I believe that Whewell had begun to construct the basic outlines of his view by 1831, which is earlier than some other commentators claim. For example, Geoffrey Cantor, following David McNally, dates it from the summer of 1834. See Geoffrey Cantor, "Between Rationalism and Romanticism," p. 70, and D. H. McNally, "Science and the Divine Order," p. 115.

28. See especially the earliest draft of *The Philosophy of Inductive Sciences,* dated June 28, 1830, pp. 98, 104, where his view of induction sounds much like the conception expressed later by Mill in his *System of Logic,* which was strongly criticized by Whewell (WP R.18.17 f. 12).

29. William Whewell, "Modern Science—Inductive Philosophy," p. 379.

30. William Whewell to Richard Jones, February 11, 1831, WP Add.Ms.c.51 f. 98.

observation or pure reason—Induction according to me and Aristotle (admire my modesty) *is the whole process by which the intellect gets a general principle from observing particulars or individuals and in that process both observation and pure reason have a part*—when observation has collected the facts abstraction . . . seizes on the law or principle and then the inductive process is compleat in all its parts.[31]

In this passage Jones suggested to Whewell a way to synthesize the empirical and the ideal in his epistemology: namely, by seeing induction as an act that includes *both* observation *and* pure reason. Whewell seems to have been inspired by this characterization of induction. He soon began to describe induction as a *process* involving observation and reason. The rational element of induction was provided by certain "conceptions of the mind." In a notebook dated 1831–32, Whewell for the first time characterized induction as involving such conceptions: "Induction supposes a power of clearly representing phenomena by the mind as subordinate to the conceptions of space, number, etc."[32] In another notebook of this period he described the importance of distinctly conceiving conceptions in order to make proper inductions,[33] and added what appears to be his first list of conceptions regulating different sciences.[34]

In the summer of 1832 Whewell was reading the *Exposition de la Doctrine de Saint-Simon* upon the suggestion of Jones.[35] In his reading notes he commented that "there is in the view of the nature of science here given a good deal which . . . falls in with my views," including a passage reading "the arrangement of facts suppose general conceptions."[36] Whewell elaborated on this point of agreement in a letter to Jones: "There are as you say several right notions about the character of science—one in which they have hit on in the same way which I have used for nearly the same thing. The conceptions which must exist in the mind in order to get by induction a law from a collection of facts; and the impossibility of inducting or even of collecting with-

---

31. Richard Jones to William Whewell, March 7, 1831, WP Add.Ms.c.52 f. 26; emphasis added.
32. William Whewell, notebook, WP R.18.17 f. 13, p. 54.
33. William Whewell, notebook, WP R.18.17 f. 15, p. 2.
34. Ibid., p. 47, in an entry dated July 2, 1831.
35. In January, Jones had written to Whewell, "But oh! The Saint-Simonians!! You must read and ponder and wonder and profit—L'Organisateur is, I am bold to say, one of the half-dozen *most* extraordinary and interesting books in the world—learned—logical—powerful—feeling—good—and taming—ignorant—unreasonable—feeble, mischievous and disgusting." Moreover, he noted that the book contains "some excellent speculations on induction and much in your spirit as to the provinces of the imagination—the intellect and the senses in seizing general laws" (January 17, 1832, WP Add.Ms.c.52 f. 46).
36. William Whewell, notebook, WP R.18.17 f. 15, p. 56.

out this."[37] By 1833 Whewell had come to believe that, as he explained in a
notebook entry, "knowledge implies passive sense [that is, observation] and
active thought [that is, reason]."[38] In an address to the British Association for
the Advancement of Science in that same year, he claimed that "a combination of theory with facts . . . is requisite" in order to discover new truths.[39]
And in a letter to Jones of 1834 Whewell described his "Philosophy of Induction," claiming that "you will see that a main feature is the assertion of ideas
and facts as equally and conjointly necessary to science."[40]

By 1837, Whewell was ready to formulate more systematically this position, and to express it publicly. He did so in his "Remarks on the Logic of
Induction," which was appended to his textbook *Mechanical Euclid*. In this
essay Whewell explained that induction requires both an idea provided by the
mind and facts provided by the world.[41] A "general idea," which is not given
by the phenomena but "by the mind," is "*superinduced* upon the observed
facts."[42] In his *History of the Inductive Sciences*, published that same year,
Whewell similarly noted that both facts and ideas are requisite for the "formation of science." "Real speculative knowledge," he claimed, "demands the
combination of the two ingredients."[43] By the time he published the first edition of *The Philosophy of the Inductive Sciences* in 1840, he had worked out
more details of the position, and developed an argument situating his epistemological view as a "middle way" between stark empiricism and full-blown
rationalism.

Because of the dual or antithetical nature of knowledge, Whewell claimed,
gaining knowledge requires attention to both Ideas and Sensations: "without
our ideas, our sensations could have no connexion; without external impressions, our ideas would have no reality; and thus both ingredients of our
knowledge must exist."[44] An exclusive focus on one or the other side of the
antithesis is to be avoided. Whewell criticized both Kant and the German Idealists for their exclusive focus on the ideal or subjective element, and Locke
and his followers of the "Sensationalist school," for their exclusive focus on
the empirical, objective element.

37. William Whewell to Richard Jones, February 19, 1832, in Isaac Todhunter, *William
Whewell, D.D.*, 2:141.

38. William Whewell, notebook, WP R.18.17 f. 8, facing p. 20.

39. William Whewell, "Address," p. xx.

40. William Whewell to Richard Jones, August 5, 1834, WP Add.Ms.c.51 f. 174.

41. William Whewell, "Remarks on the Logic of Induction," p. 181.

42. Ibid., p. 178.

43. William Whewell, *History of the Inductive Sciences*, 1:5–6.

44. William Whewell, *The History of Scientific Ideas*, 1:58.

What exactly are these "ideas," which comprise the ideal or subjective element in Whewell's antithetical epistemology? He generally referred to them as "Fundamental Ideas," explaining that "I call them *Ideas*, as being something not derived from sensation, but governing sensation, and consequently giving form to our experience;—*Fundamental*, as being the foundation of knowledge, or at least of Science."[45] They are supplied by our minds in the course of our experience of the external world; they are not simply received from our observation of the world.[46] This is why Whewell claimed that the mind is an active participant in our attempts to gain knowledge of the world, not merely a passive recipient.[47]

Although these Ideas are supplied by our minds, they are such that they enable us to have real knowledge of the empirical world. They do so by connecting the facts of our experience; this occurs because the Ideas provide the general relations that really exist in the world between objects and events. These relations include Space, Time, Causation, and Resemblance, among numerous others. By enabling us to connect the facts under these relations, the Ideas provide a structure or form for the multitude of sensations we experience.[48] Thus, for example, the Idea of Space allows us to apprehend objects as existing in space, in spatial relations to each other, and at a particular distance from us. Indeed, we need these Ideas in order to be able to make sense of our sensations: "Our sensations, of themselves, without some act of the mind, such as involves what we have termed an Idea, have not form. We cannot see one object without the Idea of Space; we cannot see two without the idea of resemblance or difference; and space and difference are not sensations."[49] Every science, Whewell believed, has one or more Fundamental Ideas particular to it, which provide the structure for all the facts with which that science is concerned.[50] The Idea of Cause is especially associated with the science of Mechanics, while the Idea of Space is the Fundamental Idea of Geometry. Whewell explained further that each Fundamental Idea has certain "conceptions" included within it; these conceptions are "special

45. William Whewell, *On the Philosophy of Discovery*, p. 336.
46. Whewell explained that the Fundamental Ideas are "not a consequence of experience, but a result of the particular constitution and activity of the mind, which is independent of all experience in its origin, though constantly combined with experience in its exercise" (*The History of Scientific Ideas*, 1:91). See also William Whewell, *On the Philosophy of Discovery*, p. 336.
47. William Whewell, *On the Philosophy of Discovery*, p. 218.
48. See William Whewell, *The Philosophy of Inductive Sciences*, 1:25.
49. William Whewell, *The History of Scientific Ideas*, 1:40.
50. William Whewell, *Novum Organon Renovatum*, p. 137, and *The History of Scientific Ideas*, 2:3.

modifications" of the Idea applied to particular types of circumstances.[51] For example, the conception of force is a modification of the Idea of Cause, applied to the particular case of motion.[52]

Whewell and the "Germano-Coleridgian" School

Thus far, this discussion of the Fundamental Ideas may suggest that they are similar to Kant's forms of intuition and categories, and there are indeed some similarities. Because of this, numerous commentators argue that Whewell's epistemology was derived from his reading of Kant, or perhaps a view of Kant refracted through the writings of S. T. Coleridge and the Coleridgian circle at Cambridge, which included Whewell's close friends Rose and Hare.[53] Mill suggested this reading of Whewell when he called his antagonist a member of the "Germano-Coleridgian school." However, although there are some similarities between his view of the Fundamental Antithesis and certain notions of Kant's and Coleridge's, there are important differences, which are frequently overlooked by commentators on Whewell.[54]

On the one hand, Whewell did read and appreciate Kant, in a time when German philosophy was not terribly popular in England.[55] He knew German well enough to publish translations of German prose and poetry[56] and, if a

51. William Whewell, *Novum Organon Renovatum*, p. 187. Whewell was not always consistent in maintaining the distinction between Ideas and conceptions. For instance, in this work he referred to both the "Idea of Number" (p. 54) as well as the "conception of number" (pp. 50 and 56).

52. See William Whewell, *The History of Scientific Ideas*, 1:184–85 and 236.

53. On readings of Whewell as a "British Kantian," see Robert Butts, "Necessary Truth in Whewell's Theory of Science"; Gerd Buchdahl, "Deductivist versus Inductivist Approaches in the Philosophy of Science as Illustrated by Some Controversies between Whewell and Mill"; Michael Ruse, *The Darwinian Revolution*, p. 58; and Silvestro Marcucci, *L'"Idealismo" Scientifico di William Whewell*. Robert Preyer has argued for a strong link between the views of Whewell and Coleridge (see his "The Romantic Tide Reaches Trinity"). Geoffrey Cantor claims that "by the end of 1838 [Whewell] had adopted a Coleridgian notion of intuition" ("Between Rationalism and Romanticism," p. 77). Susan Cannon also suggests a link between Whewell and Coleridge by claiming that for Whewell "reason is not man's highest faculty"; rather, it is his "ability to have moral sentiments, and his emotions of love, including the love of beauty" ("The Whewell-Darwin Controversy," p. 381). Most recently, Phillip Sloan has claimed that Whewell's notion of Fundamental Ideas owes a strong debt to Coleridge (see "Whewell's *Philosophy of Discovery* and the Archetype of the Vertebrate Skeleton," pp. 50–53).

54. However, as I suggest below, Whewell may have been influenced by his reading of Kant in a way that has not been noted by other commentators.

55. See René Wellek, *Immanuel Kant in England*.

56. In 1847, Whewell edited *English Hexameter Translations from Schiller, Goethe, Homer, Callinus, and Meleager*, which contained his translation of Goethe's *Herman and Dorothea*. Herschel was engaged in translations into the hexameter form as well: he translated Schiller's *Walk* and the *Iliad* (see Isaac Todhunter, *William Whewell, D.D.*, 1:286, 298). See also William Whewell,

story related by Isaac Todhunter is true, well enough to pass in Germany as a native speaker.[57] A notebook from 1825 contains reading notes of the *Critique of Pure Reason*.[58] Like Kant, Whewell was interested in examining how it is possible for us to have knowledge that has a universality and necessity that experience alone cannot give it. And his answer, like Kant's, involves certain conceptions and ideas that are in some sense a priori, because they are not derived from experience. As he wrote to Herschel, "My argument is all in a single sentence. You *must* adopt such a view of the nature of scientific truth as makes universal and necessary propositions possible; for it appears that there are such, not only in arithmetic and geometry, but in mechanics, physics and other things. I know no solution of this difficulty except by assuming *a priori* grounds."[59] Moreover, Whewell rather modestly admitted that his discussions of the Fundamental Ideas of Space and Time are mere "paraphrases" of Kant's discussion of the forms of intuition in the first *Critique*.[60] A contemporary translator of Kant claimed, in fact, that in his *Philosophy of the Inductive Sciences*, Whewell "more elegantly expressed" certain doctrines of Kant.[61] These most basic Ideas of Space and Time (in some works, Whewell includes the Idea of Cause among these) do function in Whewell's epistemology as similar to Kant's forms of intuition as being conditions of our having any knowledge of the world; indeed, Whewell referred to them as "conditions of experience"[62] and "necessary conditions of knowledge."[63] For instance, on his view, experiencing objects as having form, posi-

---

Verse Translations from the German, Including Burger's "Lenore," Schiller's "Song of the Bell," *and Other Poems.*

57. Isaac Todhunter quotes a letter from the scientist Alexander von Humboldt, who bemoaned the fact that he had "lost the pleasure" of seeing Whewell in Potsdam because he had told his servant to admit an English gentleman who would present himself, and Whewell was turned away because "you have spoken German like an inhabitant of the country!" (see *William Whewell, D.D.*, 1:411)

58. See William Whewell, notebook, WP R.18.9 f. 13, p. 19.

59. William Whewell to John Herschel, April 22, 1841, in Isaac Todhunter, *William Whewell, D.D.*, 2:298.

60. William Whewell, *On the Philosophy of Discovery*, p. 335.

61. Whewell's library contains a translation of the *Kritik der reinen Vernunft* by Francis Hayward (1848) inscribed "The Rev. Dr. Whewell/With the Translator's respects." In Kant's preface to the second edition, the translator added a note referring to Whewell. Kant wrote, "If the intuition must regulate itself according to the property of the objects, I do not see how one can know anything with regard to it, *a priori*, but if the object regulates itself, (as objects of the senses,) according to the property of our faculty of intuition, I can very well represent to myself this possibility" (pp. xxv–xxvi). The added note reads, "See Whewell's *Philosophy of the Sciences* . . . where this is more freely translated and more elegantly expressed" (p. xxvi).

62. William Whewell, *The History of Scientific Ideas*, 1:268.

63. William Whewell, "Demonstration That All Matter Is Heavy," p. 530.

tion, and magnitude requires the Idea of Space.[64] Because of these similarities, Whewell was accused by his contemporaries (not all of whom had themselves studied Kant) of trying to import Kant into British philosophy: in his
review of the *Philosophy* even Whewell's friend Augustus DeMorgan expressed surprise that "the doctrines of Kant and Transcendental Philosophy
are now promulgated in the university which educated Locke."[65] The logician
H. L. Mansel—who was the closest thing to a British Kantian in those days—
thought Whewell had not gone far enough, criticizing him for his "stumble
on the threshold of Critical Philosophy."[66]

However, this was a threshold that Whewell did not intend to cross. There
are important differences between Kant's transcendental philosophy and
Whewell's antithetical epistemology. Whewell did not follow Kant in distinguishing between the a priori components of knowledge provided by intuition (*Sinnlichkeit*), the Understanding (*Verstand*), and the faculty of Reason
(*Vernunft*). Thus Whewell drew no distinction between "precepts," or forms
of intuition, such as Space and Time, the categories, or forms of thought,
in which Kant included the concepts of Cause and Substance, and the transcendental ideas of reason. Further, many of Whewell's Fundamental Ideas
function not as conditions of experience but as conditions for having knowledge within their respective sciences: although it is certainly possible to have
experience of the world without having a distinct Idea of, say, Chemical
Affinity, we could not have any knowledge of certain chemical processes
without it.[67] Unlike Kant, Whewell did not attempt to give an exhaustive list
of these Fundamental Ideas; rather, he believed that there are others which
will emerge in the course of the development of science. Moreover, and perhaps most important for his philosophy of science, Whewell rejected Kant's
claim that we can only have knowledge of our "categorized experience." The
Fundamental Ideas, on Whewell's view, accurately represent objective features of the world, independent of the processes of the mind, and we can use
these Ideas in order to have knowledge of these objective features.[68] Indeed,

64. See William Whewell, "Remarks on a Review of the *Philosophy of the Inductive Sciences*"
(letter to John Herschel, April 11, 1844), p. 487.

65. Augustus DeMorgan, "Review of the *Philosophy of the Inductive Sciences*," p. 707.

66. H. L. Mansel, *Prolegomena Logica*, p. 258. Mansel's teacher, William Hamilton, more
scathingly accused Whewell of "recently dipping into the Kantian philosophy" ("Review of [Whewell's] *Thoughts on the Study of Mathematics as Part of a Liberal Education*," p. 417).

67. See William Whewell, *The History of Scientific Ideas*, 2:3 and *On the Philosophy of Discovery*, p. 349.

68. In 1865 the American philosopher and logician Charles Peirce expressed a similar view of
Whewell's divergence from Kant: "Dr. Whewell has usually been considered a Kantian. Up to a certain point this is true. He accepts Kant's division of the matter and form of our knowledge and also

he criticized Kant for viewing external reality as a "dim and unknown region."[69] His justification for the presence of these concepts in our minds takes a very different form from Kant's transcendental argument. For Kant, the categories are justified because they make experience possible. For Whewell, though the categories *do* make experience (of certain kinds) possible, the Ideas are justified by their origin in the mind of a divine creator. And finally, the type of necessity that Whewell claimed is derived from the Ideas is very different from Kant's notion of the synthetic a priori.

Thus we should take seriously Whewell's frequent denials that his epistemology is identical or even particularly similar to Kant's. In his response to DeMorgan's review, a privately published pamphlet, Whewell noted that his critic had gone too far in associating him with the Kantian philosophy: "It might have occurred to him, . . . that by the very circumstance of classing many other ideas with those of space and time, I entirely removed myself from the Kantian point of view."[70] In his reply to Mansel, later published as part of the *Philosophy of Discovery*, Whewell clearly noted that he never intended to follow Kant's view, and pointed out ways in which his philosophy differed from Kant's.[71] What he particularly admired of Kant's work was his having shown the untenable nature of the account of knowledge given by the Lockean school.[72] Whewell's dislike of Locke began early, even before his 1825 reading of the first *Critique:* in 1814 he wrote to his former headmaster, the Reverend George Morland, that while reading Locke he "grew out of humour with [him]."[73] In his chapter "The Influence of German Philosophy in Britain" in the final edition of the *Philosophy of the Inductive Sciences*, Whewell set up a tripartite division showing himself to be in between the idealism of the

---

his theory of space and time but he seems to have cast away from the doctrine of the limits of our knowledge which is the essence of the critical philosophy" ("Lecture on the Theories of Whewell, Mill, and Comte," p. 205).

69. William Whewell, *On the Philosophy of Discovery*, p. 312. At B312 of the first *Critique*, Kant noted that our understanding "limits sensibility by applying the term 'noumena' to things-in-themselves, i.e. to things not regarded as appearances. But in doing so it at the same time sets limits to itself, recognizing that it cannot know these noumena through any of the categories, and that it must therefore think of them only under the title of an unknown something." As commentators have noted, it is unclear to what extent Kant is justified in asserting even the existence of things in themselves; indeed, it seems to conflict with his claim that the noumenon is "a merely *limiting concept*, the function of which is to curb the pretensions of sensibility" (A255/B311). See, for example, C. D. Broad, *Kant, an Introduction*, pp. 201–4.

70. William Whewell, "Remarks on the Review of the *Philosophy of the Inductive Sciences*" (reply to DeMorgan)," p. 4.

71. See William Whewell, *On the Philosophy of Discovery*, pp. 335–46.

72. Ibid., p. 308.

73. William Whewell to George Morland, April 1814, in Isaac Todhunter, *William Whewell, D.D.*, 2:2.

Germans and the strict empiricism of Mill, standing in here for the Lockean school. Speaking of himself in the third person, he wrote, "Kant considers that Space and Time are conditions of perception, and hence sources of necessary and universal truth. Dr. Whewell agrees with Kant [only] in placing in the mind certain sources of necessary truth; he calls these Fundamental Ideas, and reckons, besides Space and Time, others, as Cause, Likeness, Substance, and several more. Mr. Mill, the most recent and able expounder of the opposite doctrine, derives all truths from Observation, and denies that there is such a separate source of truth as Ideas."[74] Whewell was, as he insisted, seeking a "middle way" between the excesses of Locke and Kant.

It is noteworthy that while working out his theory, Whewell made no references to Kant and used no technical Kantian terminology such as "forms of intuition," "categories," and "analytic and synthetic truths" in his letters to Jones, his notebooks, or his early drafts of the *Philosophy*.[75] This suggests that Kant was not explicitly on Whewell's mind as he strove to develop an epistemology that could incorporate both empirical and ideal elements of knowledge. While there are, as we have seen, some elements of Whewell's view that are similar to Kant's epistemology, it is not necessarily the case that Whewell even derived these elements directly from his reading of Kant in 1825. He certainly appreciated Kant's critique of the Sensationalist school, and took note of his ideal solution. But his reading of Dugald Stewart, Thomas Reid, and other Scottish philosophers similarly impressed Whewell, and may well have first exposed him to the type of ideal solution given by Kant.[76] Whewell expressed an appreciation of the work of the Scottish philosophers as early as the 1814 letter to Morland cited earlier.[77] Years later, when he published the *Philosophy*, he applauded the "intelligent metaphysicians in Scotland," including "Reid, Beattie, Dugald Stewart, and Thomas Brown," for ar-

74. William Whewell, *On the Philosophy of Discovery*, p. 336. Gerd Buchdahl claims that Whewell shared Kant's "idealist" or "subjectivist" approach to the Ideas, and therefore also to the knowledge we gain using them ("Deductivist versus Inductivist Approaches in the Philosophy of Science as Illustrated by Some Controversies between Whewell and Mill"). Robert Butts maintains that Whewell "owes his theory of science to Kant," which is clearly overstated ("Induction as Unification," p. 278). Menachem Fisch has a more nuanced view, claiming that there are "Kantian undertones" in Whewell's epistemology, but less than generally supposed (see *William Whewell, Philosopher of Science*, p. 105).

75. Menachem Fisch agrees with the importance of this point. See *William Whewell, Philosopher of Science*, p. 105.

76. Pietro Corsi has suggested, for example, that Whewell's earliest views of space and time as conditions of experience may have been sparked initially by his reading of Dugald Stewart's *Philosophical Essays*, prior to his reading of Kant (see *Science and Religion*, p. 155).

77. William Whewell to George Morland, June 15, 1814, in Isaac Todhunter, *William Whewell, D.D.*, 2:6.

guing against the "sensationalism" of Hume and Locke.[78] He noted that their shared view, "according to which the Reason or Understanding is the source of certain simple ideas, such as Identity, Causation, Equality, which ideas are necessarily involved in the intuitive judgments which we form, when we recognize fundamental truths of science, approaches very near in effect to [my] doctrine . . . of fundamental ideas."[79] Stewart had argued, for example, that every act of observation involves an "interpretation of nature." In his *Elements of the Philosophy of the Human Mind*, he explained that "without theory (or in other words, without general principles inferred from a sagacious comparison of a variety of phenomena), experience is a blind and useless guide."[80] It seems likely that Stewart was at least as much an influence upon the development of Whewell's epistemology as Kant.[81]

And what of the influence of Coleridge upon Whewell's antithetical epistemology? Commentators have noted that the network of friendships at Trinity included classical and German scholars influenced by Coleridge and German philosophy such as Hare, Rose, Connop Thirlwall, F. D. Maurice, and John Sterling, and men of science such as George Peacock, G. B. Airy, Herschel, Adam Sedgwick, and Whewell.[82] However, friendship in these cases did not imply philosophical agreement. Whewell was especially close to Hare and Rose, but he did not join them in their enthusiasm for the Romantic movement in general or its specific views of education, history, poetry, or philosophy. He particularly objected to the way the Romantics set up an opposition between reason and feeling or "enthusiasm," elevating the latter over the

78. See William Whewell, *On the Philosophy of Discovery*, p. 214 and *The Philosophy of the Inductive Sciences*, 2:309.

79. William Whewell, *On the Philosophy of Discovery*, p. 215; see also *The Philosophy of the Inductive Sciences*, 2:310–11.

80. Dugald Stewart, *Elements of the Philosophy of the Human Mind*, 2:444.

81. By 1847, Whewell had come to believe that Kant's notion of cause was closer to his own than that given by Thomas Brown and the "Scotch philosophers." "According to the Scotch philosophers, the cause and the effect are two things, connected in our minds by a law of our nature. But this view requires us to suppose that we can conceive the law to be absent, and the course of events to be unconnected. . . .The Kantian doctrine, on the other hand, teaches us that we cannot imagine events liberated from the connexion of cause and effect: this connexion is a condition of our conceiving any real occurrences." Hence the Scottish philosophers only account for universality, Kant for necessity (see *The Philosophy of the Inductive Sciences*, 1:173–74). But the necessity Whewell wanted to account for is not a Kantian synthetic a priori necessity, as I show below. In his younger days, before reading Kant, Whewell had had a more positive assessment of Brown's notion of causality; in a letter to Hugh James Rose in 1822 Whewell asked, "Have you read Brown's books? They are dashing, and on some material points strongly wrong, but about cause and effect he has an admirable clearness of view and happiness of illustration" (September 24, 1822, WP R.2.99 f. 17).

82. See Robert Preyer, "The Romantic Tide Reaches Trinity," and Geoffrey Cantor, "Between Rationalism and Romanticism," especially p. 77. Richard Yeo has also pointed to the importance of this connection in a different context in his *Defining Science*, pp. 65–71.

former. He argued this point with Rose: "Why will you not see that in speculative matters, though Reason may go wrong if not guided by our better
affections, you cannot do without her? All your efforts not to reason at all
will only end in your reasoning very ill. . . . Finding that Reason alone cannot invent a satisfactory system of morals or politics, are you not quarrelling
with her altogether, and adopting opinions *because* they are irrational?"[83] In
a notebook entry from 1816 to 1817 containing reading notes on Coleridge's
On the Constitution of the Church and State according to the Idea of Each,
Lay Sermons, Whewell accused Coleridge of trying to "make reason commit
suicide" by ceding ground to enthusiasm.[84]

Moreover, Whewell associated Coleridge with the views of Plato: both
were guilty of overlooking one side of the Fundamental Antithesis, the side
of facts.[85] Whewell believed that Plato and his followers had a problem similar to that of Kant and the German idealists.[86] Although the followers of Plato
avoided the error of the Sensationalists, namely believing that "facts alone
are valuable" as the materials of knowledge, they fell into the opposite error
of "despising or neglecting" facts.[87] But, unlike Kant, Plato and his followers
held that the Ideas have a real existence outside our minds, and that our true
knowledge is only knowledge of the Ideas, not of objects in the world or even

83. William Whewell to Hugh James Rose, December 29, 1823, in Janet Stair Douglas, *The Life
and Selections from the Correspondence of William Whewell, D.D.,* p. 95.

84. See a notebook entry from 1816 to 1817, containing reading notes on Coleridge's *On the
Constitution of the Church and State according to the Idea of Each, Lay Sermons.* Quoted in Isaac
Todhunter, *William Whewell, D.D.,* 1:349.

85. Whewell—like Rose and most others at the time—considered Coleridge a Platonist rather
than a Kantian. This was the most common interpretation of Coleridge in Whewell's time,
notwithstanding Coleridge's own dubious claim that he had in fact *anticipated* the ideas of Kant
and Schelling. See G. N. G. Orsini, *Coleridge and German Idealism,* p. 265. For instance, Robert
Chenevix Trench, afterward Dean of Westminister, described his fellow members of the Apostles
in 1830 as "that gallant band of Platonico-Wordsworthian-Coleridgean-anti-Utilitarians" (quoted
in Robert Preyer, "The Romantic Tide Reaches Trinity," p. 45). Indeed, in a notebook entry Whewell expressed doubt that Coleridge "fully understands Kant" (July 25, 1827, WP R.18.9 f. 4; quoted
in Richard Yeo, "William Whewell's Philosophy of Knowledge and Its Reception," p. 192). Although Coleridge visited Germany in 1798–99, he did not begin studying Kant intensively until
1801. At this time he introduced positive comments about Kant's philosophy into his writings
(G. N. G. Orsini, *Coleridge and German Idealism,* p. 52). By fifteen years later Coleridge was incorporating notions of Kant into his work, but often with very different meanings than Kant's (p. 78).
His fullest and most explicit discussion of Kant appeared in a 600-page manuscript for a "Treatise
on Logic," which was written between 1822 and 1823 and which consisted mainly of an exposition
of Kant's Transcendental Analytic. However, this was not published until the twentieth century
(p. 246).

86. On the question of whether Plato or Kant exerted a greater influence upon Whewell, see
Silvestro Marcucci, "William Whewell" and *L'"Idealismo" Scientifico di William Whewell,* and
Robert Butts's criticisms of Marcucci in his "Professor Marcucci on Whewell's Idealism." Both writ-
ers overemphasize the role of ideas in Whewell's epistemology, at the expense of the role of facts.

87. William Whewell, *On the Philosophy of Discovery,* p. 11.

of our experience of these objects.[88] Plato, Whewell complained, considered the study of Ideas the only method of pursuing knowledge.[89] Like Plato, Coleridge was guilty of the "tendency to exalt Ideas above Facts,—to find a Reality which is more real than Phenomena,—to take hold of a permanent Truth which is more true than truths of observation. . . . He tries to separate the poles of the Fundamental Antithesis, which, however antithetical, are inseparable."[90] Because of this he developed an epistemology opposed to Whewell's.

At first glance, however, there are similarities between Whewell's epistemology and the view expressed by Coleridge in his *Preliminary Treatise on Method* (1817), which served as the general introduction of the *Encyclopedia Metropolitana*. Unlike some Cambridge admirers of Coleridge, such as Whewell's friend Rose, Coleridge himself did not disparage natural science,[91] and intended his treatise on method to illustrate the application of his philosophy to the scientific study of nature.[92] Coleridge noted that discoveries of truth are not made by accident, but by the distinct presentation of an *Idea*. He claimed that the science of Electricity had progressed more rapidly than that of Magnetism because the former contained a clear Idea of Polarity, while the latter had no clear regulative Idea.[93] Coleridge described the "perfect" scientific method as involving the placing of particulars under a general conception, which becomes their "connective and bond of unity."[94] So far, this is not opposed to Whewell's view of the role of Fundamental Ideas in our knowledge. Yet, as Whewell perceived, even in this work Coleridge separated the poles of the Fundamental Antithesis by asserting an absolute division between the ideal and the empirical parts of knowledge. Coleridge drew a distinction between "Metaphysical" and "Physical" Ideas, explaining that "Metaphysical Ideas, or those which relate to the essence of things as pos-

88. See ibid., p. 17.

89. Ibid., p. 8.

90. Ibid., pp. 424–25. In his 1856 essay "Of the Intellectual Powers according to Plato," Whewell complained that Coleridge presented English writers in a false light in order to make their views seem to accord with Plato's; here Whewell seems to be referring to Coleridge's claim, in the *Preliminary Treatise on Method*, that Bacon was the "British Plato."

91. For a recent and definitive dismissal of the common assumption that German Romantic thinkers were uniformly opposed to science and reason, see Robert J. Richards, *The Romantic Conception of Life*.

92. See Kathleen Coburn, "Coleridge." Although, to my knowledge, Whewell never referred explicitly to Coleridge's *Preliminary Treatise* in his letters to Rose or in his notebooks, he surely must have read it, given his interest in scientific method and given that Rose was one of the editors of the *Encyclopedia Metropolitana*.

93. S. T. Coleridge, *Preliminary Treatise on Method*, see pp. 17–21.

94. Ibid., p. 54.

sible, are of the highest class. . . . Physical Ideas are those which we mean
to express, when we speak of the *nature* of a thing actually existing and cog-
nizable by our faculties."[95] Metaphysical Ideas, then, have nothing to do with
empirical experience, while Physical Ideas have only to do with it. In his
Aids to Reflection and *The Friend*, Coleridge asserted further that there are
two distinct faculties, the Understanding and the Reason, claiming that
these different faculties are responsible for our having knowledge of different
types.[96] The Understanding is the "conception of the Sensuous, or the faculty
by which we generalize and arrange the phenomena of perception"; it is the
faculty that deals with our perceptions of material objects, which we gain
through the senses.[97] The faculty of Reason is "the organ of the Super-
sensuous" in that it does not depend on the senses. Rather, Reason is the
source of necessary and universal principles of mathematics and science, as
well as the laws of thought. Understanding is the faculty that leads us to Theories, which, according to Coleridge, describe relations that are only contingent, being "the result of observation." Reason is the faculty that leads us to
Laws, which he claimed describe necessary relations between things.[98] Thus
Coleridge maintained that the Ptolemean and the Newtonian systems were
developed by different faculties: the Ptolemean by the Understanding, the
Newtonian by Reason.[99]

Whewell pointedly criticized Coleridge's view of the discoveries of
Ptolemy and Newton, calling it "altogether false and baseless." He noted that
"the Ptolemaic and the Newtonian system do not proceed from different faculties of the mind, but from the same power, exercised more and more completely."[100] As we have seen, Whewell believed that *all* knowledge requires
the use of both sensations and conceptions; he explained that "there is in
science no faculty which judges according to sense without doing more; and
no creative or suggestive faculty which must not submit to have its creations
and suggestions tested by the phenomena."[101] Further, Whewell objected to
Coleridge's denigration of the Understanding by associating it with the

95. Ibid., p. 20.
96. S. T. Coleridge, *Aids to Reflection*, in *Collected Works*, 9:252n. See also *The Friend*, in *Collected Works*, 4:158.
97. S. T. Coleridge, *The Friend*, in *Collected Works*, 4:156.
98. S. T. Coleridge, *Preliminary Treatise on Method*, p. 21. On Whewell's view, this distinction
between laws and theories is invalid because, as we will see below, no fundamental division exists
between necessary truth and empirical truth.
99. S. T. Coleridge, *Aids to Reflection*, in *Collected Works*, 9:252–53.
100. William Whewell, *Lectures on the History of Moral Philosophy, 2nd edition, with Additional Lectures on the History of Moral Philosophy*, p. 122.
101. Ibid., p. 123.

instinct of animals. (In *Aids to Reflection* Coleridge compared this faculty to the instinct of bees and ants.) In this way Coleridge made even more explicit his disdain for the faculty of perception and the empirical facts with which it deals in the attainment of human knowledge.[102] After completing one of his papers on Plato, Whewell sent it to DeMorgan, writing "I hope you will think that in the paper I send you I have demolished the Coleridgean account of *Reason* and *Understanding.*"[103]

Thus Whewell's notion of the Fundamental Antithesis was neither derived from nor greatly influenced by his reading of either Kant or Coleridge, nor was it shaped by his friendships with members of the Coleridgian circle at Cambridge. Mill's accusation that Whewell belongs to the "Germano-Coleridgian" school in epistemology indicated that he did not appreciate the bridge that Whewell had built between apriorism and empiricism. Mill focused his attention—and his attacks—solely on the a priori side of Whewell's epistemology, as we will see in chapter 2.

Discoverers' Induction

Once he had developed his antithetical epistemology, Whewell was able to construct an inductive methodology that accounted for the discovery of laws while incorporating both empirical and a priori elements. His first explicit, lengthy discussion of this inductive method is found in the 1840 first edition of *The Philosophy of the Inductive Sciences;* his view of it remained essentially unaltered through his publication of the second edition of the *Philosophy* in 1847, and the third and final edition, which appeared as three separate works between 1858 and 1860 (*The History of Scientific Ideas, Novum Organon Renovatum,* and *On the Philosophy of Discovery*).[104] Whewell called his induction Discoverers' Induction and claimed that it is used to discover laws of two types: "Laws of Phenomena" and "Laws of Causes." Laws of phenomena are, as their name suggests, laws that express *what* takes place, while laws of causes explain *why* it does. Laws of phenomena include Kepler's first law of planetary motion (that is, the orbits of all the planets

---

102. See S. T. Coleridge, *Aids to Reflection, Collected Works,* 9:243–45, and William Whewell, *Lectures on the History of Moral Philosophy, 2nd edition, with Additional Lectures on the History of Moral Philosophy,* pp. 123–26. See also Whewell's notes on Coleridge's *Aids to Reflection* (WP R.G.13 f. 39, p. 21; quoted in T. R. Levere, "Samuel Taylor Coleridge on Nature and Reason," pp. 1686–89).

103. William Whewell to Augustus DeMorgan, February 2, 1858, WP O.15.47 f. 23.

104. One change in his view, which I shall discuss below, is that by the early 1850s, Whewell had come to believe that our intellectual development consists in our approximating closer and closer to the Ideas that exist in the Divine Mind.

describe elliptical curves). As an example of a causal law, Whewell often cited Newton's law of universal gravitation, which is taken to explain why Kepler's first law is true.[105] Phenomenal laws must be discovered before their associated causal laws can be found. Thus Whewell claimed that since the phenomenal laws of reflection, refraction, and dispersion had been discovered, the causal law for optical phenomena could be sought.[106]

In describing the inductive process by which laws are discovered, Whewell began by noting in the *Philosophy* that the standard view of induction holds that it is "the process by which we collect a General Proposition from a number of Particular Cases."[107] However, he rejected this overly narrow notion of induction, casting induction in the light of the Fundamental Antithesis by arguing that in scientific discovery, it is not the case that "the general proposition results from a mere juxta-position [sic] of the cases" (that is, from simple enumeration of instances).[108] Rather, Whewell explained that "there is a New Element added to the combination [of instances] by the very act of thought by which they were combined."[109] As Jones had suggested years before, induction was described by Whewell in the various editions of the *Philosophy* as an "act of the intellect" that includes both observation and reasoning.[110] In the *Philosophy*, Whewell coined the term "colligation" to describe this "act of thought."

Colligation, Whewell explained, is the mental operation of bringing together a number of empirical facts by "superinducing" upon them a fundamental conception that unites them and renders them capable of being expressed by a general law. The conception provides the "true bond of Unity," tying the phenomena together by providing a property shared by the known members of a class. (Note that in the case of causal laws, this can be the property of sharing the same cause.)[111] We have already seen that in his 1831 review of Herschel's *Preliminary Discourse*, Whewell used a metaphor sug-

---

105. See William Whewell, *Novum Organon Renovatum*, p. 118.

106. Ibid., p. 119.

107. William Whewell, *The Philosophy of the Inductive Sciences*, 2:48.

108. Ibid.; emphasis added.

109. Ibid.

110. See William Whewell, *Novum Organon Renovatum*, p. 36, and compare to letter from Richard Jones, March 7, 1831, WP Add.Ms.c.52 f. 26.

111. William Whewell, *The Philosophy of the Inductive Sciences*, 2:46. In his review of the *Novum Organon Renovatum*, Augustus De Morgan similarly interpreted Whewell's view of colligation. He defined induction as reasoning in which "the general conclusion [is] established by bringing in, separately all its particulars"; and he noted that Whewell's discoverers' induction involves "finding out what to bring them into" ("Review of Whewell's *Novum Organum* [sic] *Renovatum*," p. 44).

gesting that the facts of nature are like pearls, and the conception is the string upon which the pearls can be threaded. Without the string we have nothing but an "indiscriminate" heap of pearls, while with the string we have something ordered and beautiful.

In order to colligate facts with a conception, we must first have suitable facts. Whewell noted that it is necessary to have "already obtained a supply of definite and certain Facts, free from obscurity and doubt."[112] It is useful at this point to consider precisely what Whewell meant by a "fact." In some places he oversimplified the situation, as when he claimed that the colligation of facts involves establishing a connection "among the phenomena which are presented to our senses."[113] This suggests that "facts" are simply observed phenomena. However, the situation is complicated for Whewell by his antithetical epistemology. As we have seen, his view of the nature of knowledge entails that all observation is mediated by our Ideas and conceptions; thus there can be no conception-free observation. Whewell explained,

> An activity of the mind . . . according to certain Ideas, is requisite in all our knowledge of external objects. We see objects, of various solid forms, and at various distances from us. But we do not thus perceive them by sensation alone. Our visual impressions cannot, of themselves, convey to us a knowledge of solid form, or of distance from us. Such knowledge is inferred from what we see:—inferred by conceiving the objects as existing in space, and by applying to them the Idea of Space.[114]

One consequence of this, Whewell noted, is that we cannot easily separate in our perceptions of things the element that our mind contributes and the element that comes from outside our mind.[115] Indeed, it is impossible to do so.[116] Nevertheless, we can attempt to make explicit the Ideas and conceptions that are involved in our perception of specific phenomena. Once we do so, we can use those facts that have reference to the more "exact Ideas" of Space, Time, and Cause, or the conceptions associated with these Ideas, such as Position, Weight, and Number, as the "foundation of Science."[117] This is use-

---

112. William Whewell, *The Philosophy of the Inductive Sciences*, 2:26.
113. Ibid., 2:36. Whewell was being sloppy here, as the passage follows the pages in which he described the need to decompose facts.
114. William Whewell, *The History of Scientific Ideas*, 1:32.
115. William Whewell, *The Philosophy of the Inductive Sciences*, 2:27.
116. Ibid., 2:30.
117. Ibid., 2:30–32.

ful because these Ideas and conceptions are particularly definite and precise.[118] The operation by which we separate complex facts into more simple
facts exhibiting the relations of Space, Time and Cause, is called the *Decom-
position of Facts.*[119] Whewell described it as a method of "render[ing] observation certain and exact."[120] He outlined a number of "Methods of Observation" that can be used in this process.[121]

Once we have decomposed facts, we can examine them with regard to
other Ideas and conceptions. According to Whewell, we can bind together the
facts by applying to them a "clear and appropriate" conception; although
such a conception may not be as exact and precise as the conceptions associated with Space, Time, and Cause, it must still be clear enough to be capable
of giving "distinct and definite results."[122] Conceptions used in colligation
must be not only clear, but also "appropriate" to the facts involved. As
Whewell pointed out, attention is generally not given to this aspect of discovery. He noted that "the defect which prevents discoveries may be the want
of suitable ideas, and not the want of observed facts."[123] An "appropriate" or
"suitable" conception is one that expresses a property or cause shared by the
facts which it is used to unify. As Whewell put it, conceptions must be "modifications of that Fundamental Idea, by which the phenomena can really be
interpreted."[124] Often scientists apply an inappropriate conception to a set
of facts, as when astronomers prior to Kepler applied the conception of epi-
cycles to colligate planetary motions. This was not the appropriate conception, because planetary motions do not, in fact, share the property of following in epicyclic orbits. Scientific discoveries are made not merely when
accurate observations are obtained, as was the case after Tycho Brahe's observations of the orbit of Mars, but when in addition to accurate observations
the appropriate conception is used, as when Kepler applied the conception of
an ellipse rather than that of the epicycle. Whewell observed that finding this
correct conception is often the most difficult part of discovery, the part that

118. Ibid., 2:38.
119. Ibid., 2:33.
120. Ibid., 2:35.
121. Ibid., 2:337–58. Whewell's interest in methods of observation was not merely theoretical
but also practical. In the mid-1830s he organized a vast international project of tidal observations
for the Royal Society, which awarded him a gold medal in recognition of his efforts in 1837. As part
of this work Whewell wrote a *Memorandum and Directions for Tide Observations.* For more on
Whewell's work on tidal science, see my "Whewell and the Scientists."
122. William Whewell, *The Philosophy of the Inductive Sciences,* 2:39.
123. William Whewell, "Remarks on the Review of the *Philosophy of the Inductive Sciences*"
(reply to DeMorgan), p. 7.
124. See William Whewell, *Novum Organon Renovatum,* p. 30.

gives it its "scientific value." As he noted, Tycho Brahe already had access to the "facts," but only Kepler was able to determine that these facts could be correctly colligated by the conception of an ellipse. Another example used by Whewell is Aristotle's inability to account for the mechanical forces in the lever because he attempted to use the conception of a circle to colligate the known facts regarding the proportion of the weights that balance on a lever. Archimedes later showed that this was an inappropriate conception, and instead used the idea of pressure to colligate these facts.[125] The problem with other notions of induction—most notably that of Whately—is that they omit precisely this crucial and difficult step of finding the appropriate conception (or, in Whately's characterization, the minor premise).

How does a scientist discover the appropriate conception with which to colligate a group of facts? Whewell believed this to be an essentially rational process, but one that is not susceptible to being put in algorithmic form. What is most important to this process is that the discoverer's mind contains a clarified or "explicated" form of the appropriate conception. The explication of conceptions is necessary because, although Whewell claimed that the Ideas and their conceptions are provided by our minds, they cannot be used in their innate form. Indeed, in an 1841 paper read before the Cambridge Philosophical Society, "Demonstration That All Matter Is Heavy," he denied that his Fundamental Ideas were innate ideas in the typical sense of the term. Unlike innate ideas, the Fundamental Ideas are not "self-evident at our first contemplation of them."[126] Later, Whewell introduced the term *germs* to describe the original form of the conceptions in our minds. In the third edition of *The Philosophy of the Inductive Sciences*, Whewell explained that "the Ideas, the germ of them at least, were in the human mind before [experience]; but by the progress of scientific thought they are unfolded into clearness and distinctness."[127]

Interestingly, Whewell began to use a metaphor—that of the germ theory of organic origins—which Kant had used in his earlier writings, but then rejected by the time of the B edition of the first *Critique*. The theory of the "pre-existence of germs," or the *Keim* theory, held that the germs of all future generations were contained within the present one, and that these germs are "unfolded" in the parental organism. This theory was given its most explicit

125. See William Whewell, *History of the Inductive Sciences*, 1:71–72.
126. William Whewell, "Demonstration That All Matter Is Heavy," p. 530.
127. William Whewell, *On the Philosophy of Discovery*, p. 373. See also William Whewell, *Novum Organon Renovatum*, pp. 30–49.

formulation in the mid-eighteenth century by Charles Bonnet, and was supported by Georges Cuvier.[128] In his early writings Kant defended the *Keim* theory and applied it to his writings on anthropology and epistemology.[129] In his anthropology lectures of 1775–76, he noted that "innate to human nature are germs which develop and can achieve the perfection for which they are determined. . . . Who[ever] has seen a savage Indian or Greenlander, should he indeed believe that there is a germ innate to this same [being] to become just such a man in accordance with Parisian fashion. . . . ? He has, however, the same germs as a civilized human being, only they are not yet developed."[130]

Kant related the *Keim* theory to his epistemology in the first edition of the first *Critique*, where he claimed that the pure concepts of the Understanding are to be "followed to their first *Keime und Anlagen* in the human understanding where they lie prepared until developed on the occasion of experience and through one and the same understanding, freed from the attending empirical conditions, are displayed in their purity."[131] However, Kant soon shifted his position somewhat, probably upon the influence of the philosopher Johann Gottfried von Herder. Herder held the opposing "epigenesis" view, on which there was no need for a preformed *Keim* to account for organic generation, only organic matter acted upon by dynamic living forces. In his review of Herder's *Ideen zur Philosophie der Geschichte der Menschheit* (1785), Kant took a middle position between the pre-existence of germs and the epigenesis theory. Around this time he started revising the first *Critique*. In the B edition, Kant used the notion of the "epigenesis of reason" to account for how the categories could be determinate on nature while not being derived from it. At B166 he wrote, "There are only two ways in which we can account for a *necessary* agreement of experience with the concepts of its objects: either experience makes these concepts possible, or these concepts make experience possible."[132] Kant rejected the first of these possibilities, because the categories are a priori, that is, independent of experience, so to ascribe to them an empirical origin "would be a sort of *generatio aequiv-*

128. In his *Considérations sur les corps organisés* (1762), Charles Bonnet argued that God created a multitude of germs, each containing or encapsulating an embryonic organism. These embryonic organisms in turn possess yet smaller individuals, down through a whole series of encapsulated organisms, enough to populate the world until the Second Coming (see Robert J. Richards, "Kant and Blumenbach on the *Bildungstrieb*," p. 15).

129. See Phillip Sloan, "Preforming the Categories."

130. Immanuel Kant, *Die Vorlesung des Wintersemesters 1775/76, Kant's Gesammelte Schriften*, Prussian Academy Edition (Berlin: Reimer, 1911), 25:694, translated by Felicitas Munzel; quoted in ibid., p. 241.

131. Immanuel Kant, *Kritik der reinen Vernunft*, A66, quoted (and translated) in Phillip Sloan, "Preforming the Categories," p. 241.

132. Immanuel Kant, *Critique of Pure Reason* B166, p. 174.

oca." But the second is "a system . . . of the *epigenesis* of reason—namely, that the categories contain, on the side of the understanding, the grounds of the possibility of all experience in general."[133] This is the position Kant advocated.[134] He mentioned, but rejected, a third "middle way," which was a version of the *Keim* theory he had earlier endorsed. This middle way holds that the categories are neither derived from experience nor are self-thought a priori principles, but are "implanted in us from the first moment of our existence, and so ordered by our Creator that their employment is in complete harmony with the laws of nature in accordance with which experience proceeds—a kind of *preformation-system* of pure reason."[135]

By the mid-1850s, Whewell had begun to endorse a version of the middle way rejected by Kant. We have seen that he faced a problem similar to Kant's, namely, to account for why it is that the Fundamental Ideas existing in our minds can allow us to have knowledge of our experience. Indeed, his problem was even more difficult, since on his view the Fundamental Ideas allow us to have knowledge of not only our experience of nature but also the true structure of the world. Whewell is a realist about our knowledge of the world, as he reassured Herschel (who had questioned his commitment to realism in his review of the *Philosophy* and the *History*).[136] Our Idea of Space, he explained, is what enables us to conceive things as existing in space with spatial characteristics; but the reason we conceive things as so existing in space is that "they do so exist."[137] Objects in the world really do exist in spatial relations to each other, and with spatial characteristics of size and shape. Whewell came to argue that our Fundamental Ideas correspond to the world because

133. Ibid., B167.

134. As Phillip Sloan points out, the version of epigenesis supported by Kant at this stage was a limited one. He did not believe that formless matter was acted upon by vital forces; rather, he held that there was some a priori structure to this matter. Later, however, under the influence of Johan Friedrich Blumenbach, Kant rejected completely even this aspect of the germ theory, and more fully accepted the epigenesis view. This is the position he used in the *Critique of Judgment*, though he seemed to continue to rely on some kind of preformed *Keim* in the moral realm (see Sloan, "Preforming the Categories," pp. 246–50, and also Robert J. Richards, "Kant and Blumenbach on the *Bildungstrieb*").

135. Immanuel Kant, *Critique of Pure Reason*, pp. 174–75. For Kant, the decisive argument against this middle way was that it would sacrifice the necessity (that is, logical necessity) of the categories. But for Whewell this was not a problem, as he did not argue that the Fundamental Ideas or the necessary truths that flowed from them were logically necessary, as I show below.

136. When a person has knowledge about something, Whewell elsewhere claimed, "we conceive that he knows it because it is true, not that it is true because he knows it" ("On the Fundamental Antithesis of Philosophy," p. 479). I thus disagree with Gerd Buchdahl's reading of Whewell as a conventionalist about knowledge in his "Deductivist versus Inductivist Approaches in the Philosophy of Science as Illustrated by Some Controversies between Whewell and Mill."

137. William Whewell, "Remarks on a Review of the *Philosophy of the Inductive Sciences*" (letter to John Herschel, April 11, 1844), p. 488.

both Ideas and world have a common origin in a divine creator. Our minds contain the germs of these Ideas because God "implanted" these germs within our minds. Thus, although Whewell was not motivated to write about science by his theological interests, as I argued in the introduction, it is certainly the case that his mature view of science rested upon a theological foundation.

This theological foundation for Whewell's epistemology was inspired by the scientific work of his childhood friend, the comparative anatomist Richard Owen.[138] In the early to mid-1840s, Owen developed his theory of the archetypal vertebrate.[139] The theory was first published in 1846 in his *Anatomy of Fishes* and developed further in *On the Archetype and Homologies of the Vertebrate Skeleton* (1848) and *On the Nature of Limbs* (1849). Owen argued that individual vertebrate animals could be seen as modified instantiations of patterns or archetypical forms that existed in the Divine Mind. This "unity of plan" view allowed Owen to explain homologies, that is, similar structures that have different purposes, such as the wing of a bird and the forelimb of a quadruped. According to Owen, the similarity of structures used in diverse ways was a consequence of their being variations on the vertebrate archetype. Such homologies did not, therefore, cast doubt on the claim that the creator of the universe designed each of the structures within it. Similarly, structures without apparent purpose, such as male nipples, did not contradict the claim that all of creation was designed; rather, they were the result of the application of general archetypes.[140]

Whewell used his friend's theory in two ways. In *Of the Plurality of Worlds*, first published in 1853, Whewell argued against the widespread view that intelligent life existed on other planets, and against the mainstream position that the argument from design entailed the plurality position.[141] He claimed that the existence of a vast, mostly unpopulated universe was not a sign of waste, and hence did not controvert the design of the universe by a

138. In his "Whewell's *Philosophy of Discovery* and the Archetype of the Vertebrate Skeleton," Phillip Sloan argues that Whewell's philosophy influenced Richard Owen's development of his archetype theory, in that Owen's concept of the archetype was both empirical and transcendental (see pp. 58–59).

139. For more details regarding the development of Owen's theory and its appearances in his lectures in the early 1840s, before its publication, see ibid., pp. 55–57.

140. For more on Owen, see Nicolaas Rupke, *Richard Owen.*

141. Twenty years earlier, in his Bridgewater Treatise, Whewell had supported the plurality position, using it as further evidence of design in the universe. But even in 1833 Whewell was not strongly wed to the plurality view. By 1853, he was vehemently against the position, for reasons I discuss in my "'Gegen alle vernunftbegabten Bewohner anderer Welten.'"

beneficent and powerful God. Rather, he argued, uninhabited celestial bodies were products of an overall pattern used by God in designing the universe. While Whewell resisted following Owen in claiming that the unity of plan *proves* design, he did hold that it allows us to see that apparent divergences from adaptation and direct purpose cannot be used as evidence against design.[142] The second way in which Whewell used Owen's archetype theory was to provide a justification for his epistemology.

This justification received its most complete exposition in five chapters originally written for the *Plurality*, which Whewell deleted at the very last moment before publication.[143] (He took this drastic step on the advice of his friend Sir James Stephen, who had been Regius Professor of History at Cambridge since 1849, and who was concerned that these chapters were too "metaphysical" for the general reader.) He added a discussion of this justification in the last volume of the final edition of *The Philosophy of the Inductive Sciences: On the Philosophy of Discovery*, published in 1860; presumably, the intended audience for this book could be expected to follow metaphysical expositions.[144] Whewell argued there that the Divine Mind contains many archetypical ideas, in accordance with which the universe was created. As he put it, God "exemplified" in his creation certain Ideas existing in his mind, by creating the universe in accordance with these Ideas.[145] For example, God exemplified an Idea of Space in his universe by creating all physical objects as having spatial characteristics, and as existing in spatial relations to each other. The Idea of Space, then, became on Whewell's view an archetypical Divine Idea similar to the Idea of the vertebrate skeleton: it was no less embodied in the physical world. In creating us in his own image, God implanted us with the germs of the Divine Ideas. Whewell claimed that "our Ideas are given to us by the same power which made the world." We can know the world because God has created us with the Ideas needed to know it. More precisely,

142. See William Whewell, *Of the Plurality of Worlds*, p. 213. In the third edition of his *History of the Inductive Sciences*, published in 1857, Whewell adjusted his earlier claims about the importance of teleological explanations in biological science in accordance with Owen's "unity of plan" theory.

143. Whewell wrote to Richard Jones, "I am myself disposed to believe that my book is well written, because I wrote it with pleasure and facility. When you come to the *hard* chapters I expect you would be staggered—the more you ought to admire me for my self denial in cutting them out after they were written—indeed, after they were printed" (December 30, 1853, WP Add.Ms.c.51 f. 278).

144. See William Whewell, *On the Philosophy of Discovery*, chaps. 30 and 31.

145. Ibid., pp. 371–79, and *Of the Plurality of Worlds*, printer's proofs including five chapters not included in the published version, WP Adv.c.16 f. 27, p. 277. These chapters are included in the edition of the *Plurality of Worlds* edited by Michael Ruse.

our minds are created with the "germs" from which the Ideas can develop.[146] God, then, has implanted within us the germs that need to be unfolded into Ideas representing those archetypical Divine Ideas upon which the universe was created. Gaining a clear view of these Ideas enables us to have knowledge of both the natural world and its Creator.

Because the Fundamental Ideas exist innately within us only in the form of germs that must be unfolded, the explication of Ideas and conceptions is a crucial part of science. Explication of conceptions is a necessarily social process, proceeding by discussion and debate among scientists; it is not a process that can take place solely within the mind of an individual genius. This idea was represented pictorially by Whewell's "signature" engraving, which appeared on the title page of his major works: an image of one hand passing a torch to another, accompanied by the Greek saying "Those who carry the light pass it to one another."[147] Whewell noted in the *Philosophy* that disputes concerning different kinds and measures of Force were important in the progress of mechanics, and the conception of the Atomic Constitution of bodies was currently being debated by chemists.[148] He explained that by arguing in favor of a particular meaning of a conception, scientists are forced to clarify and make more explicit what they really mean. This is beneficial, whether or not the original expression prevails. If it does not, then it is replaced by a more accurate or more clear expression of the conception. If it does, the original expression has been improved by the scientists' efforts. Thus, like Bacon, Whewell argued that "the tendency of all such controversy is to diffuse truth and to dispel error. Truth is consistent, and can bear the tug of war; Error is incoherent, and falls to pieces in the struggle."[149] He claimed that the explication of conceptions is a "necessary part of the inductive movement."[150] Indeed, a large part of the history of science is the "history of scientific ideas," that is, the history of their explication and subsequent use as colligating concepts.[151] This is why, even though Whewell worried about

146. William Whewell, *On the Philosophy of Discovery*, p. 373.
147. Herschel had earlier noted that "knowledge can neither be cultivated nor adequately enjoyed by a few" (*Preliminary Discourse on the Study of Natural Philosophy*, p. 69).
148. William Whewell, *The Philosophy of the Inductive Sciences*, 2:6–7.
149. Ibid., 2:7. Indeed, Whewell ended his *Plurality of Worlds* with the explicitly Baconian exhortation for nations to bind together in cooperation, in order to extend "man's intellectual empire" (see *Of the Plurality of Worlds*, p. 423). He noted that "if the nations of the earth were to employ, for the promotion of human knowledge, a small fraction only of the means, the wealth, the ingenuity, the energy, the combination, which they have employed in every age, for the destruction of human life and of human means of enjoyment; we might soon find that what we hitherto know, is little compared with what man has the power of knowing" (p. 424).
150. William Whewell, *Novum Organon Renovatum*, p. vii.
151. William Whewell, *The History of Scientific Ideas*, 1:16.

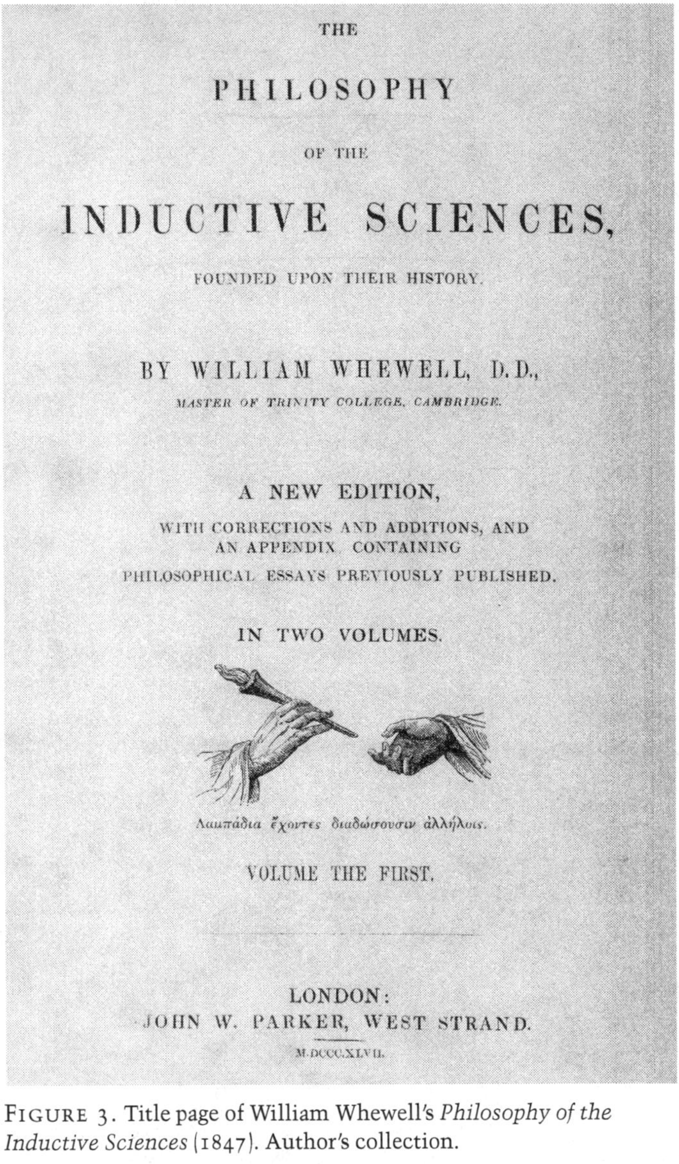

THE

PHILOSOPHY

OF THE

INDUCTIVE SCIENCES,

FOUNDED UPON THEIR HISTORY.

BY WILLIAM WHEWELL, D.D.,

MASTER OF TRINITY COLLEGE, CAMBRIDGE.

A NEW EDITION,

WITH CORRECTIONS AND ADDITIONS, AND
AN APPENDIX CONTAINING
PHILOSOPHICAL ESSAYS PREVIOUSLY PUBLISHED.

IN TWO VOLUMES.

Λαμπάδια ἔχοντες διαδώσουσιν ἀλλήλοις.

VOLUME THE FIRST.

LONDON:
JOHN W. PARKER, WEST STRAND.
M.DCCC.XLVII.

FIGURE 3. Title page of William Whewell's *Philosophy of the
Inductive Sciences* (1847). Author's collection.

his philosophy being considered too metaphysical by Jones and Herschel,
ultimately he believed that metaphysics is a necessary part of science at all
stages; he explained that "the explication or . . . the clarification of men's
ideas. . . . [is] the metaphysical aspect of each of the physical sciences."[152]

152. William Whewell, *Novum Organon Renovatum*, p. vii.

Disagreeing with Auguste Comte, who had claimed that at its highest level of advancement the intellect becomes "scientific" by purging itself of metaphysics, Whewell argued that successful discoverers differ from "barren speculators," not by rejecting metaphysics, but by employing "*good* metaphysics" rather than bad.[153]

Explication of conceptions is, then, a process that occurs by discussion and debate among groups of scientists. Yet scientific discoveries are, Whewell noted, generally made by individuals. Why is it that particular individuals are able to discover the appropriate conception to apply to a set of facts? Whewell pointed to a facility for "invention," the quality of "genius"; yet he strongly denied that there is anything "accidental" about scientific discoveries. He explicitly opposed the view popularized by David Brewster in his *Life of Newton* (1831) that the discovery of the law of universal gravitation was a "happy accident." (In his review of Whewell's *History of the Inductive Sciences*, Brewster criticized him in turn, claiming that, contrary to Whewell's assertions, "It cannot, we think, be questioned that many of the finest discoveries in science have been the result of pure accident.")[154] Whewell stressed that discovery is always preceded by much intellectual and scientific preparation. Thus he noted that Kepler was able to discover the elliptical orbit of Mars because his mind contained the clear and precise Idea of Space, and the conception of an ellipse derived from this Idea. Whewell explained that "to supply this conception, required a special preparation, and a special activity in the mind of the discoverer. . . . To discover such a connection, the mind must be conversant with certain relations of space, and with certain kinds of figures."[155] That Kepler's mind was more "conversant" with the long-explicated conceptions derived from the Idea of Space, such as that of the ellipse, explains why he was able to recognize that the ellipse was the appropriate conception with which to colligate the observed points of the Martian

153. Ibid.

154. David Brewster, "On the History of the Inductive Sciences," p. 121. Brewster's review was so nasty that Whewell felt obliged to write a letter to the editor of the *Edinburgh Review,* objecting to Brewster's reference to Cambridge as "the cloisters of antiquated institutions, through whose iron bars the light of knowledge and liberty has not been able to penetrate" (see William Whewell, letter to editor of the *Edinburgh Review,* WP 289.c.80.84 f. 14). In a letter to Richard Jones, Whewell claimed that Brewster's article is merely "an angry remonstrance in favour of his rights unjustly withheld"—that is, because Whewell did not give Brewster enough credit for his contributions to science. Whewell's publisher John Parker called the review Brewster's "Scotch grumble" for its suggestion that Whewell slighted the contributions of all Scottish men of science. As Whewell noted, Brewster conveniently overlooked Whewell's glowing remarks about the contributions of J. D. Forbes, Brewster's Scottish "rival." (Whewell to Jones, September 6, 1837, Add.Ms.c.51 f. 209).

155. William Whewell, *Of Induction,* pp. 28–29.

orbit, while Brahe and his assistant Logomontanus (who had made the most precise observations) were not.[156]

Another useful quality for the discoverer to possess is a certain facility in generating a number of possible options for the appropriate conception—often a necessary step before the appropriate conception can be ascertained and applied to the facts. Whewell sometimes used the terminology of "guesses" to describe this stage of the discovery process. In part because of this, most twentieth-century commentators, for the most part, have incorrectly viewed his methodology as similar to the "method of hypothesis" (or, as it is now known, the "hypothetico-deductive" method). On that view, no rational inference is required to arrive at a hypothesis; its formation is generally described as a "guess."[157] However, Whewell claimed that the *application* or the *selection* of the appropriate conception, in Kepler's case and in all cases of discovery, is not a matter of guesswork. He described this process as being one in which "trains of hypotheses are called up and pass rapidly in review; and the *judgment* makes its choice from the varied group."[158] Thus, even though at a certain point in his investigation (once he had inferred that the Martian orbit was some type of oval) Kepler called up in his mind "nineteen hypotheses" of possible ovals, his choice of the appropriate ellipse conception was based on "calculations," and hence on a rational process—and, certainly, his conclusion that the orbit was oval involved much rational inference.

Nor is the selection of an appropriate conception a matter of mere observation. Whewell claimed that choosing the appropriate conception requires more than this: "there is a special process in the mind, in addition to the mere observation of facts, which is necessary."[159] This "special process of the mind" is inference. In order to colligate facts with the appropriate conception, "we infer more than we see."[160] Whewell claimed that in the case of choosing the conception of force to colligate the observed motions of a needle toward

156. For a more detailed discussion of how Kepler's discovery conformed to Whewell's methodology, see my "Discoverers' Induction."

157. See, for example, Robert Butts, "Pragmatism in Theories of Induction in the Victorian Era"; Larry Laudan, "Why Was the Logic of Discovery Abandoned?" and "William Whewell on the Consilience of Inductions"; Gerd Buchdahl, "Deductivist versus Inductivist Approaches in the Philosophy of Science as Illustrated by Some Controversies between Whewell and Mill"; Peter Achinstein, "Inference to the Best Explanation"; N. R. Hanson, *Patterns of Discovery*; Carl Hempel, *Philosophy of Natural Science*; Michael Ruse, "Darwin's Debt to Philosophy"; Ilkka Niiniluoto, "Notes on Popper as a Follower of Whewell and Peirce"; and John Wettersten, "Discussion: William Whewell." For the most recent example, see Richard Yeo, "Whewell, William."

158. William Whewell, *The Philosophy of the Inductive Sciences*, 2:42; emphasis added.

159. William Whewell, *Of Induction*, p. 40. See also "Modern Science—Inductive Philosophy," p. 379, and *On the Philosophy of Discovery*, pp. 256–57.

160. William Whewell, *The History of Scientific Ideas*, 1:46.

a magnet, an inference, or an "interpretative act" of the mind, is required.[161] Inference is required before we see that force is the appropriate conception with which to colligate magnetic phenomena. Since the selection of the appropriate conception with which to colligate the facts involves inference, Whewell noted that discoveries are made "*not* by any capricious conjecture of arbitrary selection."[162]

Indeed, selecting the appropriate conception typically requires not just one inference, but a series of inferences. This is why Whewell claimed that discoverers' induction is a process involving a "train of researches."[163] He explicitly rejected limiting inductive discovery to enumerative inference, writing that "induction by mere enumeration can hardly be called induction."[164] Clearly, as we have seen, enumerative inference—by which we infer from our observation of many black crows that, probably, all crows are black—could not, in most cases, account for the discovery of the appropriate conception with which to colligate the data. Rather, Whewell allowed that any form of valid inference could be used, and throughout his writings he especially stressed the power of analogical reasoning. He extolled the importance of analogical inference in two reviews from the early thirties, his review of Herschel's *Preliminary Discourse* and his discussion of the second volume of Charles Lyell's *Principles of Geology.*[165] Decades later, in his *Of the Plurality of Worlds*, he more systematically discussed the importance of defining the varying degrees of precision and relevance of different kinds of analogies. Not everyone was pleased with this liberal notion of inductive reasoning. Perhaps making a snide reference to Whewell's origin as the son of a carpenter, DeMorgan complained, in his 1847 logic text, about certain writers using the term "induction" as including "the use of the whole box of [logical] tools."[166]

We have seen how Whewell's discoverers' induction allows the inference to an unobservable elliptical *shape* for the orbit of Mars. His view also allows for the postulation of unobservable or theoretical *entities*, such as light waves moving in a luminiferous ether. Whewell often used the example of the wave theory of light to illustrate a well-formed theory (even though, as we

161. See ibid.,1:31, 45.

162. Ibid., 1:29.

163. William Whewell, *History of the Inductive Sciences*, 1:326. See also Andrew Lugg, "History, Discovery and Induction."

164. William Whewell, "Criticism of Aristotle's Account of Induction," p. 451.

165. See William Whewell, "Modern Science—Inductive Philosophy," p. 385, and "Lyell's *Principles of Geology*, Volume 2," p. 110. See also *The History of the Inductive Sciences*, 1st ed., 2:391.

166. Augustus DeMorgan, *Formal Logic; or, The Calculus of Inference*, p. 216.

will see in chapter 3, he admitted that this theory was not yet conclusively confirmed). I will follow his example and use the wave theory in order to illustrate the ability of discoverers' induction to lead to inferences to unobservable entities. In the next chapter we will see that Mill rejected these kinds of inferences.

In his *History of the Inductive Sciences*, Whewell explained that Optics is one of the "Secondary Mechanical Sciences" (along with Acoustics and Thermotics), because in these sciences facts are not presented to the senses as modifications of position and motion, but as secondary qualities. Moreover, the phenomena are reduced to their laws in a "secondary" manner, by treating them as the "operation of a *medium* interposed between the object and the organ of sense."[167] Thus in these sciences laws are discovered by colligating the facts with the appropriate conception of such a medium. In the case of Acoustics, this was accomplished much earlier, as it was known from the time of Aristotle that the medium of communication of sound was air. Aristotle conceived of air as acting like a rigid body, for example a staff, transmitting an impulse from one end to the other.[168] Later, this conception of air as the medium of transmission was conceived more correctly as being like a liquid, transmitting sound by circular waves. Yet it was not understood precisely how this worked, until Newton introduced the conception of an undulation or wave, which Whewell described as being an "exact and rigorous conception."[169]

Developing a theory of the nature of light involved the selection of an appropriate conception or Idea of a medium through which we perceive objects. As Whewell noted, it had long been accepted that there were two possible hypotheses regarding the nature of light: the particle, or emission, theory, and the wave theory. He described these as "two hypotheses of the nature of the luminiferous medium . . . the one representing light as Matter emitted from the luminous body, the other, as Undulations propagated through a fluid."[170] Humphrey Lloyd, in a well-known review article of 1834 (to which Whewell referred in his *History*),[171] showed that this claim can be inferred from two observed facts: (1) light travels from one point in space to another with a finite velocity; and (2) in nature one observes this type of finite motion occurring either by the motion of a body or by the motion of vibrations or undulations propagated through a medium. The inference involved is enumer-

167. William Whewell, *History of the Inductive Sciences*, 2:233.
168. William Whewell, *The Philosophy of the Inductive Sciences*, 1:308–9.
169. Ibid., 1:310.
170. Ibid., 1:315.
171. William Whewell, *History of the Inductive Sciences*, 2:369n.

ative: from the observed fact that known types of finite motion occur in only two different ways, it is inferred that the observed finite motion of light occurs in one of these two ways—that is, either by the motion of bodies (light particles) or by vibrations through a fluid medium (light waves in the ether). Wave theorists then used eliminative reasoning to infer that the wave theory is probably true, by showing that the particle theory is probably false.[172] Because inference was used to conclude the probable existence of an unobservable ether, Whewell believed that the wave theorists were arguing in accordance with his methodology.

It will now be useful to discuss briefly a case in which Whewell *rejected* the postulation of an unobservable entity on the grounds that the case failed to satisfy the requirements of discoverers' induction. He did not accept the atomic hypothesis, that is, the hypothesis that all bodies are composed of indivisible, unobservable atoms, even though accepting it would enable us to "explain the occurrence of definite and multiple proportions" (Dalton's law), and even though the hypothesis has other true empirical consequences as well.[173] He had two reasons for not accepting the atomic hypothesis: first, because the chemical facts do not yield "any *inference* with regard to the existence of certain smallest possible particles," and second, because the facts do not enable us to eliminate a second possible hypothesis. Whewell explained that "the assumption of *indivisible* particles, smaller than the smallest observable, which combine, particle with particle, will explain the phenomena; but the assumption of particles bearing this proportion, but *not* possessing the property of indivisibility, will explain the phenomena at least equally well."[174] Scientists were unable to make inferences from the observations to the atomic theory, including those inferences that could eliminate alternative competing hypotheses. Thus Whewell's inductive methodology led to the rejection of unobservable atoms as the cause of "definite and multiple proportions" (pending further evidence), because it was not at that time possible to make well-supported inferences to these unobservable entities.[175]

---

172. Augustin Fresnel, for example, made this kind of argument in his famous "Memoir on the Diffraction of Light" of 1818. Whewell discusses the importance of Fresnel's work on diffraction in his *History of the Inductive Sciences*, 2:326–27. See also Peter Achinstein, *Particles and Waves*, chap. 3.

173. William Whewell, *The History of Scientific Ideas*, 2:49.

174. Ibid., 2:49–50.

175. This point provides further support for my claim that Whewell's discoverers' induction is not a form of hypothetico-deductivism.

Renovating Bacon

We have seen that Whewell explicitly framed his reformist project in Baconian terms. Yet it might seem that his discoverers' induction, because of the antithetical epistemology underlying it, is inherently opposed to Bacon's inductive method. Herschel, who shared Whewell's desire to renovate Bacon, thought that Whewell's epistemology vitiated his Baconianism, and accused him of having abandoned their shared project. Later commentators have read Whewell in a similar way, as having rejected inductivism for some kind of hypothetical methodology by the time he published *The Philosophy of the Inductive Sciences* (if not sooner).[176] However, as we have seen, this is not the case. In the *Philosophy*, Whewell clearly endorsed an inductive view of discovery. Moreover, he continued to support an inductive methodology in later works. In *Of Induction* (1849), his work responding to Mill's criticisms of him in *System of Logic*, Whewell noted, in opposition to Mill's characterization of Kepler, that new hypotheses are properly "collected from the facts," and not merely guessed.[177] In his *Of the Plurality of Worlds*, published in 1853, he strongly criticized the "bold assumptions," both "arbitrary and fanciful," that led some people to claim there was good reason to believe in the existence of intelligent life on other worlds. Since we had no inductive evidence for this, Whewell claimed, we should not engage in such "conjectures" and "speculations."[178] In his 1857 review article of the Spedding, Ellis, and Heath edition of the collected works of Bacon, Whewell continued to praise Bacon for emphasizing the gradual successive generalization that Whewell believed characterized the historical progress of science. He also stressed this in the third edition of his *History of the Inductive Sciences*, published the same year. In his "Additions" to the first volume, Whewell explained that "laborious observation, narrow and modest inference, caution, slow and gradual advance, limited knowledge, are all unwelcome efforts and restraints to the mind of man, when his speculative spirit is once roused: yet they are the necessary conditions of all advance in the Inductive Sciences." He criticized the "bold guesses and fanciful reasonings of man unchecked by doubt

176. Richard Yeo has argued that by the time of writing his review of Herschel's *Preliminary Discourse*, Whewell had already abandoned an early endorsement of inductivism in favor of a hypothetical method (see *Defining Science*, pp. 98–99). Menachem Fisch similarly argues that Whewell abandoned his commitment to Baconianism at an early stage of his career (see *William Whewell, Philosopher of Science*, p. 61).

177. See William Whewell, *Of Induction*, p. 17.

178. William Whewell, *Of the Plurality of Worlds*, pp. 41, 122, 141, and passim.

or fear of failure."[179] And, to take one final example, in *On the Philosophy of Discovery*, published in 1860, Whewell referred to the belief that "the discovery of laws and causes of phenomena is a loose hap-hazard sort of guessing," and claimed that this type of view "appears to me to be a misapprehension of the whole nature of science."[180] Indeed, as time went on Whewell had more and more reason to be opposed to the use of hypotheses in science. In his two reviews of Charles Lyell's *Principles of Geology* from the early 1830s, he attempted to "calm down" speculations in that field, which was beginning to separate itself from Revelation, by stressing the importance of a gradual inductive progress.[181] In later decades, as we will see in chapter 3, Whewell cautioned against illegitimate evolutionary hypotheses.

It is also notable—especially for someone so concerned with the role of language in science—that Whewell consistently continued to call himself an inductivist, and his philosophy an inductive one. It seems unlikely that Whewell, who argued that scientific terminology "fixes" discoveries by connoting definitions of explicated conceptions, would continue to characterize his view as inductive if he had decided it was really hypothetical or deductive instead.

Thus I believe that Whewell never abandoned his commitment to some form of Baconian inductivism, although he certainly believed that his own view had improved upon Bacon's. I will here show that there are several central aspects of Whewell's discoverers' induction that mirror Bacon's inductive methodology.[182] Whewell echoed Bacon in claiming that hypotheses are invented by inference from the data. Moreover, he agreed with Bacon's emphasis upon the "gradual and continuous ascent" to hypotheses.[183] He claimed

179. William Whewell, *History of the Inductive Sciences*, 1:339–40.

180. William Whewell, *On the Philosophy of Discovery*, p. 274.

181. For instance, he criticized Lyell's uniformitarianism—which rejected the need for miracles in explaining the origin of organic life on earth—by noting that "his theory must speedily fall back into the abyss of pure fantasies and guesses from which he has evoked it in vain: leaving us the conviction, that many less sweeping propositions and intermediate generalizations must be firmly established on innumerable observations, before we can hope to ascend to one general theorem, including . . . all terrene changes throughout endless ages" (William Whewell, "Lyell's *Principles of Geology*, Volume 1," p. 204).

182. More details of the comparison between Bacon and Whewell can be found in my "Renovating the *Novum Organon*."

183. In the third edition of the *Philosophy*, Whewell agreed with Bacon's insistence "upon a *graduated and successive induction*, as opposed to a hasty transit from special facts to the highest generalizations" (*On the Philosophy of Discovery*, p. 130; see also pp. 131 and 145). In his *History* Whewell similarly noted that the inductive sciences proceed by "successive steps": they are not "formed by a single act. . . . On the contrary, they consist in a long-continued advance; a series of charges; a repeated progress from one principle to another" (*History of the Inductive Sciences*, 1:7–8).

that this is where Bacon's importance and originality lie, not in the claim that knowledge must be sought in experience, which many others prior to Bacon had professed.[184]

Whewell's conception of science followed the gradualism of Bacon in two respects. First, the method of inventing hypotheses from data involves a "connected and gradual process" of inference, namely, discoverers' induction.[185] The second way in which Whewell followed Bacon's gradualism concerns his view of the history of science. Whewell claimed that the progress of science over time is slowly cumulative, in the sense that theories of progressively greater generality are derived from less general theories once those are proved to be true. In this way, Whewell noted, Kepler's law of planetary motion, invented by discoverers' induction, was used by Newton to derive a law of greater generality, the universal inverse-square law of gravitation.[186] The development of the theory of electromagnetism is another case in the history of science pointed to by Whewell as exemplifying the Baconian process of gradual generalizing.[187] His own "Inductive Tables" are meant to illustrate the successive generalization that occurs over the history of science. In his earlier works Whewell even called them Inductive Pyramids, echoing Bacon's use of the term *Pyramids of Knowledge* to describe the process of successive generalization in the sciences.[188] Whewell's Inductive Tables can be constructed only for sciences that have achieved a large degree of generality, that is, where phenomenal laws have been subsumed under causal laws of greater generality, and these causal laws are themselves seen as instances of laws of even greater generality. According to Whewell, the two sciences

184. William Whewell, "Spedding's Complete Edition of the Works of Bacon," p. 158. Whewell thus rejected the French tradition (exemplified in the writings of Étienne de Condillac) of adopting Bacon as the "patron saint of the Enlightenment" by claiming that he *was* the first to emphasize the origin of all knowledge in sense experience (see Richard Yeo, *Defining Science*, p. 182).

185. Richard Yeo claims to find a "tension" between Whewell's endorsement of Bacon's gradualism and his claim that, in Yeo's words, there is an "inductive *leap* from disconnected facts to unifying theory" (*Defining Science*, p. 13). Ernan McMullin has made a similar point to me in personal correspondence. However, I do not believe that there is any incompatibility between these claims. The same moment of discovery is described psychologically, as feeling like a "mental leap" to the scientist who suddenly sees the facts under a new point of view, and logically, as being a rational move supported (but not entailed) by the data. Thus Whewell explained in a notebook entry, "Imagination which jumps from something near to something far remote is strongly opposed to the inductive spirit which never takes an unsupported step.—What then are brilliant discoveries? They always imply much knowledge of subordinate generalizations—" (WP R.18.17 f. 12, p. xx).

186. William Whewell, *On the Philosophy of Discovery*, p. 182; see also Richard Yeo, "An Idol of the Marketplace," p. 273.

187. William Whewell, "Spedding's Complete Edition of the Works of Bacon," pp. 159–60.

188. Whewell explicitly drew this comparison between Bacon's pyramids and his own Inductive Tables in his *On the Philosophy of Discovery*, p. 132.

able to support the formal structure of the Inductive Tables thus far are astronomy and (to a certain extent) optics. So central to Whewell's discoverers' induction is its gradualism that Whewell claimed these tables present the "Logic of Induction, that is, the formal conditions of the soundness of our reasoning from the facts."[189] He suggested by such comments that the soundness of our reasoning from facts to law is the greater the more gradual is the generalization, and that this soundness is therefore exhibited in the formal structure of the tables.[190]

Whewell's discoverers' induction shares another important feature of Bacon's method of interpretation: namely, the claim that the inference from data to hypothesis is not limited to inductive generalization. We have seen that, on Whewell's view, the selection of the appropriate conception with which to colligate the data involves an inferential process, and that this process can involve any type of inference. Bacon too argued that mere inductive generalization (enumerative induction) was not sufficient; he criticized the "logic of the schoolmen" as being "incompetent."[191] Nor was it enough merely to add eliminative induction. Since this aspect of Bacon's thought is not widely recognized, it deserves some detail here.

In seeking "forms of simple natures"—that is, underlying causes for the simple properties of things, such as the underlying cause of heat—Bacon believed that the natural philosopher must construct "Natural Histories" consisting of tables of presences, absences, and variations of the simple nature under investigation. The investigator must first collect a table of presences, which lists bodies that are dissimilar from one another except for sharing the nature under investigation. Bacon's table of presences for his own "Investigation into Heat" included the rays of the sun, boiling water, flame, hairy animals, certain herbs, and others.[192] Once the table of presences is constructed, the investigator must collect a table of absences. This table consists of instances that are each similar to one of the instances of presence but differ in lacking the nature in question.[193] In the case of heat, the table of absences constructed by Bacon included rays of the moon, stars, and comets, which are instances similar in many respects to the rays of the sun, but which

189. Ibid., p. 207.

190. See also ibid., p. 134, and *Novum Organon Renovatum*, p. 115; and letter to Augustus DeMorgan, January 18, 1859, in Isaac Todhunter, *William Whewell, D.D.*, 2:416–17. Indeed, Whewell criticized Bacon for not being sufficiently gradualist in his depiction of science. See William Whewell, *On the Philosophy of Discovery*, pp. 136–37.

191. Francis Bacon, *Works*, 3:387; see also 4:410.

192. See ibid., 4:127–29.

193. See ibid., 4:129–37.

lack the quality of heat. The third type of table required is a table of degrees or variations. In constructing such a table the investigator describes instances in which the nature varies to a greater or lesser degree. Bacon's table of variations for heat included circumstances under which animal bodies increase in heat.[194]

The construction of the table of presences implicitly involves enumerative induction, since it contains general instances. (These instances are types, not tokens, in the philosophical jargon.) But what is more, the construction of the table of absences involves analogical reasoning. Both positive and negative analogies are employed in determining instances to include in this table. For example, rays of the sun and rays of the moon are known to share several properties, including those of being rays of light and emanating from celestial bodies. These shared properties comprise the *positive* analogy between rays of the sun and rays of the moon. But sun rays and moon rays disagree in some qualities as well, including the fact that rays of the sun feel warm to us and rays of the moon do not. This *negative* analogy suggests that the rays of the moon do not share with rays of the sun the quality of heat. (This is not entailed, however, for it is possible that the heat of moon rays might just be of a lesser degree than that of sun rays, below the threshold for our sensation of heat.) This suggestion can be tested by the use of another type of analogical reasoning, which Bacon called learned experience (*experientia literata*).[195] What this means is that the investigator argues by analogy from a situation in which a particular kind of experiment was fruitful, to another, similar situation in which the same kind of experiment may also be fruitful (in the sense of yielding informative results). Thus, in the case of rays of the sun and rays of the moon, it is a known experimental result that passing the rays of the sun through a magnifying lens so intensifies heat that the rays can set combustible material on fire.[196] Learned experience suggests to us that we might attempt an experiment passing the rays of the moon through a magnifying lens, to see if heat is thus "actuated to any degree."[197] If it is not,

---

194. Ibid., 4:139.

195. Interestingly, this notion resembles closely a legal concept with which Bacon would certainly have been familiar, that of legal reasoning. See Rose-Mary Sargent, *The Diffident Naturalist*, p. 46, and Barbara Shapiro, "Sir Francis Bacon and the Mid-17th Century Movement for Law Reform." In seventeenth-century law, "legal reasoning" meant the kind of general experience with the law that would allow a good lawyer to discern similarities and differences between past and present cases, in order to decide whether it was reasonable to apply the precedents of past cases to the present one. Bacon used this concept in his discussion of scientific method, explaining that learned experience "proceeds from one experiment to another" (Francis Bacon, *Works*, 4:413).

196. Francis Bacon, *Works*, 4:414. For another example of learned experience see 4:131.

197. Ibid., 4:414.

this would seem to confirm the conclusion that rays of the moon do not share the quality of heat with rays of the sun, and thus that moon rays belong on the table of absences and not the table of presences in the investigation of heat.

After the three tables have been compiled, the second step of Bacon's method occurs. Eliminative inference is used on the instances in the tables to rule out possible forms that are not both necessary and sufficient for the appearance of the given nature. Every nature or quality that is not common to all instances of presence of heat, or that is present even when heat is absent, or that does not vary as the quantity of heat varies, is excluded.[198] Bacon recognized that the exclusion could only be complete if the tables were complete and correct, including every instance of presence, every corresponding instance of absence, and every case of variation. He seemed to believe that this is an ideal toward which science should aim, but which is impossible in its early stages and may always remain so.[199] Meanwhile, the natural philosopher may perform an incomplete exclusion, followed by an "Indulgence of the Understanding." This results in the provisional postulation of a form, which Bacon termed the First Vintage or Commencement of the Interpretation.[200] The First Vintage of Heat, Bacon concluded, is this: "Heat is motion, expansive, restrained, and acting in its strife upon the smaller particles of bodies. But the expansion is thus modified; in while it expands all ways, it has at the same time an inclination upwards. And the struggle in the particles is modified also; it is not sluggish, but hurried and with violence."[201]

However, it is obvious that any method of exclusion based on *observed* instances of presence, absence, and variation could not result in even the provisional acceptance of an *unobservable* form, such as the motion of unobservable particles. How then is the natural philosopher to postulate an unobservable form? Bacon did not endorse pure guesswork—the postulated form must be obtained, as he put it, "on the strength both of the instances given in the tables, and of any others it may meet with elsewhere."[202] He thus suggested that the investigator must infer a form on the basis of the evidence remaining after the exclusion.[203] In his section discussing what he called

198. See ibid., 4:147–49.

199. Ibid., 4:149. This is why it is clear that Bacon does not think that, in practice, his method can reach certainty.

200. Ibid., 4:149.

201. Ibid., 4:154–55; emphasis removed.

202. Ibid., 4:149.

203. Mary Horton calls this move a kind of nonrational "intuition"; thus she claims, contrary to my position, that Bacon's method of induction "is fundamentally and explicitly an alogical process" (see "In Defense of Francis Bacon," pp. 245–55 and 262–63).

"Prerogative Instances," Bacon explained that analogical inference is "employed, when things not directly perceptible are brought within reach of the sense, not by perceptible operations of the imperceptible body itself, but by observation of some cognate body which is perceptible."[204] Further, he noted, "information is to be derived from [such instances] in the absence of instances proper."[205] Since there are no "instances proper"—that is, observed instances—in the case of the search for forms, it appears that discovering forms requires analogical reasoning. In reaching the First Vintage of heat, Bacon was therefore making an analogical inference from instances of heat in which motion is observed to co-occur on the macro level, such as in the instances of flame, boiling water, and rays of the sun, to the existence of imperceptible motion on the micro level, both in these cases and in instances of heat where motion is not observed at the macro level, such as in the cases of certain herbs. Even cases in which motion does not seem obvious at the macro level can be seen as analogous to his proposed form of heat: the heated iron does expand, as do gases under pressure. While Bacon was far from explicit on this point, seeing him as endorsing the use of analogical reasoning here is consistent both with his result and with his remarks about analogy in his discussion of Prerogative Instances.[206] Thus Bacon—like Whewell later—allowed enumerative, eliminative, and, significantly, analogical reasoning as part of his method of discovery.

It has sometimes been claimed that Bacon believed the First Vintage to be infallibly certain, not merely provisional. This was a popular interpretation of his view among nineteenth-century commentators in England, perhaps

204. Francis Bacon, *Works*, 4:203.
205. Ibid., 4:204.
206. John Maynard Keynes recognized the importance of analogical reasoning in Bacon's method in his *A Treatise on Probability*, p. 268. For more on analogical reasoning in Bacon's method, see Mary Hesse, "Francis Bacon's Philosophy of Science," p. 127; Lisa Jardine, *Francis Bacon*, p. 144; Ernan McMullin, "Conceptions of Science in the Scientific Revolution," pp. 53–54; Katherine Park et al., "Bacon, Galileo and Descartes on Imagination and Analogy," pp. 297–99; and Antonio Perez-Ramos, *Francis Bacon's Idea of Science and the Maker's Knowledge Tradition*, p. 293, and "Francis Bacon and the Disputations of the Learned," pp. 584–85. Although Whewell himself did not explicitly mention the use of analogical inference in Bacon's method, it is likely that his view of Bacon was influenced by that of his young friend and protégé J. D. Forbes, who wrote an essay on the importance of analogical inference in Bacon's induction a few years before meeting Whewell in 1831("On the Inductive Philosophy of Bacon, His Genius, and Atchievements" [sic], in "Moral Philosophy Essays," 1827–1828, Item 4, Box V–VI, in correspondence and papers of J. D. Forbes at St. Andrews University Library; quoted in Richard Olson, *Scottish Philosophy and British Physics, 1750–1880*, pp. 225–26). A later paper, "On the Refraction and Polarization of Heat," which Whewell discussed in letters to Forbes, argues for the importance of analogical reasoning in scientific discovery. See letters from Forbes to Whewell in J. C. Shairp et al., *Life and Letters of James David Forbes, FRS*, pp. 115–17, and from Whewell to Forbes in Isaac Todhunter, *William Whewell, D.D.*, 2:203–4.

because Robert Ellis expressed the opinion in his general preface to the *Works*.[207] However, there is much textual evidence that this was not Bacon's view at all, as both Whewell and Herschel recognized.[208] Bacon claimed that the First Vintage must be tested in order to be confirmed, disconfirmed, or corrected. This step is especially crucial given his caution regarding analogical inference, that it is "useful, but is less certain" than other forms of reasoning.[209] He explained in the *Valerius Terminus* that an axiom discovered by the method of interpretation can be confirmed when it leads to or predicts consequences that are not yet known to be true but are found to be so afterwards.[210] He made similar claims in *De Augmentis* and the *Novum Organum*.[211] Moreover, when Bacon noted the goal of the interpretation of nature as being the creation of "works," he claimed that he intended works to function as "pledges of truth" of the interpretation—that is, as evidence that the interpretation is indeed a "true model of the world." Thus he wrote

207. See Robert Ellis, "General Preface to the Philosophical Works," p. 24. In his review of this edition of the *Works*, Augustus DeMorgan repeated Ellis's accusation (see "The Works of Francis Bacon," p. 368). The author of the entry on Bacon in the ninth edition of the *Encyclopaedia Britannica* expressed this view as well (R. Adamson, "Bacon," p. 187). On the other hand, prior to Ellis's preface, this view of Bacon was not universal in Britain. Robert Small, in his 1804 commentary on Kepler's *Astronomia Nova*, saw Bacon rather differently, as in this passage assimilating Kepler's methodology to Bacon's: "This work [the *Astronomia Nova*] . . . exhibited, even prior to the publication of Bacon's *Novum Organum* . . . [the] legitimate connection between theory and experiment; of experiments suggested by theory, and of theory submitted without prejudice to the test and decision of experiments" (*An Account of the Astronomical Discoveries of Kepler*, p. 2).

208. See William Whewell, *On the Philosophy of Discovery*, p. 139, and John Herschel, *Preliminary Discourse on the Study of Natural Philosophy*, pp. 104–74.

209. See Francis Bacon, *Works*, 4:203.

210. Bacon wrote that "the discovery of new works and active directions not known before is the only trial to be accepted of; and yet not that neither, in case where one particular giveth light to another; *but [only] where particulars induce an axiom or observation*, which axiom found out discovereth and designeth new particulars. . . . The nature of this trial is not only upon the point, whether the knowledge be profitable or no, but even upon the point whether the knowledge be true or no; not because you may always conclude that the Axiom which discovereth new instances is true, but contrariwise you may safely conclude that if it discover not any new instance it is in vain and untrue" (*Works*, 3:242; emphasis added). It is incorrect to read this passage as expressive of the hypothetico-deductive position, as Mary Horton does ("In Defense of Francis Bacon," p. 256). Antonio Perez-Ramos calls this aspect of Bacon's method "hypothetical," but in doing so he uses the term differently than other twentieth-century writers, who equate "hypotheses" with "bold conjectures" or "guesses" that may be reached nonrationally. Perez-Ramos applies the term "hypothesis," rather, to statements that have been reached through a gradual, rational process, but are corrigible (see *Francis Bacon's Idea of Science and the Maker's Knowledge Tradition*, pp. 256–7n). This use of the term is consistent with the usage up through the nineteenth century in Britain. More recently, Stephen Gaukroger reads this passage as evidence of Bacon's emphasis on the production of works, even though it seems that Bacon is denying this explicitly (see *Francis Bacon and the Transformation of Early-Modern Philosophy*, pp. 155–56).

211. Francis Bacon, *Works*, 4:413; see also p. 343. For a similar claim in the *Novum Organum*, see 4:98.

that "works themselves are of greater value as pledges of truth than as contributing to the comforts of life."[212]

Whewell, like Bacon, believed that hypotheses obtained inductively must
be tested by their empirical consequences. He described three confirmation
tests: prediction, consilience, and coherence. (I will discuss these tests in
chapter 3.) Also like Bacon, he believed that science could yield knowledge
about the unobservable part of the natural world. Bacon claimed that his
interpretation of nature could discover the unobservable forms of simple
natures. Whewell, as we have seen, intended his method to allow for the discovery of hypotheses referring to unobservable entities and properties. For
example, he claimed that Augustin-Jean Fresnel and the other wave theorists
had good inductive grounds for postulating the existence of unobservable
light waves in an unobservable ether, and scoffed at Mill's rejection of the
wave theory on the basis of its postulation of these unobservable entities.

Although I have shown that Whewell correctly regarded his discoverers'
induction as following in the tradition of Bacon's inductivism, he certainly
also recognized that their views were not identical. From the time of his
earliest notebooks on induction, he expressed the need to improve upon Bacon's inductivism. In 1836, he wrote to Herschel that the *Novum Organum*
"requires both to be accommodated to the present state of thought and
knowledge, and to have its vast vacuities gradually supplied."[213] Indeed, Whewell believed himself to be renovating Bacon's method. This renovation takes
the form of grafting his antithetical epistemology onto Bacon's empirical
methodology. However, this does not constitute a contradiction of Bacon's
view; for, as we will see, there are seeds of an antithetical epistemology already in Bacon's works, as Whewell often pointed out.

Whewell's epistemology entails that certain ideal conceptions, as well as
facts, are necessary materials of knowledge. We have already seen two ways
in which conceptions are crucially involved in the discovery of empirical
laws according to Whewell. First, conceptions are involved in the very process of perception; Whewell claimed that all perception is "conception-
laden." Second, conceptions are necessary to form theories from facts in the
process of colligation. The appropriate conception must be superinduced
upon, or applied to, the facts in order to bring the facts together under a general law. But there is a third way in which Whewell emphasized the role of

212. Ibid., 4:110. See also C. J. Ducasse, "Francis Bacon's Philosophy of Science," p. 296n12,
and Lisa Jardine, "*Experientia Literata* or *Novum Organum*?" p. 57.
213. William Whewell to John Herschel, April 9, 1836, in Isaac Todhunter, *William Whewell,
D.D.*, 2:234.

conceptions in science. He noted that some conceptual framework is necessary in order to guide the collection of empirical data. In his address to the
British Association meeting in 1833, Whewell claimed that "it has of late
been common to assert that *facts* alone are valuable in science. . . . [But] it is
only through some view or other of the *connexion* and *relation* of facts, that
we know what circumstances we ought to notice and record."[214] That is, we
cannot and do not collect facts blindly, without some theory or conception
guiding our choices for what to include and exclude from the collection of
data.[215]

At times, Whewell argued that Bacon did not adequately take note of the
antithetical nature of knowledge; he did not "give due weight or attention to
the ideal element in our knowledge."[216] Other times, Whewell claimed that
Bacon did not ignore the conceptual side of knowledge altogether. The problem is that Bacon never was able to complete his task of reforming philosophy: "if he had completed his scheme, [he] would probably have given due
attention to Ideas, no less than to Facts, as an element of our knowledge."[217]
It has been suggested that Whewell more or less invented this reading of Bacon
in order to "detach" the British inductive tradition from French positivism.[218]
However, this conceptual element *can* be found in Bacon's writings in various
ways, though not as explicitly as it is developed in Whewell's philosophy.

Bacon claimed, for instance, that he "established for ever a true and last-
ing marriage between the empirical and the rational faculty."[219] He elaborated
on this marriage in his famous aphorism urging the scientist to emulate
the bee:

> Those who have handled sciences have been either men of experiment or
> men of dogmas. The men of experiment are like the ant; they only collect
> and use: the reasoners resemble spiders, who make cobwebs out of their
> own substance. But the bee takes the middle course; it gathers its mate-
> rial from the flowers of the garden and of the field, but transforms and

214. William Whewell, "Address," p. xx.
215. For example, Whewell explained that "the laws of the tides have been in a great measure
determined by observations in all parts of the globe, *because* theory pointed out what was to be
observed. In like manner the facts of terrestrial magnetism were ascertained with a tolerable com-
pleteness by extended observations, *then*, and then only, when a most recondite and profound
branch of mathematics had pointed out what was to be observed, and most ingenious instruments
had been devised by men of science for observing" (*On the Philosophy of Discovery*, p. 155).
216. Ibid., p. 135.
217. Ibid., p. 136.
218. Richard Yeo, *Defining Science*, p. 247. On this point see also Antonio Perez-Ramos, *Francis Bacon's Idea of Science and the Maker's Knowledge Tradition*, p. 26.
219. Francis Bacon, *Works*, 4:19.

digests it by a power of its own. Not unlike this is the true business of philosophy; for it neither relies solely or chiefly on the powers of the mind, nor does it take the matter which it gathers from natural history and mechanical experiments and lay it up in the memory whole, as it finds it; but lays it up in the understanding altered and digested. Therefore from a closer and purer league between these two faculties, the experimental and the rational, (such as has never yet been made) much may be hoped.[220]

Notice that the "blind" gathering of facts that Whewell criticized is characteristic of the "men of experiment," or the "empirics," whose methods Bacon rejected in the above passage.[221] Moreover, Bacon pointed to the importance of the conceptual side of knowledge—in a way similar to Whewell—when he cautioned that forms cannot be discovered until conceptions or "notions" are clarified.[222] Such comments evoke Whewell's claim that laws cannot be discovered until conceptions are clarified or explicated. Whewell was aware that Bacon made this point; indeed he chastised Bacon for ignoring his own advice. Whewell complained that in his investigation into heat, "his collection of instances is very loosely brought together; for he includes in his list the *hot* taste of aromatic plants, the *caustic* effects of acids, and many other facts which cannot be ascribed to heat without a studious laxity in the use of the word."[223]

To be sure, there is one unresolvable difference in their respective views on the role of conceptions in science, and that concerns Whewell's claim that perception itself is conception-laden. Bacon famously warned against the imposition of our internal concepts upon the external world. He admonished that "all depends on keeping the eye steadily fixed upon the facts of nature and so receiving their images simply as they are. For God forbid that we should

220. Ibid., 4:92–93; see also Paolo Rossi, "Ants, Spiders, Epistemologists," p. 255.

221. Elsewhere, Bacon similarly noted that "experience, when it wanders in its own track, is . . . mere groping in the dark, and confounds men rather than instructs them" (*Works*, 4:95). In another passage he complained that "the manner of making experiments which men now use is blind and stupid . . . wandering and straying as they do with no settled course, and taking counsel only from things as they fall out, they fetch a wide circuit and meet with many matters, but make little progress" (ibid., 4:70; see also 4:81 and Paolo Rossi, "Bacon's Idea of Science" and "Ants, Spiders, Epistemologists," p. 250). In a footnote added to *De Augmentis*, James Spedding rejected the proposal that Bacon intended fact-collecting to occur in the absence of any theory (Francis Bacon, *Works*, 1:623n1). Spedding suggested that Bacon's notion of *experientia literata* was meant to provide a provisional theory of the collection of facts. (Lisa Jardine agrees with this point in her "*Experientia Literata or Novum Organum?*" pp. 60, 63n15.) See also Rose-Mary Sargent, *The Diffident Naturalist*, p. 34.

222. See Francis Bacon, *Works*, 4:49–50 and 4:61–2.

223. William Whewell, *On the Philosophy of Discovery*, p. 139.

give out a dream of our imagination for a pattern of the world."[224] Bacon claimed that in constructing our natural histories we must record phenomena that correspond as much as possible to pure, non-conception-laden observations; moreover, he seemed to believe that the correspondence can be quite high.[225] His "new way," the method of interpretation of nature, is intended to begin "directly from the simple sensuous perception."[226] Human-derived conceptions cannot, on his view, aid us in understanding a God-made world.[227] In contrast with this view, Whewell believed, as we have seen, that our conceptions aid us in perceiving and understanding the created world precisely because they correspond in some degree to Ideas in the Divine Mind. Whewell rejected the straightforwardly empiricist epistemology of Bacon for the same reasons he opposed the views of Locke and the "Sensationalist school." He replaced Bacon's purely empiricist epistemology with his own antithetical epistemology. Yet there are elements of Bacon's view that allow for the importance of the conceptual, as well as the empirical, side of knowledge. Whewell had reason to believe that this epistemological alteration was more of an organic extension of Bacon's philosophy than an outright rejection of it.

Because Bacon did not adequately cultivate the conceptual side of knowledge, Whewell claimed, he was led into another error: namely, the false notion that there can be a purely mechanical method of discovery.[228] Bacon began his work with the claim that what was new about his *Novum Organum* was the realization that "the entire work of the understanding [must] be commenced afresh, and the mind itself be from the very outset not left to its own course, but guided at every step; and the business be done as if by machinery."[229] He compared his task to that of providing a compass for the purpose of enabling any person to draw a perfect circle.[230] Such comments have been taken to suggest that Bacon sought to develop what Robert Hooke later called a "philosophical algebra." Yet, on Whewell's view, "no maxims can be given which inevitably lead to discovery. No precepts will elevate a man of ordinary endowments to the level of a man of genius."[231] As we have seen, Whewell's

224. Francis Bacon, *Works*, 4:32–33.
225. See Lisa Jardine, *Francis Bacon*, p. 135. There seems, however, to be a conflict between this optimism and his claim that the Idols of the Cave and of the Tribe cause us to "distort and discolor" our sensory experience of nature.
226. Francis Bacon, *Works*, 4:40.
227. Ibid., 4:110.
228. William Whewell, *On the Philosophy of Discovery*, p. 138.
229. Francis Bacon, *Works*, 4:40.
230. Ibid., 4:62–63.
231. William Whewell, *Novum Organon Renovatum*, p. 94.

antithetical epistemology entails the need for the existence of clear conceptions in the mind of the discoverer; and while there are methods to aid in the
clarification and selection of the appropriate conception, there is no algo-
rithm or mechanical method for this process. This is why invention requires
"genius." Whewell criticized Bacon for believing that a discovery method
could "supersede" genius.[232]

However, it is not clear that Bacon's goal actually was to give a "philosophical algebra." Many of the elements of his method of interpretation cannot be reduced to a mechanical rule. The construction of the tables of presence, absence, and variation cannot be; Bacon himself described *experientia
literata,* which is a crucial tool in their construction, as "rather a sagacity,
and a kind of hunting by sense, than a science."[233] Moreover, after the more
or less mechanical exclusion is performed, it is still not obvious what the
true form is. One needs to postulate a provisional form, seemingly by analogical reasoning, for which Bacon gives us no set of mechanical rules. His
comments about creating a discovery "machine" can perhaps be read as expressing the intention to create a machine that we may use to supplement
our creative rationality, not to supplant it, just as a compass aids the hand
in drawing a circle but does not render the hand itself unnecessary.[234] Such a
view is suggested by the following passage in the preface to the *Novum Organum:* "Certainly if in things mechanical men had set to work with their
naked hands, without help or force of instruments, just as in things intellectual they have set to work with little else than the naked forces of the understanding, very small would the matters have been which . . . they could have
attempted or accomplished."[235]

This is not so far from Whewell's own view. That Whewell denied the possibility of a mechanical method of discovery does not entail his rejection of

232. William Whewell, "Spedding's Complete Edition of the Works of Bacon," p. 158. The
author of the Bacon article in the 1875 edition of the *Encyclopaedia Britannica* agreed with this
criticism of Bacon, arguing that "the minds of various investigators can never be reduced to the
same dead mechanical level" (R. Adamson, "Bacon," p. 187). On the other hand, in his presidential
address to the British Association for the Advancement of Science in 1859, Prince Albert endorsed
this reading of Bacon by praising the goal of a discovery machine: he described inductive method
as "reasoning upwards from the meanest fact established, and making every step sure before going
one beyond it, like the engineer in his approaches to a fortress. We thus gain ultimately a roadway,
a ladder by which even a child may, almost without knowing it, ascend to the summit of truth"
(quoted in Donald Benson, "Facts and Constructs," p. 299). Whewell criticized Hooke for following Bacon in this desire for a mechanical discovery method (*On the Philosophy of Discovery,*
p. 171).

233. Francis Bacon, *Works,* 4:421.

234. See also Ernan McMullin, "Conceptions of Science in the Scientific Revolution," p. 83.

235. Francis Bacon, *Works,* 4:40.

a rational discovery method. His method does not contain rules that are universally applicable, but it does offer rational "instruments" for aiding in discovery, although invention and genius cannot be dispensed with, it is still possible to "analyse and methodize the process of discovery" to some extent.[236] Thus, in a letter to DeMorgan, Whewell noted that it is possible to have an "*Art* of Discovery"; he quipped, "if I had £20,000 a year which might be devoted to the making of discoveries, I am sure that some might be made."[237] Indeed, Whewell began the *Novum Organon Renovatum* with the claim that there may be "an effectual and substantial method of Scientific Discovery." Thus regarding the issue of a "discovery machine," Bacon's view does not need to be renovated by Whewell as much as he believed.

It is interesting that Whewell and Herschel, who as schoolmates both wished to renovate and publicize Bacon's inductivism, had very different views of what Baconian induction entailed. (They agreed, however, in rejecting the dismissive view of Bacon promulgated by T. B. Macaulay in his 1837 essay "Lord Bacon.")[238] In his review of Herschel's *Preliminary Discourse*, Whewell accused his friend of having omitted to relay Bacon's "condemnation of the method of *anticipation*, as opposed to that of gradual induction; a judgment indeed which of itself almost conveys the whole spirit and character of his philosophy."[239] What Whewell considered expressive of "the whole spirit and character" of Bacon's inductive method is his injunction against hasty anticipations and his endorsement of gradual generalizations. And Herschel, in his *Preliminary Discourse*, did ignore or at least underplay the importance of this aspect of Bacon's inductivism. He claimed that one of the legitimate methods for discovering a law was "by forming at once a bold hypothesis, particularizing the law, and trying the truth of it by following out its consequences and comparing them with facts."[240] He characterized Dalton's discovery as involving only "the contemplation of a few instances, with-

236. William Whewell, *The Philosophy of the Inductive Sciences*, 1st ed., 2:186–87. See also Richard Yeo, *Defining Science*, p. 164.

237. William Whewell to Augustus DeMorgan, February 14, 1859, in Isaac Todhunter, *William Whewell, D.D.*, 2:416.

238. Macaulay's essay was an attack on Bacon, both personally and philosophically. Belittling Bacon's claim of having put forth a "New Organ" of scientific reasoning, Macaulay claimed that induction was merely an instinctive mode of reasoning practiced by all humans, "even the very child at the breast," who uses induction to expect milk from the mother and not from the father ("Lord Bacon," p. 194). Bacon had, in fact, explicitly rejected this type of criticism, as when he explained that "better things. . .are to be expected from man's reason and industry and direction and fixed application, than from accident and animal instinct and the like" (Francis Bacon, *Works*, 4:99).

239. William Whewell, "Modern Science—Inductive Philosophy," p. 399.

240. John Herschel, *Preliminary Discourse on the Study of Natural Philosophy*, p. 199.

out passing through subordinate stages of painful inductive ascent by the intermedium of subordinate laws."[241] Whewell was worried that his friend might seem to be encouraging a spirit of "gratuitous theorizing" by not cautioning against anticipatory leaps to hypotheses. To some extent, Herschel *was* guilty of this. In a letter to Whewell some years later he wrote, "I remember it was a saying often in my father's mouth 'Hypotheses fingo' in reference to Newton's 'Hypotheses *non* fingo' and certainly it is this facility of framing hypotheses if accompanied with an equal facility of abandoning them which is the happiest pattern of mind for theoretical speculation."[242] This is where Whewell disagreed, and where he saw his friend as having strayed from the path of their "Master," Bacon.[243]

Nine years later, Herschel had the chance to return the charge of anti-Baconianism. When Whewell's *Philosophy of the Inductive Sciences* was first published, Herschel wrote to him: "Your book is a tough one—when I ruminate it chapter by chapter I chew the cud of both sweet and bitter fancies—you are too a priori rather for me—as soon as one has worked one's way up to a general law you come cranking in and tell me it is a Fundamental Idea innate in everybody's mind."[244] In his lengthy review of Whewell's *History of the Inductive Sciences* and *Philosophy of the Inductive Sciences* for the *Quarterly Review*, Herschel claimed that Whewell was the one who had strayed from the path of their "Master." What he objected to in Whewell's book was the ideal or conceptual element of Whewell's Fundamental Antithesis. While Whewell had criticized Herschel for departing from Bacon's *methodology*, Herschel criticized Whewell for departing from his *epistemology*. In his review Herschel described Whewell's position as that of the "high *a priori*

241. Ibid., pp. 305–6. Whewell remarked that "such language we cannot but think is liable to be mistaken. If Mr. Dalton had *guessed* the law to be true from a few instances, and done no more, he would have been the first person whose name has been permanently connected with the history of science in virtue of such an unexamined simplification." He then quoted from Adam Sedgwick's recent address to the Geological Society: "'The records of mankind offer no single instance of any great physical truth anticipated by mere guesses and conjectures'" (William Whewell, "Modern Science—Inductive Philosophy," p. 401).

242. John Herschel to William Whewell, August 20, 1837, WP Add.Ms.a.207 f. 30.

243. Herschel's characterization of Bacon's philosophy followed, in important respects, Dugald Stewart's interpretation of Bacon's view. Stewart had argued that Bacon not only tolerated, but *required*, the use of hypotheses (see Salim Rashid, "Dugald Stewart, 'Baconian' Methodology, and Political Economy," p. 255). Pietro Corsi has pointed out to me that Stewart himself gave a copy of his *Elements* to the astronomer William Herschel, who later passed it—along with his willingness to frame hypotheses—to his son John. Although Herschel was more open to hypothetical "leaps" than was Whewell, he required inductive constraints on hypotheses, and was therefore not a proponent of what is currently known as "hypothetico-deductivism," the position that it is acceptable to make nonrational guesses to hypotheses.

244. John Herschel to William Whewell, August 6, 1840, WP Add.Ms.a.207 f. 45.

Pegasus," in contrast with the inductive philosopher, a "plain matter of fact roadster":

> The high *a priori* Pegasus . . . is a noble and generous steed who bounds over obstacles which confine the plain matter of fact roadster to tardier paths and a longer circuit. There is no denying to this philosophy, for one of its distinguishing characters, a *verve* and energy which a merely tentative and empirical one must draw from foreign sources, from a solemn and earnest feeling of duty and devotion, in its followers, and a firm reliance on the ultimate sufficiency of its resources to accomplish every purpose which Providence has destined it to attain.[245]

Herschel, then, believed that he, not Whewell, was the true inductivist (the "merely tentative and empirical one"), the legitimate heir of Bacon. He claimed to be following Bacon in denying an a priori origin to ideas or conceptions used in forming theories. In his review, he rejected Whewell's claim of an a priori source for the certainty and necessity of the axioms of statics. "What then," Herschel asked rhetorically, ". . . is the origin of our certainty of the axiom? We reply, simple experience."[246] In the next chapter we will see that Mill followed Herschel in this criticism of Whewell.

The "Ultimate Problem" of Philosophy

Whewell not only wished to describe how scientists discover empirical laws. He also sought to account for how scientists come to know necessary truths. This was a problem that had worried him since 1819, while he was completing his first textbook on mechanics. By the time he published *The Philosophy of the Inductive Sciences* in 1840, Whewell believed he had solved what he referred to in a letter to Herschel as the "ultimate problem" of philosophy of science, by showing that necessary truths can emerge in the course of empirical science.[247] This is a rather striking claim and has not been well understood by other commentators, including Mill.[248] Comprehending Whew-

245. John Herschel, "Whewell on Inductive Sciences," p. 223.
246. Ibid., p. 218.
247. William Whewell, "Remarks on a Review of *The Philosophy of the Inductive Sciences*" (letter to John Herschel, April 11, 1844), p. 489.
248. For modern attempts to characterize Whewell's notion of necessity, see Robert Butts, "Necessary Truth in Whewell's Theory of Science" and "On Walsh's Reading of Whewell's View of Necessity"; H. T. Walsh, "Whewell on Necessity"; Menachem Fisch, *William Whewell, Philosopher of Science* and "Necessary and Contingent Truth in William Whewell's Antithetical Theory of Knowledge"; and Margaret Morrison, "Whewell on the Ultimate Problem of Philosophy."

ell's solution requires viewing it in the context of his antithetical episte-mology.[249]

One way in which Whewell described the antithetical nature of knowl-edge was by claiming that "there is no fixed and permanent line" to be drawn between the empirical and ideal elements of knowledge.[250] For example, as we have seen, sensations cannot be entirely differentiated from Ideas. Whewell claimed further that there is no permanent line to be drawn be-tween fact and theory; facts are joined together by the use of an Idea to form a theory, but a true theory is itself a fact, and can be used to form theories of even greater generality.[251] The fact/theory distinction is only relative, then, because where we draw the line between them changes as we discover that our theories are true (and thus that they are "facts"). Whewell implied that the same relation holds for the pair "experiential truth and necessary truth."[252] Although there is no "fixed and permanent line" between experi-ential and necessary truths, he did allow that they can be distinguished philosophically, for the purposes of understanding them (as can the pairs "fact and theory" and "sensation and Idea"). Experiential truths are laws of nature that are knowable only empirically.[253] Necessary truths, for Whewell, are propositions expressing laws that "can be seen to be true by a pure act of thought."[254] That is, they are knowable a priori without any experience. Fur-ther, experiential truths are recognized by us as being contingent; they are such that "for anything which we can see, might have been otherwise."[255] Necessary truths, by contrast, are those "of which we cannot distinctly con-ceive the contrary."[256] These distinctions between experiential and necessary truths are epistemic, grounded upon how we come to know a truth, and whether we can conceive of its contrary. But there is also a nonepistemic dis-tinction: Whewell claimed that necessary truths "must be true"—whether

---

249. This discussion follows closely the more detailed exposition in my "It's *All* Necessarily So."
250. William Whewell, *The History of Scientific Ideas*, 1:23.
251. William Whewell, "On the Fundamental Antithesis of Philosophy," p. 467; see also *On the Philosophy of Discovery*, p. 305. Whewell characterized the relation between facts and theories in this way as early as his 1830 book on architecture. In this work, he referred to theories as "gen-eral facts," and expressed the hope that his architectural descriptions would be intelligible to "those who prefer facts to theories, that is, particular facts to general ones." See *Architectural Notes on German Churches, with Remarks on the Origin of Gothic Architecture*, p. 40.
252. Whewell explained of the set of pairs expressive of the antithesis that "the same remarks apply to it under its various forms" (William Whewell, "On the Fundamental Antithesis of Philos-ophy," p. 465).
253. William Whewell, *The History of Scientific Ideas*, 1:26.
254. Ibid., 1:60.
255. Ibid., 1:25.
256. William Whewell, "On the Fundamental Antithesis of Philosophy," p. 463, in italics.

we recognize this or not.[257] Moreover, he suggested that the epistemic criteria are reliable tests for this nonepistemic necessity. If a general proposition
satisfies the epistemic criterion for necessary truth, then we can be certain
that it must be true. Note that Whewell did not imply the converse. That is,
a proposition might be necessary in the nonepistemic sense even if it does
not meet the epistemic criteria (only, in that case, we would not know it was
a necessary truth).

Like facts and theories, however, experiential and necessary truths "are
not marked by separate and prominent features of difference, but only by
their present opposition, which is only a transient relation."[258] There is,
Whewell claimed, merely a temporary division between truths that are experiential and those that are necessary. He believed that science consists in
a process called the "idealization of facts," whereby experiential truths are
"transferred to the side" of necessary truths.[259] The same proposition moves
from one side of the Fundamental Antithesis to the other—hence the divid-
ing line between them is "transient." Whewell claimed that by this process,
"a posteriori truths become a priori truths."[260] Truths that are first knowable
only empirically become knowable a priori. Self-evident truths, then, *become*
self-evident. In order to understand how this can occur, it is important to
grasp the relation between necessary truths and the Fundamental Ideas.

Whewell believed that necessary truths, or the "axioms" of science, can
be known a priori from the Fundamental Ideas, because they are "necessary
consequences" of these Ideas.[261] As we have seen, every science is organized
by one or more Fundamental Ideas. Each Fundamental Idea has several
axioms that follow from it. Axioms are necessary consequences of an Idea in
the sense that they express the meaning of the Idea. Whewell explained that
the axioms "in expressing the primary developments of a Fundamental Idea,
do in fact express the Idea, so far as its expression in words forms part of
our science."[262] For example, one of the three axioms of the Idea of Cause
is "every event must have a cause"; and Whewell noted that "this axiom expresses, to a certain extent [because it is only one of the axioms] our Idea of
Cause."[263] The connection between an Idea and its axioms seems to be that

257. William Whewell, *The History of Scientific Ideas*, 1:25–26.
258. William Whewell, *On the Philosophy of Discovery*, p. 305.
259. See ibid., p. 303.
260. Ibid., pp. 357–58.
261. William Whewell, *The History of Scientific Ideas*, 1:99.
262. Ibid., 1:75; see also I:58 and William Whewell, *Novum Organon Renovatum*, p. 13.
263. William Whewell, *The History of Scientific Ideas*, 1:185.

the meaning of the axiom is contained in the meaning of the Idea, and expresses nothing but what is already contained in the Idea. The proposition "Causes are such that every event has a cause" is therefore analogous to the proposition "Bachelors are never-married men," where the predicate (never-married men) expresses only what is already contained in the subject (bachelors).[264] (Kant called these analytic judgments.) As we have seen, Whewell argued that a crucial part of science is the "explication" of Ideas and their conceptions. By explicating Ideas scientists gain an explicit, clarified view of the meaning of the Idea—it becomes "distinct." Once an Idea is distinct enough that its meaning is understood, the scientist can see that the axioms are necessary consequences of the Idea, by virtue of the fact that they express part of this meaning.[265] For example, when the scientist's mind contains an explicated form of the Idea of Cause, he will know that it must be true that every event has a cause.

The two epistemic criteria Whewell gave for necessary truths follow from this understanding of the relation between the Ideas and the axioms. Once the meaning of the concept "bachelor" is understood, no empirical knowledge is required to know the truth of the proposition "Bachelors are never-married men": this proposition follows from the meaning of the concept, and hence can be known a priori. (On the other hand, if the meaning of "bachelor" is not understood, the truth of this proposition will not be knowable a priori.) Further, once the meaning of "bachelor" is understood, it will not be possible to conceive of someone who is both a bachelor and married. Similarly, Whewell claimed, only someone with a distinct understanding of the Idea of Space can know a priori that two straight lines cannot enclose a space; moreover, a person who does know this a priori will be unable to conceive of two lines that are straight but yet contradict this necessary truth.[266]

This indicates yet another way that Whewell's view differs from that of Kant: Whewell's notion of necessary truth is quite different from Kant's con-

264. Whewell's claim that the Ideas "cannot be fixed in words" might seem to vitiate my claim. However, what Whewell meant is that the Ideas cannot be given complete verbal definitions, in part because it is unlikely that we will ever come to know all the axioms and definitions of conceptions that constitute the full meaning of an Idea (William Whewell, *The History of Scientific Ideas*, 1:75). We can, nevertheless, fix *part of* the meaning of an Idea in words. This is consistent with Whewell's several remarks suggesting that the axioms are "deduced" from the Ideas (see ibid., 1:64, and *On the Philosophy of Discovery*, p. 355).

265. This is why the ability to recognize the necessity of axioms is one test for the distinctness of our Ideas. See William Whewell, *The History of Scientific Ideas*, 1:101, and *Thoughts on the Study of Mathematics as Part of a Liberal Education*.

266. For the contrary claims of other commentators, and my arguments against these claims, see my "It's *All* Necessarily So."

ception of the synthetic a priori.[267] It is clear that Whewell disagreed with Kant's claim that necessary truths of mathematics are synthetic a priori. Kant explained of the truth "5 + 7 = 12" that "this concept of 12 is by no means already thought in merely thinking this union of 7 and 5; and I may analyze my concept of such a possible sum as long as I please, still I shall never find the 12 in it."[268] That is, for Kant, "5 + 7 = 12" is not an analytic truth such that by knowing the meaning of "5 + 7" we can know the truth and necessity of "5 + 7 = 12." Whewell was certainly aware of this claim of Kant's; in one of his notebooks he transcribed this discussion of the synthetic a priori nature of mathematical principles from the first *Critique*.[269] In published work, nevertheless, Whewell presented the opposing view. Using the similar example "7 + 8 = 15," Whewell claimed that "we refer to our conceptions of seven, of eight, and of addition, and as soon as we possess the conceptions distinctly, we see that the sum must be 15."[270] Merely by knowing the meanings of *seven, eight,* and *addition,* we see that it follows necessarily that "7 + 8 = 15." Hence, for Whewell, mathematical truths (like all necessary truths) are analytic and not synthetic in Kant's sense. Mansel duly complained that

Dr. Whewell lays too much stress on *clearness* and *distinctness* of conceptions as the basis of the axiomatic truths of science. But the clearness and distinctness of any conception can only enable us more accurately to unfold the virtual contents of the concept itself; it cannot enable us to add *a priori* any new attribute. In other words, the increased clearness and distinctness of a conception may enable us to multiply to any extent our analytic judgments, but cannot add a single synthetical one.[271]

Whewell's view of mathematical truth also differs from the position of William Hamilton and Dugald Stewart. They had argued that the axioms of mathematics are conventional or "hypothetical," because axioms are deduc-

267. This is contrary to the claims of Robert Butts, "Necessary Truth in Whewell's Theory of Science" (see p. 167); Menachem Fisch, *William Whewell, Philosopher of Science,* p. 157; John Metcalfe, "Whewell's Developmental Psychologism," p. 125; and M. R. Stoll, *Whewell's Philosophy of Induction,* p. 40.
268. See Immanuel Kant, *Critique of Pure Reason,* B15, p. 53.
269. William Whewell, notebook, WP R.18.19 f. 13, p. 10. Although the entry is undated, based on the date of an earlier entry it appears to date from the mid-1820s.
270. William Whewell, "On the Fundamental Antithesis of Philosophy," p. 471. Whewell made the same claim for the axioms of geometry: "the *meaning of the terms being understood,* and the proof being gone through, the truth of the proposition must be assented to" (p. 462; emphasis added).
271. H. L. Mansel, *Prolegomena Logica,* p. 258.

tive consequences of definitions that are not themselves necessary.[272] Mathematical truths such as "Two straight lines cannot enclose a space" are necessarily true (within Euclidean geometry) only because of how Euclidean geometry defines "straight line" (that is, "A straight line is that which lies evenly between its extreme points"). The axiom would not even be true, let alone necessary, if our geometry defined "straight line" as one that lies unevenly between its extreme points. What is more, their view suggests that we could have chosen to define "straight line" differently. But Whewell rejected this merely hypothetical necessity. On his view, the necessary truths that follow from definitions are necessary because they are deductive consequences of definitions which are themselves necessary. Mathematical definitions are not merely conventional.[273] Rather, they are "descriptive."[274] According to Whewell's view the definitions of geometry and arithmetic describe the properties of certain mathematical conceptions such as point, line, circle, and number. These conceptions are necessary consequences of the Ideas of Space and Time.[275] Like the axioms, the conceptions of an Idea are "included" in the meaning of the Idea. (Hence their definitions express part of its meaning.)[276] When we verbalize the definition of "straight line," we are not conventionally *assigning* the properties of straight lines, but are expressing or *describing* what these properties really are.[277] We could not correctly define "straight line" in any other way.

Like most British thinkers prior to the late 1870s, Whewell considered geometrical definitions descriptive not only of mathematical conceptions that exist in our minds, but of physical reality as well; geometry was thus

272. See William Whewell, *Mechanical Euclid*, p. 151, and *The History of Scientific Ideas*, 1:107–8. Stewart had applied this account of truth to science as well, claiming that the laws deduced with the aid of axioms were purely hypothetical (see Pietro Corsi, *Science and Religion*, p. 157).

273. See William Whewell, *Novum Organon Renovatum*, pp. 36 and 39, *The History of Scientific Ideas*, 1:74–75, and Whewell's letter to Frederic Myers (September 6, 1845) in Janet Stair Douglas, *The Life and Selections from the Correspondence of William Whewell, D.D.*, p. 327. For William Hamilton's view see his "Review of [Whewell's] *Thoughts on the Study of Mathematics as Part of a Liberal Education*." Hamilton's position led him to deny Whewell's claim that mathematics is an important subject of study for university students; according to Hamilton, mathematics is too purely formal to be of use in developing basic reasoning skills.

274. See Joan Richards, *Mathematical Visions*, pp. 22–23.

275. William Whewell, *Novum Organon Renovatum*, pp. 30–31, and *The History of Scientific Ideas*, 1:74.

276. William Whewell, *The History of Scientific Ideas*, 1:75.

277. Whewell claimed that "a definition, to be admissible, must necessarily refer to and agree with some conception which we can distinctly frame in our thoughts." See "Remarks on Mathematical Reasoning and on the Logic of Induction," in *Mechanical Euclid*, pp. 153–55. Like the axioms, these definitions are recognized by us as being correct and necessary when our Ideas are adequately distinct.

held to have an ontological foundation. As we have seen, Whewell's Funda-mental Ideas correspond to the structure of the physical world. The Idea of Space that conforms to the geometry of physical space is (according to clas-sical Newtonian physics) *Euclidean*.[278] On Whewell's view, only definitions that follow from this Idea of Space can serve as the source of necessary truths of geometry. This is why he claimed that it is not the case that a necessary truth of mathematics "*merely* expresses what we mean by our words"; rather, it expresses a truth about some fundamental feature of physical real-ity as well.[279]

To understand this assertion, we must recall Whewell's theological justi-fication for the claim that the Ideas existing in our minds correspond to the nature of physical reality. As we have seen, this justification is based on the notion that our Ideas and the world share a divine creator. In *On the Philos-ophy of Discovery*, Whewell explicitly asked how it is possible that proposi-tions we know a priori are informative about, and indeed necessarily true of, the physical world. Answering this question, he claimed, required asking another: "how did things come to be as they are?"[280] Whewell's response to this query is that God created the physical universe in accordance with cer-tain of his "Divine Ideas." For example, God made the world such that it cor-responds to the Idea of Cause partially expressed by the axiom "Every event has a cause." Hence in the universe every event conforms to this Idea, not only by having a cause but by being such that it could not have occurred with-out one. Whewell's necessary truths are not logically necessary, in the sense of being true in all possible worlds.[281] Rather, he believed, God could have chosen to create the world in accordance with different Ideas, in which case

---

278. It was not until the late 1870s—ten years or so after Whewell's death—that British thinkers began to question the assumption that the nature of geometry is defined by Euclidean geometry (see Joan Richards, *Mathematical Visions*, chap. 2). But we may still speculate upon what Whewell's reaction would have been to non-Euclidean geometries. Since the structure of physical space was still considered Euclidean, Whewell would have maintained that only Euclidean geom-etry is "science," in the sense of containing necessary truths that conform to the nature of physi-cal reality. However, he would have needed to address the question of how geometers, who seem to have a distinct Idea of Euclidean Space, can nevertheless distinctly conceive propositions contrary to the Euclidean axioms.

279. William Whewell, *The History of Scientific Ideas*, 1:59; emphasis added. By denying the conventionality of mathematical truths, Whewell did not thereby deny their analyticity, as some commentators have claimed. Analyticity need not be limited to truths of conventional definition, but can apply to definitions that describe the true nature of physical reality. See Robert Butts, "Nec-essary Truth in Whewell's Theory of Science," p. 167; and Menachem Fisch, *William Whewell, Philosopher of Science*, pp. 155–57.

280. William Whewell, *On the Philosophy of Discovery*, pp. 354–55 and p. 358.

281. My view here differs from that of Michael Ruse in his "William Whewell and the Argu-ment from Design," pp. 251–52.

different axioms would be necessary truths. Even the axioms of mathematics are not logically necessary: "the propositions of space and number and the like, must be supposed to be what they are by an act of the Divine Mind," that is, by the act of God choosing one set of Ideas over another.[282] Given the Ideas God did choose, the axioms are necessarily true of the world, because they follow necessarily from the meanings of these Ideas. But this is not a view of hypothetical truth in the sense that Hamilton and Stewart proposed for mathematics. We do not conventionally assign meanings to the Ideas; now that the world has been created, only one set of meanings is possible.

We are at last in a position to understand what Whewell meant by claiming that a proposition that is at first knowable only empirically can become knowable a priori—that is, how it is possible to "idealize the facts." Since necessary truths follow necessarily from the meaning of our Ideas, the a priori intuition of necessary truths is possible only once our Ideas are distinct: when this is the case, we apprehend that an axiom is necessarily true because the meaning of the axiom is, in fact, contained in the meaning of a Fundamental Idea to which the universe necessarily conforms (given God's choice of Ideas to use as archetypes in creating it). But if our Idea is not distinct, we do not—and indeed cannot—apprehend this. I have already discussed the explication of conceptions and Ideas that occurs in the course of empirical science, and by which Ideas are made distinct. Once an Idea is distinct—once we understand its meaning—truths that we may have discovered empirically are seen actually to follow a priori from the meaning of the Idea. The experiential truth becomes knowable a priori from the now-understood meaning of the Idea; the experiential truth has been "idealized" into a necessary truth. The a priori intuition of necessary truths is "progressive," then, because our Ideas must be explicated before it is possible for us to know their axioms a priori. This is why Whewell claimed that necessary truths become *knowable* a priori and not merely *known* a priori. Thus, for example, he argued that "though the discovery of the First Law of Motion was made, historically speaking, by means of experiment, we have now attained a point of view in which we see that it might have been certainly known to be true independently of experience."[283] The First Law of Motion is a necessary truth that has undergone the process of idealization: though it was first knowable only empirically, it has become knowable a priori.[284]

We have seen that, by the process of idealization of facts, experiential

---

282. See printer's proofs of *On the Plurality of Worlds*, WP Adv.c.16 f. 27, chap. 12, p. 276.

283. William Whewell, *The Philosophy of the Inductive Sciences*, 2:221.

284. Whewell's view of necessary truth altered between the writing of his 1834 "On the Nature of the Truth of the Laws of Motion" and his 1840 *Philosophy of the Inductive Sciences*. In the ear-

truths come to satisfy the criteria of necessary truths. That is, they become knowable a priori from a distinct Idea, and it becomes impossible (for those who have the Idea in its distinct form) to conceive clearly their contraries. But recall that these epistemic criteria are intended to be reliable tests of a deeper kind of necessity. If a law satisfies these epistemic criteria, then we know that the law "must be true." Yet it is not the case that laws change their status regarding this nonepistemic necessity. These truths that *become* necessary in the epistemic sense are *always* necessary in the nonepistemic sense. As discussed earlier, the nonepistemic sense in which an axiom must be true is that it follows as a necessary consequence of one of the Divine Ideas used by God in creating the world. Since God created the world to conform to a particular Idea of Cause, the axioms that express the meaning of this Idea, and the necessary truths that are a priori derivable from these axioms, must be true of the objects and events of the world; and this is so even if we have not explicated our Idea of Cause enough to see their necessity.

Thus, through the idealization of facts, truths become necessary truths in the epistemic sense, yet were always necessary in the nonepistemic sense. There is a rather interesting and important consequence of this understanding of the idealization of facts. Recall that this discussion began with the Fundamental Antithesis, according to which no fixed line divides experiential and necessary truths. We now see how it is that the line we draw between them, like that between fact and theory, is a relative one, based upon epistemic distinctions that change as our Ideas become more distinct. As we explicate our Ideas, we recognize empirical truths to be necessary consequences of these Ideas; and the truths are thus transferred from the empirical to the necessary side of the antithesis. But since there is no firm division between these two classes of truths, any experiential truth can, in principle, become knowable a priori (and therefore it will become impossible to conceive distinctly its contrary). Since satisfying the epistemic criteria is a reliable test for truths that are necessary in the nonepistemic sense, it follows

---

lier work, he claimed that the laws of motion were experiential truths, while in the later one he characterized the first law of motion as a necessary truth. In 1834 he argued that the reason to be cautious about regarding the laws of motion as necessary truths is that "we know that, historically speaking, men did at first suppose the laws of motion to be different from what they are now proved to be," and this would be "impossible" if the laws were necessary and "self-evident" ("On the Nature of the Truth of the Laws of Motion," p. 573; see also *Astronomy and General Physics, Considered with Reference to Natural Theology*, p. 179). By 1840, however, Whewell had come to hold his mature view that—as he put it later—the intuition of a priori truths requires "a certain growth and development of the human mind," and thus that there is no contradiction in supposing a law to be a necessary truth even if previous science did not recognize it as such (see *On the Philosophy of Discovery*, p. 347). Those who have focused only on the 1834 essay have consequently misunderstood Whewell's mature position.

that every experiential truth is, in fact, necessary in this sense. That is, every law of nature is a necessary truth, by virtue of following a priori from some Idea used by God in creating the universe.

Whewell's view thus destroys the line traditionally drawn between laws of nature and the axiomatic propositions of the pure sciences of mathematics; mathematical truth is granted no special status. Mansel seems to have had this consequence in mind when he argued against Whewell that the difference between a priori principles and empirical laws "is not one of degree, but of kind; and the separation between the two classes is such that no conceivable progress of science can ever convert the one into the other."[285] For Whewell, there is no such separation. By virtue of their connection to the Divine Ideas, the laws of nature have the same rigorous necessity as geometrical axioms.[286] Moreover, the axioms of geometry and arithmetic are themselves laws of nature, "established by the Creator of the Universe."[287] In principle, then, it is possible to idealize all experiential truths into necessary truths knowable a priori. Hence, Whewell claimed it was possible (again, and significantly, only in principle) for all science to become purely deductive, like the mathematical sciences. Once all the axiomatic laws of a science are knowable a priori, the only task left for the scientist would be to deduce further theorems from these laws. Eventually this would mark the end of empirical science. However, there is still much work left for the empirical scientist; Whewell vehemently disagreed with Mill's claim that most remaining scientific work is deductive.[288] Moreover, it is clear that Whewell believed we will never, in fact, idealize all empirical laws. Many such laws will be seen by us only as experiential truths, as being what they are "not by virtue of any internal necessity *which we can understand.*"[289]

Whewell insisted that "Science *is* the Idealization of Facts." He claimed that even after the scientist has confirmed that he has correctly colligated a set of facts with an Idea, he must continue to explicate the Idea until it becomes clear *how* these facts are necessary consequences of the Idea, and hence why the empirical law which combines them is a necessary truth. But why should the empirical scientist be concerned with necessary truth? Whewell's position was not, like Descartes', the claim that a priori knowledge is

285. H. L. Mansel, *Prolegomena Logica*, p. 275.
286. See William Whewell, *Thoughts on the Study of Mathematics as Part of a Liberal Education*, p. 160.
287. William Whewell, *Of the Plurality of Worlds*, WP Adv.c.16 f. 27, p. 274.
288. See William Whewell, *Of Induction*, pp. 73–76.
289. William Whewell, *Astronomy and General Physics, Considered with Reference to Natural Theology*, p. 165; emphasis added.

more certain than empirical knowledge; for example, he argued that a priori mathematical laws are no more certain than empirical laws of astronomy.[290] Rather, his point can be inferred from his claim that "however far the phenomena may be idealized, there will always remain some which are not idealized. . . . And thus, in the contemplation of the universe, however much we understand, there must always be something which we do not understand."[291] Whewell here implied that without idealizing empirical laws of nature into truths knowable a priori, there can be no real understanding of the world. Scientific understanding thus requires that we know more than just that an empirical law is true, and even that it must be true: we need to know *why* it must be true. We must understand, that is, the cause of the law's truth. And we can understand this only by seeing the law's a priori connection to the Ideas of the Divine Mind, which caused the law to be what it is.

As we have seen, by the mid-1850s Whewell came to found his epistemology on a theological base. Since our Ideas are "shadows" of the Divine Ideas, to see a law as a necessary consequence of our Ideas is to see it as a consequence of the Divine Ideas exemplified in the world. Hence the more we idealize the facts, the more difficult it will be to deny God's existence. We will come to see more and more truths as the intelligible result of intentional design. We see, then, that the relation between science and theology is, for Whewell, a mutual one. Theology is invoked in providing a mechanism to resolve the "ultimate problem" of philosophy, that is, to show how it is possible for empirical science to reach necessary truths through a process of idealization of facts. And empirical science returns the favor to theology by providing what Whewell believed to be the most "profound" evidence for it. Whewell's point was not that particular scientific laws provide evidence for God's existence (though he had suggested this natural theology position in his 1833 Bridgewater Treatise), but that by coming to see empirical laws as knowable a priori by the explication of Fundamental Ideas, we come to recognize more fully the divine origin of our minds and our world.[292]

In thus placing science and religion in a mutual relation, Whewell was once again following a path taken by Bacon. In his writings, Bacon, like Whew-

---

290. William Whewell, *The Philosophy of the Inductive Sciences*, 1:46.

291. William Whewell, *On the Philosophy of Discovery*, p. 306.

292. Whewell thus proposed a more sophisticated version of natural theology just as such views were losing their grip on the educated public. As John Brooke and Pietro Corsi have pointed out, natural theology was already becoming a less uniform and less compelling philosophy even before Darwin's *Origin of Species* came on the scene. See John Brooke, "Natural Theology and the Plurality of Worlds," and Pietro Corsi, *Science and Religion*.

ell, stressed the role that natural philosophy can play in supporting religious teachings, and noted that religion in fact supported—if not demanded—the study of natural philosophy. The study of natural philosophy is an "inducement to the exaltation of the glory of God" and a "help and preservative against disbelief and error."[293] Not only does such learning enhance faith, but faith demands such learning: Bacon argued that man could regain his dominion over nature, lost in the Fall, by the study of natural philosophy.[294] In the *Novum Organum* he wrote, "For man by the fall fell at the same time from his state of innocency and from his dominion over Creation. Both of these losses, however, can even in this life be in some part repaired; the former by religion and faith, the latter by arts and sciences."[295] In this way Bacon sought to vindicate the study of science against those who claimed it to be an antireligious undertaking.[296] Indeed, he exhorted, "let no man, upon a weak conceit of sobriety or an ill-applied moderation, think or maintain that a man can search too far or be too well-studied in the book of God's word or in the book of God's works—divinity or philosophy. Rather let men endeavor an endless progress or proficiency in both."[297] Charles Darwin would later quote this passage on the frontispiece of the *Origin of Species*, pairing it with a passage from Whewell's Bridgewater Treatise.

Newton, too, had provided Whewell with another source—an even more scientifically legitimate one—for the complementary roles of science and theology.[298] Whewell's regard for Newton is well known; he considered Newton the one mainly responsible for astronomy's status as "queen of the sciences," and indeed "the only perfect science."[299] In his Bridgewater Treatise, he pointed to Newton as one of the instances who prove the rule that "inductive discoverers" are more likely to be religious believers than deductive

293. Francis Bacon, *Works*, 3:300–301.

294. On Bacon's "religious vindication" of natural philosophy, see Stephen Gaukroger, *Francis Bacon and the Transformation of Early-Modern Philosophy*, pp. 75–80.

295. Francis Bacon, *Works*, 4:247–48.

296. Bacon claimed that the Fall was not caused by Adam's illegitimate desire for knowledge of the natural world, but rather for knowledge of good and evil, which is the knowledge that he should have been content to receive by revelation from God. See ibid., 3:296.

297. Ibid., 3:268. For a discussion of Bacon, Baconianism, and millenarianism, see Charles Webster, *The Great Instauration*.

298. See David B. Wilson, "Herschel and Whewell's Versions of Newtonianism," p. 94. Vincent Kavaloski has argued convincingly that this theological element in Newton's thought provided what was considered scientifically legitimate justification for the inclusion of theistic explanations within science by nineteenth-century catastrophist geologists and special creationists (see "The Vera Causa Principle," pp. 35–36).

299. William Whewell, "Address," p. xiii.

thinkers, citing comments made by Newton in the *Opticks* and the *Mathematical Principles of Natural Philosophy* (the "*Principia*").[300] In these works and others, Newton clearly expressed his view that science and religion are not warring disciplines, but rather that science can provide support for theology. In particular, Newton stressed that we can infer the existence of God from his effects in the natural world. In the General Scholium added to the second edition of the *Principia*, he explained that "this most beautiful system of the sun, planets, and comets, could only proceed from the counsel and dominion of an intelligent and powerful Being . . . and lest the systems of the fixed stars should, by their gravity, fall on each other, he hath placed those systems at immense distances from one another."[301] "Blind metaphysical necessity" could never have produced such a world.[302] In Query 31 of the *Opticks* Newton used other examples of design, which were later echoed by Whewell in the Bridgewater Treatise, including the concentric orbits of planets and the general "Uniformity in the Planetary System," as well as the "Uniformity of Bodies" of animals, all of which, again, requires the "Wisdom and Skill of a powerful ever-living Agent."[303] The *Opticks* ends with Newton giving a theological justification for scientific knowledge, by connecting the reform of natural philosophy with the reform of moral philosophy.[304]

Whewell's predecessors at Trinity, Bacon and Newton, provided him with a model for his reform of science. Whewell's renovation, in turn, provided John Stuart Mill with a target against which to aim the reforms he outlined in his *System of Logic*.

---

300. William Whewell, *Astronomy and General Physics, Considered with Reference to Natural Theology*, pp. 238–39. Whewell concluded, "It must be evident to the reader that the succession of great philosophers through whom mankind have been led to the knowledge of the greatest of scientific truths, the law of universal gravitation, did, for their parts, see the truths which they disclosed to men in such a light that their religious feelings, their reference of the world to an intelligent Creator and Preserver, their admiration of his attributes, were exalted rather than impaired by the insight which they obtained into the structure of the universe" (p. 239).

301. Isaac Newton, *Mathematical Principles of Natural Philosophy*, 1:388–89. A longer portion of this passage is quoted in William Whewell, *Astronomy and General Physics, Considered with Reference to Natural Theology*, p. 239.

302. Isaac Newton, *Mathematical Principles of Natural Philosophy*, 1:391. Newton also expressed the argument from design in four letters to Richard Bentley in 1692, which were published in 1756. Whewell cited these letters in his *Astronomy and General Physics, with Reference to Natural Theology* (see p. 136).

303. Isaac Newton, *Opticks*, pp. 402–3.

304. Newton explained that "if natural philosophy in all its Parts, by pursuing this Method, shall at length be perfected, the Bounds of Moral Philosophy will also be enlarged. For so far as we can know by natural Philosophy what is the first Cause, what Power he has over us, and what Benefits we receive from him, so far our Duty towards him, as well as that towards one another, will appear to us by the Light of Nature" (ibid., p. 405).

CHAPTER TWO

Mill's Radicalization of Induction

The notion that truths external to the mind may be known by intuition or consciousness, independently of observations and experience, is, I am persuaded, in these times, the great intellectual support of false doctrines and bad institutions. . . . There never was such an instrument devised for consecrating all deep seated prejudices. And the chief strength of this false philosophy in morals, politics, and religion, lies in the appeal which it is accustomed to make to the evidence of mathematics and of the cognate branches of physical science. To expel it from these, is to drive it from its stronghold. . . . In attempting to clear up the real nature of the evidence of mathematical and physical truths, the "System of Logic" met the intuition philosophers on ground on which they had previously been deemed unassailable.

—*John Stuart Mill*, Autobiography

In his *Autobiography*, Mill clearly stated his motivation for writing *System of Logic:* to expel the intuitionist philosophy from its "stronghold" in physical science and mathematics. He saw this as being the crucial precondition for reforming moral and political philosophy. The *Logic*, then, is very explicitly "a reformer's book,"[1] and needs to be read as such. In his own time, even before he explicitly stated his reformist motivation, Mill's book was understood in these terms. Leslie Stephen reported that the *Logic* became "a kind of sacred book for students who claimed to be genuine Liberals," and was considered "the most important manifesto of Utilitarian philosophy."[2]

1. Alan Ryan uses this phrase in his *J. S. Mill*, p. 85.
2. See Leslie Stephen, *The English Utilitarians*, vol. 3, *John Stuart Mill*, pp. 75–76. See also John Grote, *Exploratio Philosophica*, p. 172; and, more recently, T. H. Heyck, *The Transformation*

The position Mill intended to defeat with his *Logic* was the epistemology he had earlier associated with S. T. Coleridge: the intuitionist or "a priori" philosophy, holding that there are certain truths known by the mind whose source was not experience, and which could be recognized as being necessary truths. This was the part of Coleridge's philosophy that Mill rejected unequivocally. The intuitionist epistemology led to political and social conservatism, Mill believed, because it reassured people that what they believed deeply must be true. As he noted in the *Autobiography*, his concern was that intuitionism allowed "every inveterate belief and every intense feeling" to be "its own all-sufficient voucher and justification."[3] Those in power could use their own "intense feelings" about the way things ought to be to crush any movement for reform. Mill noted that the intuitionists claimed the criterion of a necessary truth was that its contrary was "inconceivable"; he feared that this standard could be used to argue that any innovations which seemed inconceivable to the ruling classes of society must be rejected as contrary to the way things necessarily must be. Intuitionism was therefore an impediment to reform. Some years later, writing to the Austrian philosopher and philologist Theodor Gomperz (who supervised the German translation of Mill's works), Mill explained that "I consider that school of philosophy as the greatest speculative hindrance to the renovation so urgently required, of men and society; which can never be effected under the influence of a philosophy which makes opinions their own proof, and feelings their own justification."[4] Before he could undertake the reform of morality and politics, Mill believed, he must purge the minds of the public of this mistaken and dangerous philosophy. If he could demonstrate that knowledge of physical science and even mathematics did not require any a priori elements, Mill hoped, then he would have proved the superfluity of a priori axioms in morality and political phi-

---

of Intellectual Life in Victorian England, p. 42, and Oskar Kubitz, *The Development of John Stuart Mill's "System of Logic,"* p. 54. Geoffrey Scarre has disparaged this reading of Mill's motivation for writing his *Logic*, claiming that there is "no evidence" for a political motive. Mill's own admission, in the passage cited above, seems to be adequate evidence. Yet to say that Mill had a political motive for arguing against the intuitionist philosophy is not to suggest that he was not interested in logic and metaphysics; Scarre seems to suggest that to allow a political motive is to deny any other interest (*Logic and Reality in the Philosophy of John Stuart Mill*, see pp. 13–14). At the same time it is clear that logic and metaphysics were not Mill's first or primary interests, the way science and mathematics were for Whewell. As he told Theodor Gomperz, "[You] have rightly judged that, to give the cultivators of physical science the theory of their own operations, was but a small part of the object of the book" (August 19, 1854, *Collected Works* (hereafter *CW*), 14:238). See also the letter from John Stuart Mill to Auguste Comte, December 1841, *CW* 22:13–14.

3. John Stuart Mill, *Autobiography*, *CW* 1:233.

4. John Stuart Mill to Theodor Gomperz, August 19, 1854, *CW* 14:239.

losophy. Understanding this motivation for writing *System of Logic* is necessary in order to make sense of the positions Mill expressed within its pages.

In writing the *Logic* Mill focused his attack not only on the intuitionist philosophy in general, but on Whewell's philosophy in particular. Whewell is the author cited the most times in the *Logic*, and most of these references are not very flattering ones. Whewell believed, as we have seen, that there are certain elements of knowledge—what he termed Fundamental Ideas—that are in some sense a priori, and that there are necessary truths in both mathematics and natural science; on Mill's definition this was enough to categorize him as a proponent of intuitionism, even though Whewell's antithetical epistemology is more complex than Mill's characterization suggests. Yet we might ask why Mill chose Whewell as his main antagonist in this work, and not Kant, or Coleridge, or William Hamilton (who was the chosen antagonist of a later work). To understand this, it is useful to review the history of Mill's work on epistemology and scientific method.

Mill had been interested in these topics for many years. From 1825 he was a member of a "Society of Students of Mental Philosophy" that met at the home of his friend George Grote two mornings a week to discuss issues in logic and related subjects. In 1828 Mill published a notice of Richard Whately's *Elements of Logic* in the *Westminster Review*.[5] In 1831 he reviewed John Herschel's *Preliminary Discourse* for the *Literary Examiner*. The review is quite short, but extremely positive; Mill later believed that his *Logic* expressed views similar in many respects to those found in Herschel's book. As he noted in his article, Mill agreed with Herschel's claim that the methods that had been so successful in the physical sciences could be fruitfully applied in order to clear up uncertainty in moral and social philosophy.[6] Soon after writing this review, Mill commenced work on his own large-scale text on logic and the method of the physical and moral sciences. He believed himself particularly suited to write such a book. He explained to John Sterling that "the only thing that I believe I am really fit for, is the investigation of abstract truth, and the more abstract the better. If there is any science which I am capable of promoting, I think it is the science of science itself, the sci-

5. Although Mill's protégé Alexander Bain later referred to this review as "a landmark not merely in the history of his own mind, but in the history of logic," Mill did not choose to reprint it in his collection of essays, *Dissertations and Discussions* (Alexander Bain, *John Stuart Mill*, p. 36). He felt that the work was superseded by the more detailed *System of Logic*; also, by the time of writing the *Logic*, Mill no longer agreed with all the points made in the review, as we will see in chapter 5.

6. See John Stuart Mill, "Herschel's *Preliminary Discourse*," CW 22:284–87, and John Herschel, *Preliminary Discourse on the Study of Natural Philosophy*, pp. 72–74.

ence of investigation—of method."⁷ He began by writing the sections on logic (books 1 and 2) and on the moral sciences (book 6). But he was stuck for five years on the problem of inductive reasoning (book 3). Mill was able to begin writing again in 1837, upon the publication of Whewell's *History of the Inductive Sciences*. This work provided him, Mill claimed, with many examples of scientific procedure that he used to construct his own view of induction. His work on the *Logic* was finishing up in 1840, when Whewell's *Philosophy of the Inductive Sciences* was first published. Mill recounted the effect this book had on him: "During the rewriting of the *Logic*, Dr. Whewell's *Philosophy of the Inductive Sciences* made its appearance; a circumstance fortunate for me, as it gave me what I greatly desired, a full treatment of the subject by an antagonist, and enabled me to present my ideas with greater clearness and emphasis . . . in defending them against definite objections, and confronting them distinctly with an opposite theory."⁸

One reason, then, for using Whewell as the stand-in for all intuitionist philosophy was that his work had been closely studied by Mill while constructing his own view; indeed, it had been materially useful in its construction. Further, because Whewell had just published a work presenting what Mill saw as the opposing view, it made sense to set up his own book as a response to and critique of Whewell's position. That a book by the well-known Whewell (who had recently been appointed Master of Trinity) was sure to garner much attention likely added to the appeal of this strategy for Mill. Indeed, when Whewell published his *Philosophy*, the work was much noticed in the periodical press.⁹ The work sold fairly briskly: by 1843 Whewell wrote to Richard Jones that although fifteen hundred copies had been printed, "new editions of my History and Philosophy are marching upon me quite as fast as I wish."¹⁰ (To give two points of comparison, Mary Somerville's *Mechanism of the Heavens*, a popularized account of astronomy, appeared in a print run of only 750,¹¹ and—as we saw earlier—Dickens's works were considered quite successful after selling three thousand copies.) Mill surely hoped that the success of Whewell's book would cause more people to pay attention to his own. He later admitted to having expected a public response from Whewell: "What

7. John Stuart Mill to John Sterling, October 20–22, 1831, *CW* 12:78–79.

8. John Stuart Mill, *Autobiography, CW* 1:231.

9. It was, as we saw in the last chapter, promptly reviewed by Herschel (along with the *History*) in a 61-page article for the *Quarterly Review*. DeMorgan reviewed it for the *Athenaeum*, and David Brewster for the *Edinburgh Review*. Other reviews include two in the *Dublin University Magazine* (see Isaac Todhunter, *William Whewell, D.D.*, 1:130). The second edition was reviewed anonymously in the *United Services Magazine*.

10. William Whewell to Richard Jones, April 7, 1843, WP Add.Ms.c.51 f. 227.

11. See Richard Yeo, *Defining Science*, pp. 80–81.

hopes I had of exciting any immediate attention, were mainly grounded on the polemical propensities of Dr. Whewell; who, I thought . . . would probably do something to bring the book into notice, by replying . . . to the attack on his opinions."[12] But he had to wait for quite some time. A year after the *Logic* was published, Mill believed Whewell was putting out a new book responding to his criticisms; he wrote to Auguste Comte, with whom he was then corresponding, "I am told Mr. Whewell intends to refute me . . . in the book he will publish. I had always more or less counted on his taste for polemics for him to launch a useful debate between us."[13] However, Whewell did not publish his response to Mill until 1849. Mill had also hoped for a review—in this case, a positive one—from Herschel. Soon before the book appeared, Mill wrote to John Austin that "I have hopes of a review in the Quarterly, grounded on the fact that Herschel writes in it, and his review of Whewell contains so much that chimes with my comments on the same book that he would probably like to lend a helping hand to a writer on the same side as him."[14] Yet Herschel never did review Mill's book, either out of loyalty to Whewell or for other reasons.[15]

Although reviews by influential men of science (such as Whewell and Herschel) did not appear with the alacrity Mill had desired, the book became a major success, one that far outstripped the success of Whewell's *Philosophy*. Mill never understood why this happened, expressing his surprise in the *Autobiography:* "How the book came to have, for a work of the kind, so much success, and what sort of persons compose the bulk of those who have bought . . . it, I have never thoroughly understood."[16] The *System of Logic* was adopted as the standard textbook on logic at Oxford until the end of the nineteenth century. At Cambridge, the *Logic* was widely read, not only by the

12. John Stuart Mill, *Autobiography, CW* 1:231.

13. John Stuart Mill to Auguste Comte, October 5, 1844, in Oscar Haac, *The Correspondence of John Stuart Mill and Auguste Comte*, pp. 259–60.

14. John Stuart Mill to John Austin, July 7, 1842, *CW* 13:528. In a letter to Auguste Comte, Mill noted that Whewell's reply to Herschel's review was "très faible" (October 5, 1844, *CW* 13:639). In the same letter to Austin, Mill admitted that he also hoped that it would be reviewed in the *Edinburgh Review*, if not by Austin himself, then by William Hamilton—who Mill expected would review it in a manner "hostile, but intelligent." In the end, no review at all appeared in the *Edinburgh*.

15. Several years later, Herschel indicated to Mill a different reason for not writing the review. In a letter of 1845, Herschel explained, "It was at one time my intention to have reviewed your book in the same sort of spirit as I did Whewell's (i.e., pointing out what I regarded as defects with the same freedom as its merits) but want of time prevented me" (July 10, 1845, JHP:22.6.26). Perhaps, as Pietro Corsi has suggested to me, Herschel did not wish to associate himself publicly with someone, like Mill, viewed as a radical. On the other hand, in 1845 Herschel referred very positively to Mill in his presidential address to the British Association.

16. See John Stuart Mill, *Autobiography, CW* 1:231.

relatively few students taking the moral sciences tripos, for whom it was required, but also by those taking the mathematical and classical triposes, for whom it was not.[17] It went through eight editions during Mill's lifetime, including an inexpensive edition for working-class readers. Because of the great influence of this work, the view of induction presented within its pages became the standard view of induction, remaining so today. Thus it is in part due to Mill's influence that Whewell's methodology, which differs from Mill's inductive view, is generally interpreted as being noninductive.[18] Mill can be seen, in fact, as the original source for the claim that Whewell's methodology is identical to that later endorsed by twentieth-century proponents of hypothetico-deductivism: in *System of Logic* Mill wrote that "Dr. Whewell. . . . allows of no logical process in any case of induction other than . . . guessing until a guess is found which tallies with the facts; and accordingly . . . he rejects all canons of induction, because it is not by means of them that we guess."[19]

Mill's characterization of Whewell's view as noninductive stems from Mill's denial of any contribution of the mind in induction; he thought that leaving room for any such contribution was to open the door to intuitionism. In attacking Whewell's view of induction, Mill criticized his characterization of Kepler as an inductive discoverer, arguing instead that Kepler used no rational or inferential process in making his discovery of the elliptical orbit of Mars, claiming that he merely "guessed." In this chapter I examine Mill's alternative characterization of Kepler's discovery, showing that making sense of it requires seeing it in the context of his worries about Whewell's "intuitionism." Next I discuss the ultra-empiricist scientific method Mill constructed in order to avoid intuitionism, noting that he rejected the inference to any theoretical or unobservable entities. Mill also denied necessary connections of causation, and necessary truths, both in empirical science and in mathematics. In later work, he endorsed an immaterialist metaphysical view that has much in common with Berkeley's idealism. These positions are all consequences of his fear of intuitionism, based on the moral and political effects he believed followed from this philosophy. Important details of Mill's epistemology and scientific method, then, stem directly from his moral and

17. See E. H. Fox Bourne, et al., *John Stuart Mill*, p. 75.

18. For discussion and evaluation of the claim that Whewell was not an inductivist, see my "Discoverers' Induction" and "The Mill-Whewell Debate."

19. John Stuart Mill, *System of Logic*, CW 7:304. As we will see, he ignored Whewell's protestation in response that he "must remind" Mill "that I have given various methods of Induction . . . all of which I have illustrated by conspicuous examples from the History of Science" (William Whewell, *Of Induction*, p. 65).

political views (which I will examine more closely in chapter 4). I end the chapter by considering Whewell's criticisms of the "young Mill," especially those that concern their differing views on the relation between philosophy of science and history of science.

Mill against Whewell on Kepler

Whewell often praised Kepler as being an exemplary discoverer.[20] Specifically, he commended Kepler's discovery of the elliptical orbit of Mars, considering it a model use of discoverers' induction. He did not subscribe to the claim that Kepler reached this discovery by mere guesswork.[21] Whewell noted that "we have seen how long and how hard Kepler laboured, before he converted the formula for the planetary motion from an epicyclical combination, to a simple ellipse."[22] Elsewhere, discussing Kepler's discovery of his third law, Whewell explained that "Kepler always sought his formal laws by means of physical reasonings [that is, by reasoning from known facts about the physical world]."[23] Although, as we saw in the last chapter, Whewell admired Kepler for his facility in bringing before his mind nineteen possible oval curves once he had inferred that the orbit was not circular, he believed that it was by "judgment," or further inference, that Kepler chose the correct elliptical curve.

In his *System of Logic*, Mill used the example of Kepler (knowing it to be

20. In his earliest Induction notebook, Whewell wrote that Kepler "offers one of the best examples possible of the kind and great labour which is requisite in *inducting laws* from phenomena." (This notebook is undated; one that seems to have been written later is dated June 1830 [WP R.18.17 f. 10, p. 49].) Years later, in his *History*, Whewell claimed that Kepler's works "are a very instructive exhibition of the mental processes of discovery. . . . They exhibit to us the usual process of inventive minds: they rather exemplify the *rule* of genius than . . . the *exception*" (*History of the Inductive Sciences*, 1:318). In the last edition of the *Philosophy*, Whewell noted of Kepler that "we see how clearly he apprehended that *colligation of the facts* which is the main business of the practical discoverer" (*On the Philosophy of Discovery*, p. 121).

21. For a different interpretation of Whewell's view of Kepler's discovery, see Andrew Lugg, "History, Discovery, and Induction." For a reading of Kepler as having guessed his ellipse hypothesis, see Arthur Koestler, *The Watershed*, pp. 141–91. For a related interpretation, which sees Kepler as having used only a form of explanatory inference unique to discovery, see Charles Peirce, *Collected Papers*, 1:31; N. R. Hanson, *Patterns of Discovery*, p. 85; and Scott Kleiner, "A New Look at Kepler and Abductive Argument." A criticism of this latter interpretation is found in Andrew Lugg, "The Process of Discovery." Kepler himself had denied the charge of guesswork in a letter to David Fabricius: "You think that I can just invent some elegant hypothesis and pat myself on the back for embellishing it and then finally examine it with respect to observations. But you are far off. . . . I demonstrated the . . . figure [of the orbit] from the observations." Quoted in Job Kozhamthadam, *The Discovery of Kepler's Laws*, p. 106.

22. William Whewell, *Novum Organon Renovatum*, p. 201. Elsewhere, Whewell discussed Kepler's discovery of his third law, which relates the periodic times of any two planets by the radii of their orbits (*History of the Inductive Sciences*, 1:322–23).

23. William Whewell, *History of the Inductive Sciences*, 1:323.

a favorite case study of Whewell's) in order to argue that Whewell's discoverers' induction was not, in fact, a form of induction. He explained that induction "is a process of inference from the known to the unknown; and any operation involving no inference, and any process in which what seems the conclusion is no wider than the premises from which it is drawn, does not fall within the meaning of the term."[24] Mill, like Jones and Whewell, denied Whately's claim that inductive inference can be reduced to the syllogism.[25] (In his monograph on Mill's *System of Logic*, Whewell praised him for taking this position.)[26] Yet Mill argued here that Whewell's discoverers' induction did not satisfy the conditions for inductive inference. He concentrated his attack on Whewell's notion of colligation. According to Mill, what Whewell called "colligation" is nothing more than a matter of observation and description. He suggested that Kepler's colligation of the Martian orbit was similar to the assertion "All observed crows are black," in which we are simply summarizing numerous observations of black crows, and are not going beyond these observations as in an inductive inference. Mill illustrated his position with the following analogy:

> A navigator sailing in the midst of the ocean discovers land: he cannot at first, or by any one observation, determine whether it is a continent or an island; but he coasts along it, and after a few days finds himself to have sailed completely around it: he then pronounces it an island. . . . He ascertained the fact by a succession of partial observations, and then selected a general expression which summed up in two or three words the whole of what he so observed. But is there anything of the nature of an induction in this process? *Did he infer anything that had not been observed, from something else which had? Certainly not.* He had observed the whole of what the proposition asserts.[27]

According to Mill, there is "no difference in kind" between the sailor's "discovery" and Kepler's. Although he did concede that Kepler's discovery of the Martian orbit was not as simple as the sailor's, he nevertheless claimed that Kepler's hypothesis that the observed points lie on an ellipse, like the sailor's

24. John Stuart Mill, *System of Logic*, CW 7:288.

25. This rejection reflects a change of position from his 1828 review of Richard Whately's *Elements of Logic*, in which Mill endorsed Whately's view. In chapter 5 we will examine the reason for this shift.

26. William Whewell, *Of Induction*, p. 85.

27. John Stuart Mill, *System of Logic*, CW 7:292; emphasis added.

hypothesis that "this land mass is an island," was not an inference from the known facts but rather a summary of them.[28] To determine the curve defined by the observed positions of Mars, Mill insisted, "there was no other mode than that of direct observation."[29] In this way he suggested that Kepler found the property shared by the observed positions by simple curve-fitting; that is, that Kepler determined that the observed points of the orbit share the property of lying on an elliptical curve merely by plotting the observations of Mars and then "connecting the dots," as it were, in order to see what curve included them. Thus Mill accused Whewell of "confounding a mere description, by general terms, of a set of observed phenomena, with an induction from them."[30] Far from being impressed by Kepler's achievement, as Whewell was, Mill sniffed that "the only wonder" was that no one had made this discovery before, once Tycho Brahe's accurate observations had been recorded.[31]

This interpretation of Kepler's discovery is quite odd, to say the least (the philosopher N. R. Hanson once characterized it as "ludicrous").[32] For one thing, Mill's claim ignores the obvious fact that our observations are from the vantage point of the Earth, whereas determining the orbital path of Mars requires determining its path around the Sun. Observations made from the Earth do not yield an ellipse. Kepler needed to develop a theory of the Earth's motions before he could infer the true path of Mars from its Earth-observed positions.[33] Further, even if Kepler's theory of the Earth plus the observations could have yielded "connectable dots," or points of the orbit around the Sun, this itself would not have enabled Kepler to see the orbital path as being elliptical. As the logician John Venn pointed out, "The path is so nearly circular that, if it were displayed, no ordinary eye could detect that it was not a circle: the verdict of mere observation, however carefully applied to the path when

28. Ibid., 7:293. It has been noted that Mill was mistaken even in claiming that the sailor's conclusion involves no inference. His observations alone say only that he has returned to his starting point; without inference from other information the sailor cannot know whether he has gone around a land mass or the circumference of an inland sea. See C. J. Ducasse, "John Stuart Mill's *System of Logic*," p. 214.

29. John Stuart Mill, *System of Logic, CW* 7:292–93.

30. Ibid.

31. Ibid., 7:652.

32. N. R. Hanson, *Patterns of Discovery*, p. 84. Charles Peirce also criticized Mill's interpretation, writing, "Mill denies that there was any reasoning in Kepler's procedure. . . . But so to characterize Kepler's work is to betray total ignorance about it" (*Collected Works*, 1:30). See also Andrew Lugg, "The Process of Discovery."

33. See Curtis Wilson, "Kepler's Derivation of the Elliptical Path," pp. 5–10; Bruce Stephenson, *Kepler's Physical Astronomy*, pp. 49–61; and Job Kozhamthadam, *The Discovery of Kepler's Laws*, pp. 155–61.

drawn to scale, would be that it was apparently a circle, but that the sun was not quite in the center."[34] Kepler needed more than mere observations to find the ellipse.

At times Mill seemed to acknowledge that the elliptical property of Mars's orbit could not be determined by direct observation, because we are not in a privileged position to see its path around the sun. He wrote that in the case of Kepler's discovery, "the facts [were] out of the reach of being observed, in any such manner as would have enabled the senses to identify directly the path of the planet."[35] Here he admitted (rather inconsistently with his earlier assertion) that Kepler's discovery of the ellipse was not achieved by mere summary of the observations, by connecting observed data points. Nevertheless, Mill continued to reject the view that Kepler's discovery of the ellipse required any inference. He offered a second argument against Whewell's claim. Sometimes, Mill explained, a conception can be obtained from earlier experience and applied to present observations. In Kepler's case, the property of lying on an elliptical curve was "derived from his former experience," presumably from his experience of mathematical curves.[36] However, Mill neglected to explain why the discovery that this property is shared by the members of the set of observed points of Mars's orbit does not constitute an inference; it would seem that to apply a property not directly observed in the facts to these facts *is* to go beyond premises about what is observed—especially when doing so goes against two thousand years of accepted dogma regarding circular celestial motion.[37]

Mill's argument here relies on a curious counterfactual. He noted that if we had adequate visual organs, or if the planet left a visible track as it moved through the sky, and if we occupied a privileged position with which to view this path, we could directly observe the planet's orbital path.[38] He suggested that Kepler's discovery that the observed positions of Mars share a property

34. John Venn, *The Principles of Empirical or Inductive Logic*, p. 355n.

35. John Stuart Mill, *System of Logic*, CW 7:296.

36. Ibid., 7:651; see also CW 7:296.

37. Some light is shed on Mill's intention here by a comment in a letter to W. G. Ward in 1859. Mill explained that "I do not think there is any ground for the distinction you draw between the evidence of present and that of past sensations, classing the one as experience and the other as intuition. . . . Memory I take to be the present consciousness of a past sensation." Moreover, he believed (following Dugald Stewart) that "internal consciousness" is a form of experience that can be used in his experiential epistemology. So perhaps Mill held that Kepler's memory of his past sensations of elliptical curves was equivalent epistemologically to his having had a present sensation of the elliptical orbit of Mars. However, since Kepler did not have a memory of the past sensation *of the elliptical orbit*, it still seems that his application of this remembered elliptical sensation to a new case requires inference. See John Stuart Mill to W. G. Ward, November 28, 1859, CW 15:648.

38. John Stuart Mill, *System of Logic*, CW 7:297.

"derived from his former experience" did not constitute an inference because it was merely an "accident," or a contingent fact, that this property was not directly observed by him. Thus he explained that "if the path [of the planet] was visible, no one I think would dispute that to identify it with an ellipse is to describe it: and I cannot see why any difference should be made by its not being directly an object of sense."[39] Later, in book 6, Mill made a similar point.[40]

This strange argument of Mill's does not defeat Whewell's claim that Kepler's discovery of the ellipse required inference from the known to the unknown. The problem with Mill's argument is that it invalidly narrows the scope of ampliative inference. Whewell claimed that Kepler's hypothesis required an inference from the observed positions of Mars to what was unobserved—namely, the shape of an orbital path that included these positions. The shape of such a curve was not directly observed by Kepler. Mill's argument against considering this operation a type of inference from the known to the unknown is that this property would be observable *under certain conditions* (if we were at the proper viewing angle, if the planet left a visible trail).[41] But this is surely irrelevant to the question of inference. What matters is what is, in fact, observed. This is so in the case of enumerative induction as well. After all, every individual crow is observable, yet we (and Mill) still allow that the conclusion "All crows are black" can be reached only by ampliative inference if it is the case that every crow has not been observed. It is exactly because Kepler did not see the orbit's path directly and at the correct angle that he needed to make an inference to a property shared by the points of the orbit; that this property may be, under some idealized conditions, "observable" is irrelevant.[42]

Mill realized that he needed to explain what type of noninferential procedure could be used to obtain the true description of the facts when the conception connecting them is not directly observed. He argued that in such cases, a conception may be applied to a set of facts by nonrational guesswork.

---

39. Ibid., 7:296.

40. Ibid., 7:651.

41. Mill thus makes a point about observability similar to that made later, in a different context, by Bas van Fraassen, who notes that certain objects (such as the moons of Jupiter) are "observable" (by us) because we would see them if we were close enough (*The Scientific Image*, pp. 16–17). Later we will see that Mill's argument may have been influenced by his reading of Berkeley.

42. This point can be illustrated by a more mundane example. Suppose I hear my doorbell ring at 1:05 p.m. I make an inference to the conclusion that my 1:00 p.m. luncheon appointment has arrived. That I would be able to directly observe my friend standing behind the door if my door were glass instead of wood does not vitiate the need for inference under the circumstances which do, in fact, obtain.

Thus he wrote that Kepler's discovery of the ellipse involved nothing but "guessing until a guess is found which tallies with the facts."[43] He argued that Kepler merely made a series of nonrational guesses, using previously observed conceptions, until he found the conception that best fit the observed positions of Mars.[44] That is, he supplied the ellipse conception "hypothetically . . . from among the conceptions he had obtained from other portions of his experience."[45] There are, however, two problems with this claim about Kepler. First, it is clear that Kepler did not merely guess his ellipse hypothesis. As suggested earlier, and as Whewell correctly claimed, Kepler made a series of rational inferences to his discovery; even if one wanted to say, with Newton, that he "guessed" the orbit to be elliptical, clearly Kepler used inference to arrive to the oval; to a great degree, this was the more difficult and revolutionary part of his discovery. The second problem with Mill's claim is that it is inconsistent with his own claims about the role of hypotheses in science. As we will see shortly and in the next chapter, Mill argued that a hypothesis merely guessed at can have only a heuristic role in science; it cannot be proved to be true or likely merely by being found to fit the data—even if it leads, in addition, to a successful prediction of an unexpected consequence. Here, however, Mill seems to be claiming that hypotheses may be proved to be true solely by testing whether they conform to the observations.[46]

Mill's Ultra-Empiricism and the Rejection of Necessity in Mathematics

Mill's rejection of Whewell's claim that Kepler's discovery was an induction is a consequence of his desire to keep out any contribution of the mind to knowledge about the physical world. Mill wanted to show that the only materials of our knowledge are particular, observed phenomena, and that science proceeds by generalizing these pieces of data into empirical laws. Any suggestion that the mind itself provides a conception—especially an unobservable one—to these phenomena was considered by Mill a form of intuitionism, leading to conservatism of thought. The same was true for the extension of knowledge beyond the realm of the observable, for example to any unobservable entity postulated as a cause of observed phenomena, or to any claim of "necessity." Thus Mill was led to an ultra-empiricist epistemology. As he wrote in the

43. John Stuart Mill, *System of Logic*, CW 7:304.
44. See ibid., 7:297.
45. Ibid., 7:296.
46. Ibid., 7:296–97.

Coleridge essay, "Sensation, and the mind's consciousness of its own acts, are not only the exclusive sources, but the sole materials of our knowledge."[47] He thus accepted the "relativity of knowledge" principle, as he called it in his later work on Hamilton's philosophy: the view that all our knowledge is relative to—and in a sense *concerns*—our sensations.

As part of this ultra-empiricist epistemology, Mill denied necessity, both the necessary connections associated with causation, and the necessary truths of science, mathematics, and logic. According to Mill, the acceptance of necessity would entail some form of the hated intuitionism, as necessary truths or necessary connections between events cannot be known by experience alone. Rather, he argued that propositions that appear to us to be necessary truths because they seem to derive from innate ideas are actually gained empirically. This was the case, he claimed, for the "laws of thought" that constrain our thinking, such as the principles of contradiction and the excluded middle.[48] What is more, he claimed that the notion of the uniformity of nature and the principle that for every event there exists a cause—which are presuppositions of all successive inductive reasoning—originate in experience.[49] We can see how far Mill was willing to go in his desire to attack intuitionism by his denial of necessity in mathematics.[50] He forcefully rejected what he viewed as Locke's "equivocation" in allowing that mathematical knowledge is certain because it arises not from ordinary experience, but from an infallible contemplation of mental archetypes.[51]

In book 1 of the *Logic*, Mill distinguished between propositions that are "real" and those that are "merely verbal." Merely verbal propositions are those that carry no new information; the attributes connoted by the predicate are a subset of those already connoted by the subject.[52] Real propositions, on the other hand, do carry information beyond what is already stated. (They are ampliative.) Mill argued that no real proposition is a priori. He drew a corresponding distinction between real and apparent inference, claiming that an inference is merely apparent when "the proposition ostensibly inferred from another, appears on analysis to be merely a repetition of the same, or

---

47. John Stuart Mill, "Coleridge," *CW* 10:125.

48. Mill seemed to back down from this claim to some extent in his article "Grote's Aristotle," which was published in the last year of his life (see *CW* 11:499–500; for a discussion of this change, see Geoffrey Scarre, *Logic and Reality in the Philosophy of John Stuart Mill*, pp. 138–45).

49. See Geoffrey Scarre, *Logic and Reality in the Philosophy of John Stuart Mill*, pp. 80–87.

50. See Alan Ryan, *John Stuart Mill*, pp. 75ff.

51. See Geoffrey Scarre, *Logic and Reality in the Philosophy of John Stuart Mill*, p. 3.

52. John Stuart Mill, *System of Logic*, *CW* 7:112–13. See also John Skorupski, *John Stuart Mill*, chap. 3.

part of the same, assertion, which was contained in the first."[53] Progress in thought occurs only by real inferences, which go beyond what is contained in the premises. Only induction, and not deduction, consists in real inference. In book 2, Mill used this analysis of propositions and inferences to claim that mathematics contains propositions and inferences that are real and not merely verbal, and hence that mathematical propositions are a posteriori.

I will here concentrate on Mill's claims about geometrical truth, because in presenting them he explicitly argued against Whewell's position. The theorems of geometry are, according to Mill, deductively inferred from premises (the axioms and definitions of Euclidean geometry) that are themselves real propositions. Thus the theorems of geometry are only "necessary" in the sense of following deductively from the axioms and definitions. As Mill explained, "The necessity consists in reality only in this, that they follow correctly from the suppositions from which they are deduced."[54] The theorems are only necessarily true if the axioms and definitions are so. Because of this, Mill continued, "their claim to the character of necessity in any sense beyond this, as implying an evidence independent of and superior to observation and experience, must depend on the previous establishment of such a claim in favor of the definitions and axioms themselves."[55] But this claim is precisely that which Mill rejected.

Mill believed, as did Whewell, that geometrical truths are truths about the world. His view of definition, expressed in book 1 of the *Logic,* held that no proposition with informative content about the world can be deduced from a definition alone; rather, such real propositions follow from the definition *along with* an implied assumption that there exists a real thing corresponding to the definition.[56] In the case of geometry, however, this is "not strictly true," because there are no "positions without magnitude" (as Euclidean geometry defines "point"), nor any "lengths without breadth" (as it defines "line"). The common claim, as Mill noted, is that geometry deals with ideal points, lines, and figures that exist as conceptions in our minds, and out of these materials we build an a priori science. But Mill claimed that this opinion is "psychologically incorrect," because "the points, lines, circles, and squares which any one has in his mind, are . . . simply copies of the points,

53. John Stuart Mill, *System of Logic,* CW 7:158.
54. Ibid., 7:227.
55. Ibid., 7:252.
56. In book 1, Mill discussed this in a section entitled "What are called definitions of Things, are definitions of Names with an implied assumption of the existence of Things corresponding to them" (CW 7:142).

lines, circles, and squares which he has known in his experience." So, the geometrical figures we have in our minds are not perfect figures. As Mill then explained, "Neither in nature, nor in the human mind, do there exist any objects exactly corresponding to the definitions of geometry." Therefore, to take one example, the definition of a circle is "not exactly true of any circle; it is only nearly true."[57] The definitions are, rather, "hypothetical," in the sense that there is "the assumption that what is very nearly true is exactly so."[58] Thus the theorems of geometry that follow from the definitions "are so far from being necessary, that they are not even true; they purposely depart, more or less widely, from the truth."[59] His view, as he noted, here coincided with Dugald Stewart's view of the hypothetical foundation of geometry, which Mill considered "substantially correct."[60]

Where Mill claimed to differ from Stewart was on the issue of the axioms of geometry. (Here he credited Whewell for disagreeing with Stewart, though Mill did not of course believe that Whewell's view was correct either.) Mill maintained that the axioms of geometry "are true without any mixture of hypothesis."[61] Indeed, they are experimental truths, that is, generalizations from experience.[62] Mill admitted that there is no single observation or group of observations that could prove, for example, that "two straight lines cannot enclose a space." But he claimed that geometrical forms have a "capacity of being painted in the imagination with a distinctness equal to reality."[63] We can therefore "experiment" upon such forms in our minds by imagining pairs of straight lines and seeing that they never do enclose a space. And we are justified in concluding that what is true of the forms in our minds is true of the real forms in the world, because we have learned from "long-continued experience that the properties of the reality are faithfully represented in the image."[64] Thus Mill argued that the experiences of imagination could be as good (and as empirical) as the experiences of sensation.[65] We can experien-

57. John Stuart Mill, System of Logic, CW 7:225–26.

58. Ibid., 7:227n. See also John Skorupski, John Stuart Mill, p. 134.

59. John Stuart Mill, System of Logic, CW 7:227.

60. Ibid., 7:226. For more on Stewart and mathematical truth, see Pietro Corsi, "The Heritage of Dugald Stewart," especially pp. 99ff.

61. John Stuart Mill, System of Logic, CW 7:230.

62. See ibid., 7:231.

63. Ibid., 7:234.

64. Ibid., 7:234.

65. See Alan Ryan, J. S. Mill, p. 81. R. P. Anschutz has argued that Mill's view of mathematics is Platonic rather than empirical (see The Philosophy of J. S. Mill, p. 158), but this does not seem supported either by the references he makes to Mill's writings (which are concerned mainly with the importance of mathematics in learning and doing science) or by Mill's general Lockean tenor.

tially know the truth of "two straight lines cannot enclose a space" by this kind of imagined experience.[66]

What seems especially problematic here is that Mill is now claiming that the properties of geometrical forms such as straight lines that we imagine are "faithful representations" of the way the forms are in the world, even though he had earlier argued that there *are* no true lines (in terms of breadthless length) outside the mind. One possible solution to this problem is that Mill held that the definitions and axioms of geometry express "ideal limits" of geometrical forms. On this view, "straight" is definable as "the limit approached as we take lines which are progressively less crooked."[67] This suggestion is consistent with Mill's claim, in a footnote added to his discussion in 1865, that "though experience furnishes us with no lines so unimpeachably straight that two of them are incapable of inclosing the smallest space, it presents us with gradations of lines possessing less and less either of breadth or of flexure, of which series the straight line of the definition is the ideal limit." Observation shows us that as lines become narrower and straighter, the space they can enclose becomes smaller, approaching zero. Thus "the inference that if they had no breadth or flexure at all, they would inclose no space at all, is a correct inductive inference from these facts," one that in fact conforms to one of his methods of experimental inquiry, the method of concomitant variations.[68]

Next, Mill addressed the problem that he knew Whewell and others would raise: namely, that experience—even this kind of "mental experience"—can-

---

66. Mill also claimed that we use this kind of "mental experimentation" in other cases, for example, regarding our "recollection of colours or of odours" (see *System of Logic, CW* 7:235n).

67. See John Skorupski, *John Stuart Mill*, pp. 133–35.

68. John Stuart Mill, *System of Logic, CW* 7:232n. Mill similarly argued that the propositions of arithmetic are not necessary truths, but that they, too, are generalizations from experience. He claimed that arithmetic can be treated as a set of deductions from two types of real propositions: axioms and "so-called" definitions. The only axioms required are "The sums of equals are equals" and "The differences of equals are equals" (*CW* 7:258). Later, Mill more correctly characterized the axioms as "Things which are equal to the same thing are equal to one another" and "Equals added to equals make equal sums" (see *CW* 8:610, and John Skorupski, *John Stuart Mill*, p. 137). He claimed, without argument, that these are experimental truths. He spent more time arguing for his claim that what are referred to as "definitions" of numbers are generalizations from experience. Using the example "$2 + 1 = 3$," Mill claimed that our experience has shown us that "collections of objects exist, which while they impress the senses thus ooo, may be resolved into two parts, thus oo and o" (*CW* 7:257; see also Alan Ryan, *J. S. Mill*, p. 68). (Notoriously, Gottlob Frege, in his *Foundations of Arithmetic*, points out the obvious objection that we should not think the truths of mathematics to be endangered if everything were nailed down.) These truths, Mill maintained, "all rest on the evidence of sense; they are proved by showing to our eyes and our fingers that any given number of objects, ten balls for example, may by separation and re-arrangement exhibit to our senses all the different sets of numbers the sum of which is equal to ten" (*CW* 7:256). Mill claimed further that, as in geometry, there is a hypothetical element to arithmetic: he noted that "in all

not provide evidence of necessity. That is, experience can show us what does happen, but not what *must* happen.[69] Mill agreed that experience cannot give us knowledge of necessity; he argued that it need not do so, because in fact necessity does not exist. Rather, the appearance of necessity is an illusion, stemming from our own psychology.[70] Mill explained that Whewell and others argued for the existence of necessary truths because we find it inconceivable to negate certain propositions. But Mill countered that this was an unfounded basis for a claim in favor of the existence of necessary truths.

Mill concentrated his attack on Whewell's use of the "inconceivability" criterion, claiming that Whewell's whole argument for necessity rests on this. (In his later work, *An Examination of Sir William Hamilton's Philosophy*, Mill also criticized Hamilton and Herbert Spencer for their adherence to this position.)[71] In the *Logic*, Mill complained that "I cannot but wonder that so much stress should be laid on the circumstance of inconceivableness, when there is such ample evidence to show that our capacity or incapacity of conceiving a thing has very little to do with the possibility of the thing in itself, but is in truth very much an affair of accident."[72] Instead, he relied on the associationist psychology he inherited from his father to explain that there is another explanation for why we cannot conceive the contrary of certain propositions: "There is no more generally acknowledged fact in human nature, than the extreme difficulty at first felt in conceiving anything as possible, which is in contradiction to long established and familiar experience. . . . And this difficulty is a necessary result of the fundamental laws of the human mind."[73] It is only due to our habit of believing a proposition to be true that we find it "inconceivable" to consider the possibility of its falsehood. Mill used Whewell's arguments for the progressive intuition of necessity as an argument against the inconceivability criterion, claiming that Whewell's admission that the same proposition that is self-evident to some people may not be self-evident to others is equivalent to the admission that conceivability is an accidental matter based upon habits of thought.[74] This criticism ignores Whewell's claim that inconceivability of a contrary is only

propositions concerning numbers, a condition is implied, without which none of them would be true; and that condition is an assumption which may be false." That assumption is the identity relation, $1 = 1$ (p. 258). For a discussion and partial defense of Mill's claims about arithmetic, see Philip Kitcher, "Arithmetic for the Millian."

69. See John Stuart Mill, *System of Logic*, CW 7:236–37.
70. Ibid., 7:238–39.
71. See John Stuart Mill, *Examination of Sir William Hamilton's Philosophy*, CW 9:130n.
72. John Stuart Mill, *System of Logic*, CW 7:238.
73. Ibid.
74. Ibid., 7:242.

a criterion of necessity when the associated Ideas and conceptions are clearly and distinctly conceived; thus the fact that a child may be able to conceive the first law of motion as being false is no reason to reject its truth or necessity. Moreover, Mill overlooked Whewell's claim that knowledge of necessary truth often starts out as knowledge of empirical truth; thus he used Whewell's examples of necessary truths that originally required empirical evidence to argue that these truths are obviously not a priori ones. Indeed, he wrongly claimed that Whewell believed that a necessary truth must be seen as necessarily true "from the first moment" and with no experience at all.[75] Finally, Mill called a "reductio ad absurdum" of Whewell's whole doctrine the claim that we cannot conceive a world in which it should be the case that simple elements could combine in other than definite combinations; obviously, Mill contended, we can so conceive this.[76] His arguments show that he did not really grasp Whewell's inconceivability criterion, for as we saw in chapter 1, Whewell did not suggest that necessity depends on our ability to form a mental image of something; rather, for Whewell, inconceivability depends on relations of meaning between Ideas and axioms. We may form mental images of worlds in which the first law of motion is false, but this does not mean we can "distinctly conceive" that this could be so.[77] We will see in chapter 4 that this misreading of Whewell on necessity caused Mill to misinterpret not only Whewell's epistemology, but his moral philosophy as well.

Causation and Human Freedom

Another way in which Mill's ultra-empiricist rejection of necessity influenced his view of science has to do with his claims about causation.[78] He

75. Ibid., 7:231. To be fair to Mill, Whewell had suggested this view in his 1834 paper "On the Nature of the Truth of the Laws of Motion," but presented his contrary, mature view in the first edition of the *Philosophy* in 1840. See chapter 1, note 284 above.

76. See John Stuart Mill, *System of Logic, CW* 7:243–44.

77. Mill himself seemed to recognize this when he distinguished his own view of "mental experimentation" in geometry from the "imaginary looking" of the a priorist. Mill claimed that the a priorists like Whewell do not actually form any mental image of geometrical figures, but rather engage in a purely intellectual process of understanding the meaning of the propositions from which an axiom follows. Mill explained that, on their view, "it is not experience which *proves* the axiom; but that its truth is perceived *a priori*, by the constitution of the mind itself, from the first moment when the meaning of the proposition is apprehended" (*CW* 7:231; see also Geoffrey Scarre, *Logic and Reality in the Philosophy of John Stuart Mill*, p. 133).

78. Mill made clear the relation between his rejection of necessity in causation and his fight against the intuitionist school in the notes to his edition of his father's *Analysis of the Phenomena of the Human Mind* (1869), writing there that the denial of his position on causes "is perhaps the principal badge of one of the two schools which at this, as at most other times, bisect the philosophical world—the intuitional and the experiential" (*CW* 31:156–57).

made clear his rejection of the necessity of causes at the start of book 3 of the *Logic*, which he prefaced with the following quotation from Stewart's *Elements of the Philosophy of the Human Mind*:

> The highest, or rather the only proper object of physics, is to ascertain those established conjunctions of successive events, which constitute the order of the universe; to record the phenomena which it exhibits to our observations, or which it discloses to our experiments; and to refer these phenomena to their general laws.[79]

Mill followed Stewart in asserting that causation was merely the constant connection between observed events. He noted that the law of universal causation is "the main pillar of inductive science." This law asserts that every event has a cause. But what this means, according to Mill, is "but the familiar truth, that invariability of succession is found by observation to obtain between every fact in nature and some other fact which has preceded it."[80] Thus, the causal relation is defined as invariability: a cause is an antecedent that is invariably followed by a particular consequent. Mill admitted that most "metaphysicians" currently in vogue argue that the notion of cause implies something more, "a mysterious and most powerful tie" between the cause and effect. Because of this they suppose the need for "ascending higher, into the essences and inherent constitution of things, to find the true cause, the cause which is not only followed by, but actually produces, the effect." Yet Mill claimed that he was not interested in these efficient causes, but only what he called "physical" causes.[81] That is, he was not concerned with the cause that "actually produces" the effect, only with the observed antecedent that is invariably followed by an observed consequent. Mill explained that "when in the course of this inquiry I speak of the cause of any phenomenon, I do not mean a cause which is not itself a phenomenon."[82] In his later *Auguste Comte and Positivism*, Mill more explicitly defined the law of universal causation as stating that "every phenomenon has a phenomenal cause."[83]

79. In his later *Auguste Comte and Positivism*, Mill similarly noted, "We know not the essence, nor the real mode of production, of any fact, but only its relations to other facts in the way of succession or of similitude. . . . The constant resemblances . . . and the constant sequences . . . are termed their laws" (*CW* 10:265).

80. John Stuart Mill, *System of Logic*, *CW* 7:326–27.

81. Ibid., 7:326.

82. Ibid., 7:326–27. See also John Stuart Mill, *Examination of Sir William Hamilton's Philosophy*, *CW* 9:362.

83. John Stuart Mill, *Auguste Comte and Positivism*, *CW* 10:293. Mill suggested more strongly that there *are* no efficient causes, claiming that "there is . . . no other uniformity in the

However, causation cannot be characterized merely by the invariable sequence of observed antecedents and consequents, or else, for example, night would be the cause of day. Mill explained that the reason night is not the cause of day is that day only succeeds night *provided that* the sun rises above the horizon. The invariable succession of day and night, then, is "conditional" upon the invariable succession of the presence and absence of the sun over the horizon (that is, the earth's rotation).[84] Besides invariable succession, causation also requires unconditionality, in the sense of not depending upon any other antecedents. The cause of day, Mill explained, is the presence of the sun over the horizon.[85] This antecedent is invariably followed by day, and is not conditional upon any other antecedent. After *System of Logic* appeared, Mill was criticized for including the criterion of unconditionality, on the grounds that knowing whether something is an unconditional antecedent would require knowledge that goes beyond experience.[86] In later editions Mill addressed this criticism by adding a passage insisting that "it is experience itself which teaches us that one uniformity of sequence is conditional and another unconditional. . . . Twice in every twenty-four hours, when the sky is clear, we have an *experimentum crucis* that the cause of the day is the sun. We have an experimental knowledge of the sun which justifies us on experimental grounds in concluding, that if the sun were always above the horizon there would be day, though there had been no night."[87] Thus, according to Mill, it is experimentally evident that the unconditional, invariable antecedent (or cause) of day is the presence of the sun over the horizon. Although many writers use the term "necessity" to connote "unconditionality,"[88] Mill argued that what he meant by unconditionality was not a form of necessity, and could be known by experience. Consistently with

---

events of nature than that which arises from the law of causation," which, as we have seen, does not involve events that "actually produce" their effects (*System of Logic, CW* 7:577).

84. John Stuart Mill, *System of Logic, CW* 7:338–39, and *Auguste Comte and Positivism, CW* 10:293.

85. John Stuart Mill, *System of Logic, CW* 7:339.

86. This criticism was raised by John Tulloch in his *Theism: Second Burnett Prize Essay of 1855.*

87. John Stuart Mill, *System of Logic, CW* 7:340. Although he is not explicit here, it is clear that Mill believed this situation corresponded to one of his "methods of experimental inquiry," that is, the methods that he proposed for discovering causal laws. The method of difference is the only one that could qualify as an *experimentum crucis*, or a crucial experiment to find a causal law, because the method of difference is the only method that can result in knowledge of a specific causal law. (The other methods allow us to conclude merely that there is a causal relation of some type between the two events, as we will see below.)

88. Ibid., 7:339.

his rejection of intuitionism, then, Mill claimed that the law of causation does not involve any type of necessity.

Mill's rejection of necessary causation was motivated not only by his general rejection of intuitionism, but also more specifically by moral and political considerations. As part of his empiricist, naturalistic project, Mill wished to develop an inductive method that was valid for the moral sciences as well as the natural sciences. He also claimed that the law of universal causation was the foundation of any logic of induction. Thus, this law underpinned the moral, or human, sciences as well as the natural sciences. It was therefore important for Mill to make clear that applying the law of causation to human action did not conflict with human freedom. Put another way, Mill wanted to claim both that human behavior is subject to laws (including causal laws) similar to laws of nature, and also that there is freedom of will. His concern with freedom of will had mainly a political motivation. Mill (and the other radical reformers) disdained forms of Calvinist social determinism that decreed that the current hierarchical class structure and division of labor were ordained by God, and could not be altered. Mill especially detested the Calvinist view that individual fates were predestined, and could not be changed by individuals themselves.[89] Mill also rejected the claim of the socialist reformer Robert Owen and his followers that an individual's character is determined wholly by his external circumstances.[90] Such views entailed that there could be no radical self-reform of the type Mill thought necessary for the reform of society. Indeed, these views were dangerous: if people accepted the claim that their characters were fixed and unchangeable, then they would not carry out the kind of "renovation of mind" that Mill desired. Mill therefore argued strongly that individuals had control over their actions and their characters, and thus could alter themselves in the ways required for the reform of society.[91]

89. Nicholas Capaldi points out that, in opposition to this view, John Wesley, the founder of Methodism, had formulated a view of free will which allowed that individuals could perfect themselves through their own effort (see *John Stuart Mill*, p. 180).

90. G. W. Smith rather too strongly claims that Mill's entire notion of freedom was "designed specifically to deal with Owenism"; in my view, Mill's concern to allow the freedom for self-reform was more far-reaching that his worries about any one particular opposing philosophical position. (Indeed, as Vincent Guillin has pointed out to me, Mill was also interested in countering Comte's view that self-reform was impossible because individuals are endowed with unalterable capacities.) See Smith, "The Logic of J. S. Mill on Freedom," p. 538. On Mill's desire to counter Owenism, see also Alan Ryan, *J. S. Mill*, p. 86 and *John Stuart Mill*, pp. 105–6; and John Skorupski, *John Stuart Mill*, pp. 252–54.

91. In his youth, Mill had been stung by the criticism that he was merely a "'made' or manufactured man" (see *Autobiography*, CW 1:161). His overriding desire to show that men could "remake" themselves may have been fueled by this experience.

In a footnote to his discussion of the law of causation in book 3 of the *Logic*, Mill explicitly linked the notion of causation that applies in the natural world to that which applies in the moral realm.[92] He developed this connection further in his chapter "Of Liberty and Necessity" in book 6 (a chapter he believed to be one of the best in the whole work).[93] He began by explaining that the "doctrine of Necessity" affirms that the law of causality applies to human actions as it does to other phenomena. The opposing view claims that the will is not determined by antecedents, but "determines itself." Mill noted that he believed in the former of these positions, but rejected the terms in which it is often expressed and the uses to which it is often put. "Correctly conceived, the doctrine called Philosophical Necessity is simply this," he explained, "that given the motives which are present to an individual's mind, and given likewise the character and disposition of the individual, the manner in which he will act might be unerringly inferred: that if we knew the person thoroughly, and knew all the inducements which are acting upon him, we could foretell his conduct with as much certainty as we can predict any physical event."[94] Mill thus reduced philosophical necessity (or determinism, as he later called it)[95] to a matter of predictability rather than compulsion.

Mill admitted that many people reject this doctrine, believing it to imply that we are not free, that we are forced to act against our will or without control over our actions. However, this implication only follows from the incorrect conception of causation, that which holds that there is some "mysterious constraint exercised by the antecedent over the consequent." Mill denied that the position of philosophical necessity implies that our actions are compelled in any way. Mill argued that there is no difficulty in conceiving of our free volitions as having causes once the correct conception of causation—that involving only invariability of sequence—is applied. (Interestingly, Mill omitted mention of the criterion of unconditionality in this discussion.)[96] As Mill put it,

> Those who think that causes draw their effects after them by a mystical tie, are right in believing that the relation between volitions and their antecedents is of another nature. But they should go farther, and

92. See John Stuart Mill, *System of Logic*, CW 7:347n.

93. See letter from John Stuart Mill to R. B. Fox, February 14, 1843, CW 13:569.

94. John Stuart Mill, *System of Logic*, CW 8:836–37.

95. He used the term "determinism" in the *Examination of Sir William Hamilton's Philosophy*.

96. See John Stuart Mill, *System of Logic*, CW 8:837–39.

admit that this is also true of all other effects and their antecedents. If such a tie is considered to be involved in the word necessity, the doctrine is not true of human actions; but neither is it then true of inanimate objects. It would be more correct to say that matter is not bound by necessity, than that the mind is so.[97]

We do not feel our freedom to be degraded by the doctrine that "our volitions and actions are invariable consequences of our antecedent states of mind."[98] This conception of causation allows that a person's actions could be predicted if that person's character and "all the inducements acting upon" him at a given moment were known. Yet, it does not suggest that a person's actions would be compelled or necessitated by his character and circumstances; if he wished to resist a certain motive, he could do so. (His wish to resist a motive would be one further antecedent state of mind, which would need to be taken into account in order to predict correctly his action.) Mill suggested that we stop using the term "necessity" in this context, as it "involves much more than mere uniformity of sequence: it implies irresistibleness. Applied to the will, it only means that the given cause will be followed by the effect, *subject to all possibilities of counteraction by other causes:* but in common use it stands for the operation of those causes exclusively, which are supposed too powerful to be counteracted at all."[99] On Mill's view, causes, especially in the case of our volitions, are never "uncontrollable" (although we might not actually control them).[100]

The man who does something wrong is not *compelled* by any of his desires to do so; rather, he does not control his own desires properly. The importance of a moral education (which as adults we are responsible for providing to ourselves) is precisely to strengthen the desire for what is right and the aversion to what is wrong, so that we will desire and choose the right actions.[101] As Mill explained, "the difference between a bad and a good man is not that the latter acts in opposition to his strongest desires; it is that his

97. Ibid., 8:838. In his *Examination of Sir William Hamilton's Philosophy,* Mill explained that "a volition is a moral effect, which follows the corresponding moral causes as certainly and invariably as physical effects follow their physical causes. Whether it *must* do so, I acknowledge myself to be entirely ignorant, be the phenomenon moral or physical; and I condemn, accordingly, the word Necessity as applied to either case. All I know is, that it always *does*" (*CW* 9:446–47). In book 3 of the *Logic,* Mill also wrote that "our will causes our bodily actions in the same sense, and in no other, in which cold causes ice, or a spark causes an explosion of gunpowder" (*CW* 7:355).
98. John Stuart Mill, *System of Logic, CW* 8:838.
99. Ibid., 8:839.
100. Ibid., 8:840.
101. John Stuart Mill, *Examination of Sir William Hamilton's Philosophy, CW* 9:453.

desire to do right, and his aversion to doing wrong are strong enough to overcome, and in the case of perfect virtue, to silence, any other desire or aversion
which may conflict with them."[102] Mill's point is that our freedom does not
consist in being able to act against our volitions, but in being able to create
or train these volitions, by forming our characters in a certain way. Here,
then, Mill was countering the claim of Owen that "our characters are made
for us and not by us."[103] Owen had argued that individuals should not be punished for their wrong actions, because they cannot help how they act; their
actions are consequences of their characters, which are not under their control. However, Mill argued that men are responsible for their actions, because
they have the power to alter their characters.[104] What the Owenites did not
realize about man, according to Mill, is that "his character is formed by his
circumstances (including among these his particular organization); but his
own desire to mould it in a particular way is one of those circumstances, and
by no means one of the least influential."[105] As Mill put it in the *Examination*,
"Not only our conduct, but our character, is in part amenable to our will."[106]
Indeed, we have a robust "power of self-formation."[107] If we do not form our
own character, by choosing our own desires and impulses, we can be said to
have no character: "One whose desires and impulses are not his own, has no
character, no more than a steam-engine has a character."[108] (Mill was not neutral on what type of character we should form; his view will be discussed in
detail in chapter 4.)

102. Ibid., 9:452–53.

103. John Stuart Mill, *System of Logic*, CW 8:840. See also *Examination of Sir William Hamilton's Philosophy*, CW 9:453.

104. Mill did allow, of course, that there are instances in which a person may be compelled to
act, as when he is under a physical constraint, or such a violent motive that no fear of punishment
could stop him; in such cases, there should be an exemption from punishment, and this is why in
most countries people are not punished for "what they were compelled to do by immediate danger
of death" (*Examination of Sir William Hamilton's Philosophy*, CW 9:464).

105. John Stuart Mill, *System of Logic*, CW 8:840.

106. John Stuart Mill, *Examination of Sir William Hamilton's Philosophy*, CW 9:465–66. As
Nicholas Capaldi notes, the Owenites and other critics of the time believed that Mill had not really
advanced a strong argument in favor of human freedom. These critics conceded that we could
change our characters if we so wished, but claimed that this wish or desire would be the result of
prior causes, which were not under our control. Mill never addressed this concern (see *John Stuart
Mill*, pp. 181–82. See also John Skorupski, *John Stuart Mill*, p. 253). This is a larger problem for Mill
than Capaldi suggests. If it is the case that our desire to change our character is itself an effect of
causes outside our control, then this would make our desire for change a conditional antecedent,
and thus it could not be a real cause. Perhaps it is no accident that Mill did not mention the requirement of unconditionality in this discussion. Unlike Capaldi, I do not believe that the argument
Mill does advance is dependent on a metaphysical idealist position; indeed, as I show below, Mill
advanced a realist metaphysics.

107. John Stuart Mill, *System of Logic*, CW 8:842.

108. John Stuart Mill, *On Liberty*, CW 18:264.

Mill believed that our feeling of moral freedom is the feeling of being able to modify our character if we wish. Indeed, he defined moral freedom as this possibility for self-modification.[109] Moral freedom is not diminished by the true meaning of philosophical necessity (that is, the doctrine that our actions are caused) once the proper sense of causation is applied, because this notion of causation does not interfere with the ability of individuals to reform themselves. Mill's concept of causation, applied to our volitions and actions, allows us the freedom to remake ourselves in the ways required for the reform of society.

Finding Physical Causes: Mill's Methods

Having explained the type of causal laws science could seek, Mill discussed in the *Logic* five ways to discover such laws: the methods of agreement, difference, joint agreement and difference, residues, and concomitant variations. He described them as "eliminative methods," and admitted that he drew them from Herschel's adaptation of Francis Bacon's "Prerogative of Instances" in the *Preliminary Discourse.* Only the method of difference, Mill believed, can lead us to specific causal laws; the others can affirm the existence of some causal connection between phenomena, but they cannot specify the law of the connection. The method of agreement can tell us only that there is a "uniformity" between an antecedent and consequent, that they are invariably connected;[110] it cannot tell us which one is the cause and which is the effect.[111] This method cannot tell us that the antecedent is the cause of the consequent. The joint method is an "extension" of the method of agreement, and cannot be regarded as providing the proof of causation that is given by the method of difference.[112] By the method of residues we can conclude that one or more consequents are the effect of one of a number of antecedents, but we cannot be certain that one particular consequent is the effect of one particular antecedent unless we are able to perform the method of difference.[113] The method of concomitant variations yields the conclusion that the antecedent and the consequent are causally related: either the antecedent is the cause of the consequent or they share a common cause.[114] With the method of difference, however, we can discover causal laws. Mill characterized this method

109. John Stuart Mill, *System of Logic,* CW 8:841.
110. Ibid., 7:394.
111. Ibid., 7:390.
112. Ibid., 7:396.
113. Ibid., 7:397–98.
114. Ibid., 7:401.

as follows: "Whatever antecedent cannot be excluded without preventing the phenomenon, is the cause, or a condition, of that phenomenon: Whatever consequent can be excluded, with no other difference in the antecedents than the absence of a particular one, is the effect of that one."[115] In order to conduct this method we must be able to produce the consequent by means of producing the antecedent, and to exclude the consequent by excluding the antecedent.[116] For all the methods of direct induction, the phenomena purported to be the antecedent and consequent must be not merely observable, but actually observed; if we cannot produce the phenomena as we must for the method of difference, we must at least have observed and accurately measured them.[117]

According to Mill, the methods of direct induction are sufficient to discover and prove laws in cases in which a single cause acts alone to produce an effect. This includes cases in which there is a "plurality of causes," a number of different antecedent conditions that can each, individually, cause the same effect. (For example, there are many different causes of death, but often only one cause for a particular death.) When there is such a plurality of causes, Mill claimed, the method of difference is sufficient to find the cause acting in a given instance. (The method of agreement becomes less useful, however, and can only be used upon a great number of instances, in order to rule out the supposition of more than one cause operating in a given case.)[118] But the methods of experimental inquiry are not sufficient in certain cases of the "intermixture of effects," that is, when multiple causes interfere with each other to produce a single effect (which is thereby "intermixed" or intermingled causally). The most common type of the intermixture of effects, which Mill called "the composition of causes," requires an alternative method for discovering and proving the individual causes involved. His example of such instances is that of a body kept in equilibrium by two equal and contrary forces.[119] In order to discover the law governing the combined effect, it is necessary to use a method Mill calls the "deductive, or a priori" method. This three-step method "considers the causes separately, and infers the effect from the balance of the different tendencies which produce it."[120]

When the causes can be physically separated from each other, the first

115. Ibid., 7:391.
116. Ibid., 7:390; see also p. 386.
117. Ibid., 7:393.
118. Ibid., 7:439.
119. Ibid., 7:443.
120. Ibid., 7:453.

step of the deductive method is a "direct induction" to the law of each of the individual causes. As Mill put it, this step involves "a process of observation and experiment upon each cause separately." When the individual causes cannot be separated in this way, the first step of the deductive method consists in a "previous deduction," that is, the deduction of complex laws from simpler ones we have already discovered using the methods of direct induction. In neither case can this first step of the deductive method yield laws concerning unobserved causes.[121] The second step of the deductive method is "ratiocination," or deducing the effects of any given combination of these single causes;[122] this is followed by "verification," or the empirical testing of these consequences.[123] Note that the methods of direct induction, when used in cases of single (non-intermixed) causes, do not require the second two steps in order to confirm the laws reached; in the case of an "intermixture of effects," ratiocination and verification are required to ensure that we have not neglected any of the concurring causes, nor included any that are not causes of the effect. Only in these more complex cases, then, does Mill require a deductive verification step in addition to inductive discovery.

Now, Mill allowed that scientists may dispense with his methods of direct induction, that is, the first step of the deductive method, and "hypothesize" laws that refer to unobserved causes. He pointed to the important heuristic value of such hypotheses, noting that they "may be very useful by suggesting a line of investigation which may possibly terminate in obtaining real proof."[124] However, he made it clear that these hypotheses cannot be proved to be true merely by the second and third steps of the deductive method, that is, by deducing empirical consequences from the hypotheses and then verifying them. Mill appears to have been contradicting his own methodology by claiming that Kepler's discovery was made by "guessing" the ellipse hypothesis, and then testing its fit with the observed data. Indeed, he insisted here that such hypotheses cannot be proved to be anything more than mere "possibilities" until the cause they postulate is "detected," or "brought conclusively to the test of observation"; that is, until the cause is observed and the causal law or laws proved by the methods of direct induction.[125] He explained that "to entitle an hypothesis to be received as one of the truths of nature, . . .

---

121. Ibid., 7:454.
122. Ibid., 7:458–59.
123. Ibid., 7:460.
124. Ibid., 7:495.
125. Ibid., 7:499.

it must be capable of being tested by the canons of legitimate induction, and must actually have been submitted to that test."[126]

It has been claimed that Mill allowed two exceptions to this stringent rule excluding all proposed causes that are not actually observed.[127] However, these alleged exceptions actually do not allow for the inference to unobserved entities. The first apparent exception is for cases in which it is possible to eliminate even the logical possibility of any other cause, that is, when it is known that the effect is possible if and only if this particular unobserved cause exists.[128] This can be considered a deductive, rather than experimental, version of the method of difference, and is potentially of importance because it can be used to prove laws in cases for which the conditions cannot be experimentally reproduced. But Mill cautioned that this "deductive-difference" argument is not possible when an unobserved cause is assumed to exist merely in order to account for some observed effects, even if it appears to be the only logically possible cause that could account for these effects (that is, when it is a "theoretical cause," as exemplified, in Mill's view, by the "luminiferous ether" postulated by the wave theorists of light). Rather, this argument is possible only when there is what Mill called a "known cause" operating by an unknown law.[129] To illustrate this point, Mill referred to Newton's law of universal gravitation. Before Newton it was known that the Sun is the cause of planetary motion, but the causal law governing this relation was still unknown.[130] Using this "deductive difference" argument, Newton proved, according to Mill, that the known facts of planetary motion could be true if and only if the inverse-square law of gravitation were true.[131] But note that Mill allowed only causes that are already known, that is, observed;

---

126. Ibid., 7:483. He elsewhere noted that "it is indispensable that the cause suggested by the hypothesis should be in its own nature susceptible of being proved by other evidence" (*CW* 7:485). Struan Jacobs argues that Mill's mature methodological view was actually a hypothetico-deductive one, because it allows for the proof of causal laws involving unobservable causes (see "John Stuart Mill on Induction and Hypotheses," p. 76). However, Jacobs admits that Mill does not succeed in showing how such proof is possible (p. 77). He claims that what makes Mill's view a hypothetico-deductive one is Mill's allowance that laws are often initially generated by a guess, and are later found to be supported independently by his methods of eliminative induction. But this type of view is not consistent with the hypothetico-deductive view, which denies that *any* inference from the initial data to a hypothesis is necessary in order to verify the hypothesis by testing whether its consequences are true. Mill's adherence to a hypothetical method in political economy in book 6 of the *Logic* (which I discuss in chapter 5) is not evidence of a shift to hypothetico-deductivism, because for him the conclusions of political economy remain only hypothetical; they cannot be confirmed by testing their consequences. See also Jacobs, *Science and British Liberalism*, chap. 8.

127. See John Skorupski, *John Stuart Mill*, pp. 199–202.

128. John Stuart Mill, *System of Logic*, CW 7:500.

129. Ibid., 7:494.

130. Ibid., 7:490.

131. Ibid., 7:492.

in this case the cause is the Sun. He also discussed the case in which the cause is an unknown (unobserved) cause (such as Descartes' vortices) operating by known laws (the laws of rotary motion). However, he claimed that any such law cannot be confirmed until the existence of the cause is proved by "direct evidence," that is, by being observed and proved by the method of difference.[132] Thus Mill required independent evidence for the law. Until this is obtained, the hypothesis "ought not to count for more than a suspicion."[133] So his "exception" here is not to the rule against postulating unobserved causes, but to the rule that specific causal laws can be discovered only by the experimental method of difference.

Mill's second apparent exception to his rule against unobserved causes is the case of causes that are unobserved due to spatial and temporal distance from us. In the case of geological and astronomical theories, we may inductively generalize known causes and laws from our present, earthbound experience to events in distant time and space.[134] It does not appear that Mill had grounds for treating this kind of inference to unobserved causes as legitimate while denying the legitimacy of inferences to causes unobserved for other reasons besides spatial/temporal distance, such as, for instance, smallness of size.[135] In any event, as in the first "exception," this is allowed only if there is no other known cause that could have produced the effects.[136] Moreover, only causes that have been observed can be generalized to these distant events; Mill asserted that in such hypotheses there should be "no unknown substance . . . nor any unknown property or law."[137] Thus he claimed that "if we find, on and beneath the surface of our planet, masses exactly similar to deposits from water, or to results of the cooling of matter melted by fire, we may justly conclude that such has been their origin."[138] What is unobserved, then, is simply the presence of a particular instance of a known cause in a situation presumed to be similar to cases in which the cause has been observed. Mill weakened even this allowance by claiming that this kind of inductive evidence is "enough, for most purposes," suggesting that it is not enough to

---

132. Ibid., 7:499.

133. Ibid., 7:496. In the 1872 edition, Mill changed this phrase to the stronger claim that such a hypothesis "ought only to count for a more or less plausible conjecture."

134. As we will see in the next chapter, Whewell rejected the position (known, in his own terminology, as "uniformitarianism") that we must assume that only analogous causes of the same intensity were at work in the history of the earth and the universe. See William Whewell, *History of the Inductive Sciences*, 3:506–20 and *The History of Scientific Ideas*, 2:257–96.

135. I thus disagree with John Skorupski, *John Stuart Mill*, p. 210.

136. John Stuart Mill, *System of Logic*, CW 7:506.

137. Ibid., 7:507.

138. Ibid., 7:506.

prove such proposed laws true or even probable. Even in the case of these so-called exceptions, then, Mill rejected the inductive inference to unobservable causes.

Given Mill's position on unobservable entities, there are two peculiar discussions in the *Logic*. One has to do with molecules. He claimed that "we have proof by our senses of the existence of molecular motion among the particles of all heated bodies,"[139] and elsewhere that "we have positive evidence of the existence of molecular motion in these manifestations of force."[140] In these passages Mill suggested that these particles are known causes, and that we know them via our senses. However, molecular motion was *not* observed at this time; it was, rather, inferred from the motion of microscopic particles suspended in liquid or gas. (In 1828 Robert Brown had published his study of such movement, later known as Brownian motion.) To understand Mill's claim here, it is important to note that both passages were added in the final (eighth) edition published during his lifetime, in 1872. One year earlier James Clerk Maxwell's *Theory of Heat* was published. In it, Maxwell had argued that the *only* possible cause of the observed phenomena of heat transfer was motion within the radiating body, that is, molecular motion. It may have seemed to Mill, therefore, that "molecules in motion" fulfilled the requirements of the "deductive-difference" method. But these requirements could not be completely satisfied, for molecules were not, at the time, capable of being "brought to the test of observation." And indeed, there is a contradictory passage unchanged from the original edition, in which Mill agreed that it would be "an important addition to our knowledge, *if proved*, that certain motions in the particles of bodies are the conditions of the production of heat and light."[141] He suggested in this passage that such molecular motions were *not* known causes. For this reason it is likely that Mill may have seen Maxwell's work or its reviews and hastily added the initial passages without enough thought as to whether the case really satisfied the requirements of the "deductive-difference" method. As we saw, Mill believed that in these cases, until the cause is "brought to the test of observation," it remains merely hypothetical.

Another problematic discussion concerns the luminiferous ether, a theoretical entity proposed by adherents of the wave theory of light. In the first and second editions, Mill noted of the ether that "it can never be brought to the test of observation, because the ether is supposed wanting in all the properties by means of which our senses take cognizance of external phenomena.

139. Ibid., 7:505.
140. Ibid., 7:350.
141. Ibid., 8:787; my emphasis.

It can neither be seen, heard, smelt, tasted, nor touched."[142] In his response to Mill, published in 1849, Whewell ridiculed him for this comment. Perhaps because of this, Mill omitted the phrase from the third (1851) and later editions of the *Logic*. Indeed, he allowed in the later editions that the luminiferous ether is "not in its own nature entirely cut off from the possibility of direct evidence in its favor." He referred to the case of Encke's comet, for which there had been found to be a difference between the calculated and the observed times of its periodic return. It had been conjectured that this was due to a medium diffused through space that was capable of opposing resistance to motion.[143] Mill noted that if there were an accumulation of similar variances in the orbits of other celestial bodies, "the luminiferous ether would have made a considerable advance towards the character of a *vera causa*," or true cause.[144]

It may appear, then, that Mill allowed that an unobservable cause might be proved by his inductive methods. However, that this was not his intention is clear from the continuation of the passage. Even if we had such an accumulation of cases of variance of calculated and observed periodic times, Mill noted, it would still be necessary to prove that the luminiferous ether postulated by wave theorists was identical to this "cosmical agent" causing the effects on planetary motions, and this would raise "many difficulties." Moreover, in the case of "such a hypothesis," we are unable to meet the main requirement for the deductive-difference method—namely, that the hypothesis is the only one that could possibly account for the known facts.[145] Why is this so? Because in the case of an unobservable cause, there is no way to rule out the possibility of a conflicting hypothesis that also accounts for all the known phenomena. As Mill wrote, "Most thinkers of any degree of sobriety allow, that an hypothesis of this kind is not to be received as probably true because it accounts for all the known phenomena; since this is a condition sometimes fulfilled tolerably well by two conflicting hypotheses; while there are probably many others which are equally possible, but which, for want of anything analogous, our minds are unfitted to conceive."[146] Thus, in all editions of the *Logic*, Mill continued to deny that we could have even probable knowledge of a luminiferous ether as the cause of optical phenomena.

142. Ibid., 7:499.
143. Whewell more accurately described this case in the third edition of his *History of the Inductive Sciences*, 2:468–69.
144. John Stuart Mill, *System of Logic, CW* 7:499.
145. Ibid., 7:500.
146. Ibid.

Mill's Metaphysics

We have seen that in *System of Logic*, Mill endorsed an ultra-empiricist epistemology, one that rejected necessity in logic, mathematics, and causal relations. His epistemology led to a scientific methodology that denied the possibility of inductive discovery of theories postulating unobservable entities or events. This epistemology was based on his adherence to the "relativity of knowledge" principle, which held that the materials of our knowledge consist solely in our sensations. As Mill put it in the *Logic*, "Of the outward world we can know absolutely nothing, except the sensations which we experience from it."[147] Nevertheless, while holding that the materials of knowledge consist solely in our sensations, Mill's metaphysical commitment in the *Logic* appears to be a "commonsense" form of realism, namely the view that material objects exist externally to us and are the cause of our sensations. Yet later, in his book *An Examination of Sir William Hamilton's Philosophy*, published in 1865, he seems to support two different metaphysical views, both of which differ from the commonsense realism of the *Logic*. At times in the *Examination*, Mill argued for a phenomenalist immaterialism, holding that material objects do not exist except as groups of actual and possible sensations. By being immaterialist, this view obviously conflicts with commonsense realism; yet by preserving the notion of a world outside the mind, a world common to different subjects, the view can be considered a form of realism. However, other comments in this work appear to support an idealist metaphysics, denying the existence of both material objects and a mind-independent reality.

Mill's writing *An Examination of Sir William Hamilton's Philosophy* was sparked by the republication of Hamilton's *Lectures on Metaphysics and Logic*, which Mill had intended at first only to review.[148] While in the *Logic* Mill claimed to be writing a book in which proponents of different metaphysical views could "meet and join hands," because "logic is common ground,"[149] the *Examination* is more openly polemical. As he related in the *Autobiography*, in the 1860s Mill believed the time had come for a "hand to hand fight" between the philosophies of experience and intuition.[150] His more ecumenical approach in the *Logic* had not had the effect he had desired, of purging science and mathematics of intuitionism. Certainly, it had not

147. Ibid., 7:62. See also John Stuart Mill, "Coleridge," *CW* 10.
148. See John Stuart Mill, *Autobiography*, *CW* 1:270.
149. See John Stuart Mill, *System of Logic*, *CW* 7:14.
150. John Stuart Mill, *Autobiography*, *CW* 1:270.

stopped the advance of what Mill considered "intuitionism" in moral philosophy; as we will see in chapter 4, the attacks of Whewell and others on utilitarian moral philosophy continued, culminating in a debate between Mill and Whewell on this topic in the early 1850s. Moreover, it was during the 1860s that Mill reentered public life and the political arena, from which he had withdrawn some years earlier after his marriage to Harriet Taylor in 1851. He won his seat in Parliament the same year of the initial publication of the *Examination.*

Mill's renewed interest in both politics and defeating intuitionism was related to his concern over the Civil War in America. He explained that "my strongest feelings were engaged in this struggle, which, I felt from the beginning, was destined to be a turning point, for good or evil, of the course of human affairs."[151] He worried about the "perverted state of public opinion" in England, which favored the Southern slave-holding states. It was soon after writing two articles supporting the Northern states that Mill turned to his study of Hamilton. In describing his motivation for writing the *Examination,* Mill explained,

> I have long felt that the prevailing tendency to regard all the marked distinctions of human character as innate, and in the main indelible, and to ignore the irresistible proofs that by far the greater part of those differences, whether between individuals, races, or sexes, are such as not only might but naturally would be produced by differences in circumstances, is one of the chief hindrances to the rational treatment of great social questions, and one of the greatest stumbling blocks to human improvement. This tendency has its source in the intuitional metaphysics.[152]

Mill continued to see the intuitionist philosophy as the support of conservatism; indeed, in the *Examination* he often referred to Hamilton as a political conservative, although he had actually been a Whig and not a Tory.[153] Thus it is no coincidence that Mill's renewed attacks on intuitionism arose concurrently with his renewed activity in politics.

This book received the immediate and extensive attention of reviewers, to a much greater extent than had the *System of Logic.* H. L. Mansel, whose recent Bampton Lectures (published as *The Limits of Religious Thought*) were

151. Ibid., 1:266.
152. Ibid., 1:270.
153. See Alan Ryan, introduction to Mill's *Examination, CW* 9:xiii.

explicitly targeted by Mill in the *Examination*, wrote a long review that appeared a few months after the publication of the book.[154] Hamilton's former student James McCosh published a book defending the parts of Hamilton's philosophy that coincided with the Scottish commonsense philosophy, called *In Defense of Fundamental Truth*.[155] On the other hand, one of Mansel's opponents at Oxford, Mark Pattison, proclaimed that "the effect of Mr. Mill's review is the absolute annihilation of all Sir W. Hamilton's doctrines, opinions, of all he has written or taught."[156] Sales of the *Examination* exceeded even those of the *Logic:* the first edition of one thousand copies sold out almost immediately, prompting the preparation of a second edition within a few months; a third edition was published two years later, although it would have appeared sooner if Mill had not wanted to spend more time formulating responses to his many critics. A fourth edition was published in 1872, only five years after the initial publication.[157]

Hamilton (1788–1856) had been professor of logic and metaphysics at Edinburgh from 1836 until his death. In his day he was considered the most learned man alive.[158] Mill described him, in his *Autobiography*, as "the great fortress of the intuitional philosophy in this country."[159] Hamilton claimed to accept the "relativity of knowledge" principle, so called because it held that all knowledge is only of the "relative," or the phenomenal, denying that we could have knowledge of the "unconditioned," or the "thing in itself." However, as Mill complained in the *Examination*, Hamilton's commitment to this principle was half-hearted and inconsistent. At times Hamilton asserted, at other times he denied, that we could have knowledge of things in themselves, through innate principles of our thought.[160] Mill therefore put Hamilton in the category of the "intuitive philosophers" who believed in the existence of certain "ultimate" facts that are not the result of experience but rather of an original intuition, and that are therefore necessary truths. He charged that "in this Sir W. Hamilton is at one with the whole of his own section of the philosophical world; with Reid, with Stewart, with Cousin, with Whewell, and we may add, with Kant."[161]

154. In a letter to Bain, Mill called Mansel's *The Limits of Religious Thought* "detestable" and "absolutely loathsome." Mansel felt similarly about Mill; he believed that Mill's philosophy was "utterly mischievous" (see John Stuart Mill to Alexander Bain, January 7, 1863, *CW* 15:817 and 817n).

155. Alan Ryan, introduction to Mill's *Examination, CW* 9:xiii.

156. Mark Pattison, "J. S. Mill on Hamilton," p. 562.

157. John M. Robson, "Textual Introduction" to Mill's *Examination, CW* 9:lxix.

158. See Rudolf Metz, *One Hundred Years of British Philosophy,* p. 61.

159. John Stuart Mill, *Autobiography, CW* 1:270.

160. See John Stuart Mill to Alexander Bain, January 7, 1863, *CW* 15:817.

161. John Stuart Mill, *Examination of Sir William Hamilton's Philosophy, CW* 9:143.

Mill focused on Hamilton's claim that we have intuitive knowledge of the existence of an external world. Mill's argument against this claim provides a model for the types of associationist explanations he believed it possible to give for all so-called innate ideas. Mill complained that the intuitionists use a faulty "introspective method" to reach the conclusion that we have an innate idea of an external world. Instead of showing that a belief could not have been acquired and thus must be an "original fact of consciousness," these philosophers conclude it is original merely because it appears to them that "our consciousness cannot get rid of it now."[162] Mill countered, "That a belief or knowledge . . . is in our consciousness now . . . is no reason for concluding that it was there from the beginning, until we have settled the question whether it could possibly have been brought in since."[163] That it could have been brought in since is precisely what Mill's own "psychological method" claimed to demonstrate. This method constitutes the required "scientific investigation" of the mind, in order to determine whether beliefs we consider innate could actually be inferences from external sensations, which we mistakenly take to be intuitions.[164] Not surprisingly, Mill argued that our belief in an external world is an acquired belief, and is therefore not an innate idea.[165]

As Mill admitted, his argument was influenced by the views of David Hartley, whose *Observations on Man* had made a strong impression on Mill when he was young.[166] It was also much influenced by the work of James Mill, especially his *Analysis of the Phenomena of the Human Mind* (1829), which John Mill later edited with the addition of his own notes. Mill explained in the *Examination* that we acquire our belief in the existence of an external world through experience. Our numerous experiences of sensations lead us, by the mechanism of "expectation," to form the concept of "possible sensations": we have the expectation that, were we to repeat the conditions under which we have experienced a certain sensation, we would once again have a similar sensation. Moreover, we experience individual sensations as occurring in stable groups. When we experience these sensations often enough in certain groups, we are led, by psychological laws of association of ideas, to link the individual sensations together.[167] When we think of one of these

162. Ibid., 9:146.
163. Ibid., 9:140.
164. Ibid., 9:141.
165. Note that Mill therefore seemed to accept that any genuinely innate belief *would* be a necessary truth.
166. See John Stuart Mill, *Autobiography*, CW 1:71, and Geoffrey Scarre, *Logic and Reality in the Philosophy of John Stuart Mill*, pp. 173ff.
167. Following his father and David Hartley, Mill noted that there are four laws of association. The first states that "similar phenomena tend to be thought of together." The second extends the

possible sensations, we are bound by the laws of association to think of the whole group of them, and we come to believe that the group comprises one thing, a kind of "permanent substratum" underlying and causing our actual sensations. Once this occurs, Mill claimed further, we come to believe that these possible sensations are intrinsically different from our actual sensations.[168] Moreover, we come to believe that possible sensations, unlike our actual sensations, have an existence outside us, that they comprise an external world. We also have a psychological tendency, Mill explained, to posit antecedent causes for all events, so we assume that our possible sensations are the cause of our actual sensations.[169]

It is in this way that we acquire our belief in a material world independent of, and causally related to, our actual sensations. Our belief in the existence of a material, external world can therefore be explained by the psychological method as arising from our experience and certain psychological laws. We do not have an original intuition of a material, external world; rather, we make an inference to its existence. But because the inference is so habitual, we do not recognize it as an inference, and mistakenly believe it to be an innate idea.[170] Mill explicitly tied the goal of exposing this error to his political motives. In his *Autobiography*, he described how "the practical reformer has continually to demand that changes be made in things which are supported

---

tendency of association to phenomena that are experienced or conceived either simultaneously or successively. The third law of association states that as repetition of the contiguity outlined in the second law increases, so does the certainty and rapidity of the association. Moreover, the third law states that eventually, when there are no counterinstances of the contiguity, this association becomes "inseparable" or "indissoluble," in that it becomes "irresistible"; i.e., we become incapable of thinking of one thing separated from the other. Finally, the fourth law of association states that when an association has become indissoluble in this way, "not only does the idea called up by association become, in our consciousness, inseparable from the idea which suggested it, but the facts or phenomena answering to those ideas come at last to seem inseparable in existence: things which we are unable to conceive apart, appear incapable of existing apart; and the belief we have in their coexistence, though really a product of experience, seems intuitive" (*Examination of Sir William Hamilton's Philosophy, CW* 9:177–78).

168. We believe that possible sensations are different from actual sensations in the following ways: (1) possible sensations seem to be permanent, in contrast with our actual sensations, which we experience as being fleeting and impermanent; (2) possible sensations seem independent from our will, unlike actual sensations, from which we can withdraw ourselves (e.g., by leaving a room in which there is a piece of white paper); (3) possible sensations seem to be trans-subjective, or shared by everyone alike, while actual sensations appear to vary from person to person.

169. See John Stuart Mill, *Examination of Sir William Hamilton's Philosophy, CW* 9:185–86.

170. Ibid., 9:249. Mill pointed out that Hamilton had expressed the principle of "parcimony [sic]," according to which it is forbidden to consider a principle of reasoning innate when it is possible to explain it from other known causes (p. 182). Thus Hamilton was, Mill argued, violating his own precepts by attributing innateness to our belief in an external world, a belief that can be explained as arising from experience.

by powerful and widely spread feelings, or to question the apparent necessity and indefeasibleness of established facts; and it is often an indispensable part of his argument to show, how these powerful feelings had their origin, and how those facts came to seem necessary and indefeasible."[171]

Mill, then, argued that we have the tendency to extend the laws of our own experience beyond our experience. We make an inference from our experienced actual sensations to the existence of a permanent, unexperienced substratum of material substance underlying (and causing) these sensations. However, in the *Examination* Mill cautioned that his argument was meant only to *account* for our acquisition of the belief in a material, external world, not to *endorse* the belief; he pointed out that "I assume only the tendency, but not the legitimacy of the tendency" to go beyond experience in this way.[172] Indeed, he rejected the commonsense belief in the existence of a material, external world, claiming that "matter . . . may be defined [solely as] a Permanent Possibility of Sensation. . . . If I am asked, whether I believe in matter, I ask whether the questioner accepts this definition of it. If he does, I believe in matter, and so do all Berkeleians. . . . In any other sense than this, I do not."[173] However, it is not always clear what precisely is the sense in which Mill saw himself as following Berkeley. Often, he seems to endorse a phenomenalist immaterialist metaphysics, rejecting the existence of material substance but—unlike Berkeley—accepting the existence of something external to minds. At other times, however, Mill seems to go further, rejecting, with Berkeley, the existence of anything besides sensations or ideas in individual minds.

Mill explicitly endorsed Berkeley's immaterialism when he explained that Berkeley realized that our experience could be fully accounted for without need of the hypothesis of physical substance; indeed, such a substance would be, Mill noted (recalling Nicolas Malebranche's phrase), a "superfluous wheel in the machinery."[174] Yet Mill appeared, much of the time, to reject Berkeley's subjectivism. In numerous places he postulated the existence of something external to minds and in common to different perceivers. Indeed, this objectivism underlies Mill's criticisms of Berkeley for holding that what persists through time are actual sensations—thus leading Berkeley to the "weak and illogical part" of his view, the invocation of a Divine Mind, in

171. John Stuart Mill, *Autobiography*, CW 1:269.

172. John Stuart Mill, *Examination of Sir William Hamilton's Philosophy*, CW 9:187.

173. Ibid., 9:183. Mill might have claimed that we have no *legitimate* grounds for believing in a material external world, without asserting that no such world exists.

174. John Stuart Mill, "Berkeley's Life and Writings," CW 11:462.

which these sensations can exist even when they are not perceived by any other mind.[175] Mill denied that actual sensations are what exist over time and regardless of whether they are perceived by particular minds. Rather, on his view, what persists is "a potentiality of having such sensations."[176] That is, as we have seen, Mill identified matter or objects with possible sensations, rather than actual sensations existing in minds. His view implies that the permanent possibilities of sensation need not exist in any given mind in order to exist; thus they are mind-independent, and objectivity is maintained. At other times in the *Examination* Mill made this more clear by referring to matter as the "power of exciting sensations."[177] He noted that Berkeley occasionally seemed to approximate this view, pointing to the following passage from Berkeley's *A Treatise concerning the Principles of Human Knowledge* in support of this claim: "The table I write on, I say, exists, that is, I see and feel it; and if I were out of my study I should say it existed—meaning thereby that *if I was in my study I might perceive it*, or that some other spirit actually does perceive it."[178]

Another passage of Berkeley's that Mill could have cited in support of the claim that Berkeley occasionally implied the "possibilities of sensation" view is one which is interesting for another reason as well. Admitting that the motion of the Earth is not perceived by the senses, Berkeley had noted that the question of the Earth's motion amounts to this: "If we were placed in such and such circumstances, and such or such a position and distance, both from the earth and the sun, we should perceive the former to move around the choir of the planets, and appearing in all respects like one of them."[179] This recalls Mill's similar use of a counterfactual situation in order to claim that the orbit of Mars is an observable, not theoretical, property. Mill

175. Ibid., 11:464; see also Alan Ryan, *The Philosophy of John Stuart Mill*, p. 97. We might see leanings toward Platonism here, in Mill's claim that what is real are ideas existing outside of and independently of minds. But Mill rejected the suggestion that his view was similar to Plato's, by noting that Plato's ideas were not identified with "common objects of sense," the way Mill's permanent possibilities of sensation are (see John Stuart Mill, "Berkeley's Life and Writings," *CW* 11:463).

176. Mill wrote of Berkeley: "He did not see clearly that the sensations I have to-day are not the same as those I had yesterday, which are gone, never to return; but are only exactly similar; and that what has been kept in continuous existence is but a potentiality of having such sensations, or, to express it in other words, a law of uniformity in nature, by virtue of which similar sensations might and would have recurred, at any intermediate time, under similar conditions. These sensations, which I did not have,. . .are not a positive entity subsisting through time: they did not exist as sensations, but as a guaranteed belief" ("Berkeley's Life and Writings," *CW* 11:464).

177. John Stuart Mill, *Examination of Sir William Hamilton's Philosophy, CW* 9:201.

178. George Berkeley, *A Treatise concerning the Principles of Human Knowledge;* quoted in John Stuart Mill, *Examination of Sir William Hamilton's Philosophy, CW* 9:460–61.

179. George Berkeley, *A Treatise concerning the Principles of Human Knowledge*, p. 123.

claimed that *if* we were in a privileged position, and *if* the planet left a visible trail, we *would* see (or, have the visual sensation of) the planetary orbit. Using this counterfactual, Mill argued that Kepler's discovery of the ellipse was not an inference, because it was a (possible) observation—or, as we might say now, a possible sensation. It may be, then, that Mill's metaphysical position—and his interest in Berkeley—were partially responsible for his rather odd argument about Kepler's discovery.

In his discussions of Berkeley in the *Examination*, Mill thus appears to have taken a realist, though immaterialist, position. His realist stance is also exhibited in this work when he argued against the inconceivability criterion of necessity. In the *Examination*, Mill focused his attention less on Whewell and more on Herbert Spencer (later known as a champion of evolutionary theory and social Darwinism), with whom Mill had been friendly since the late 1840s, when Spencer became one of the editors of the *Economist*. While working on the *Examination*, Mill claimed that Spencer had "Out-Whewelled Whewell."[180] Spencer agreed with Mill that axioms of the sciences were learned by induction from experience, but unlike Mill he claimed that the inconceivability of its negation was the means to justify an axiom.[181] In the *Examination*, Mill argued that in general, inconceivability is often the result of strong association of ideas. However, he also claimed that even if this were not the case, and the inconceivability of certain axioms could be traced to certain capacities inherent in our minds, "this would not entitle us to infer, that what we are thus incapable of conceiving cannot exist. Such an inference would only be warrantable, if we could know *a priori* that we must have been created capable of conceiving whatever is capable of existing; that the universe of thought and that of reality . . . must have been framed in complete correspondence with one another . . . an assumption more destitute of evidence could scarcely be made, nor can one easily imagine any evidence that could prove it, unless it were revealed from above."[182] That is, Mill here suggested a distinction between the "universe of thought" and that of reality, implying that reality is something independent of our thinking of it. Here is one way that Mill's position differs from that of Berkeley, who did hold that conceivability and possibility coincide; Berkeley noted that "my conceiving

180. In a letter to Alexander Bain while working on the *Examination*, Mill complained that by claiming we cannot conceive the annihilation or diminution of force, Spencer had "Out-Whewelled Whewell" (November 22, 1863, *CW* 15:901).

181. See Geoffrey Scarre, *Logic and Reality in the Philosophy of John Stuart Mill*, p. 109, and Herbert Spencer, "Mill *versus* Hamilton."

182. John Stuart Mill, *Examination of Sir William Hamilton's Philosophy*, *CW* 9:68. Mill listed Friedrich Schelling and Hegel as those holding this incorrect view, but he might very well have included Whewell on the list, as he seemed not to have realized.

or imagining power does not extend beyond the possibility of real experience or perception."[183]

Moreover, in the chapter on logic in the *Examination*, Mill expressed a realist stance by insisting that the role of logic is to help us attain objective truths.[184] In numerous passages of this text, then, he presented a phenomenalist immaterialism consistent with realism. Yet Mill occasionally lapsed into a subjectivist idealism, suggesting that reality is mind-dependent after all, and thus that there are no objective truths, that is, truths independent of our minds. In a letter to Spencer discussing the *Examination*, he explicitly denied the objectivist interpretation of his idealism, writing that

> the ultimate elements in the analysis I hold to be themselves *states of mind*, viz.—sensations, memories of sensations, and expectations of sensation. I do not pretend to account for these, or to recognize anything in them beyond themselves and the order of their occurrence; but I do profess to analyze our other states of consciousness into them. Now I maintain that these are the only substratum I need to postulate; and that when anything else seems to be postulated, it is only because of the erroneous theory on which all our language is constructed.[185]

In a note added to the 1867 edition of the *Examination*, Mill claimed that his argument had shown that there is no ground for either of the two opinions that provide "the evidence on which Matter is believed to exist independently of our minds"—that is, the opinion that we directly perceive matter by our senses (the argument of Thomas Reid), or the intuitionist claim that we come to believe it by "an original law of our nature" (the argument of Hamilton). Instead, Mill concluded, he had shown that the notion of matter could have come to us by psychological laws, and "without being a revelation of *any* objective reality."[186] Most interestingly, in another part of the *Examination*, he suggested that our acceptance of the possibilities of sensation "as permanent and independent of us" and our "projecting them into objectivity" are

<hr>

183. George Berkeley, *A Treatise concerning the Principles of Human Knowledge*, p. 104.

184. John Stuart Mill, *Examination of Sir William Hamilton's Philosophy, CW* 9:370–71.

185. John Stuart Mill to Herbert Spencer, August 12, 1865, *CW* 16:1090; emphasis added. In a paper written after Mill's death, Spencer noted that "in upholding Realism, I had opposed in decided ways those metaphysical systems to which [Mill's] own Idealism was closely allied" ("His Moral Character," in E. H. Fox Bourne et al., *John Stuart Mill*, p. 41).

186. John Stuart Mill, *Examination of Sir William Hamilton's Philosophy, CW* 9:204n; emphasis added.

the result of a hypothesis. This hypothesis is accepted as true, because it leads to empirical consequences that turn out to be confirmed: that is, we "recognize other *sentient* beings as existing, and receive impressions from them which entirely accord with this hypothesis, [so] we accept the hypothesis as truth, and believe that the Permanent Possibilities of Sensation really are common to ourselves and other beings."[187] But we have already seen that Mill denied hypotheses—even those that lead to successful predictions—the status of empirical truths. Unless he is contradicting his own methodological dictates, he seems here to be simultaneously providing a psychological explanation of our belief in an objective (though immaterial), external world, and exposing this belief as not well founded.[188]

Mill, then, presented two conflicting metaphysical positions in the *Examination*, one an immaterial realism, the other a Berkelean idealism. He did not appear to be aware of any conflict, either between these two viewpoints or between either of these and the view underlying his discussions in the *Logic*. Rather, he believed the metaphysics of the *Examination* to be consistent with the position he supported in the *Logic*. Thus to determine which stance is the one Mill intended to take, it will be useful to decide what view is presented in the *Logic*.

An idealist metaphysics can be construed as consistent with only very little of what Mill wrote in *System of Logic*. The few comments in the text that correlate with this view were added in the 1865 edition (that is, while Mill was writing the book on Hamilton), and serve mostly to qualify some of his more strongly realist comments from the earlier editions. For example, in a footnote added to the *Logic* in 1865, Mill outlined the idealist position as that which denies the existence of a substratum existing behind and causing our sensations (or, as he put it here, as the position that "denies Noumena"). He did not argue for the truth or falsity of the position, but claimed of it merely that "I have, as a metaphysician, no quarrel."[189] In the chapter "Of Things Denoted by Names," Mill similarly added phrases that qualify some of his realist-sounding statements from the earlier editions. For example, he had originally written that "mind is the mysterious something which feels and

187. Ibid., 9:202.
188. Moreover, there is a sense in which his phenomenalism seems to lead to subjectivism. For one thing, his very use of the term *external* for the permanent possibilities of sensation is odd. Mill claimed that what is meant by their being external to our minds is not external to us spatially; rather, they are "external" in the sense of not being part of any actual sensation I may be having now. As other commentators have noted, this is a very thin sense of externality. See, e.g., Alan Ryan, *J. S. Mill*, pp. 221–23.
189. John Stuart Mill, *System of Logic*, CW 7:62–63.

thinks" and "body is the mysterious something which excites the mind to feel." In the 1865 edition, Mill revised this second clause to read, "body is *understood to be* the mysterious something which excites the mind to feel," suggesting that it is not, after all, any external thing that "excites the mind to feel."[190]

Nevertheless, the preponderance of evidence points toward Mill holding some kind of realist position when he wrote the *Logic*.[191] In his introduction to the book, he noted the centrality of logic for all the sciences, explaining that all branches of science must be bound by the relations of logic, "under the penalty of making false inferences—of drawing conclusions which are not grounded in the realities of things."[192] Later, he distinguished between ideas and reality, noting that our beliefs must have reference to "outward things"; for example, of the belief that "gold is yellow," he wrote, "What I believe, is a fact relating to the outward thing, gold, and to the impression made by that outward thing upon the human organs; not a fact relating to my conception of gold, which would be a fact in my mental history, not a fact of external nature."[193] In other words, in addition to the sensation of gold experienced by us, there is a further object that also exists, gold itself.

In a sense, the whole project of the *Logic*, which involves the discovery of objective truth by the use of appropriate patterns of reasoning, entails a realist metaphysical stance; Mill was wrong to suggest that his task in the book is "neutral" with regard to metaphysical questions. The *Logic* concerns the proper ways to use the given materials of our knowledge to infer objective truths, which entails that there are such things as truths independent of our minds and our experience. Indeed, as we will see in the following section, Mill criticized Whewell precisely for seeming to abandon a commitment to this kind of realism, by claiming that scientific truths can be constructed using conceptions that originate in our minds; Mill suggested (wrongly) that Whewell's view was a conventionalist or idealist one, rather than a realist one.[194]

Mill also clearly exhibited a realist stance in his criticisms of associa-

---

190. Ibid., 7:63; emphasis added.
191. As Leslie Stephen put it, "His whole purpose there is to show that thoughts should conform to things" (Leslie Stephen, *The English Utilitarians*, vol. 3, *John Stuart Mill*, p. 404).
192. John Stuart Mill, *System of Logic*, CW 7:10–11.
193. Ibid., 7:88.
194. Because of Mill's insistence that logic aids us in discovering objective truths, I reject the claim, made by some, that his logic was psychologistic in the sense of claiming that laws of logic merely describe patterns of human thought. If this were Mill's view, Mill would then face the problem of those asserting the inconceivability criterion of necessary truths: namely, to explain why ideas or laws of thought in our minds should accurately portray the world as it exists. For a systematic argument against the psychologistic interpretation of Mill's view, see Geoffrey Scarre, *Logic and Reality in the Philosophy of John Stuart Mill*, pp. 110–25.

tionism. Although he used the theory in the *Examination* as part of his psychological method for accounting for beliefs that seem innate, he had begun to doubt the associationist account of belief formation decades earlier. In particular, he worried that the theory did not provide a means to distinguish between belief and imagination; thus he wrote in his 1869 notes to his father's *Analysis* that "the difference between belief, and mere imagination, is the difference between recognizing something as a reality in nature, and regarding it as a mere thought of our own."[195] Belief, Mill here claimed, aims at truth of an objective kind, about "realities in nature," while imagination does not. Another way to put this is by saying that belief, and not imagination, is tightly bound up with the issue of evidence.[196] Indeed, in the *System of Logic*, Mill had already made this point by noting that evidence is not what the mind does in fact yield to, but what it ought to yield to.[197] Evidence, contrary to the suggestions of his father, should be based on causation as discovered by the empirical methods of induction, not merely casual association. Indeed, an unquestioned faith in the value of associationism could lead to exactly the kind of conservatism of thought that Mill abhorred in the intuitionists. Without making this explicit, he did note that "a mere strong association of ideas often causes a belief so intense as to be unshakable by experience or argument."[198] Another worry about associationism arose for Mill as well. In his notes to his father's work, he complained that associationism "seems to annihilate all distinction between the belief of the wise, which is regulated by evidence, and conforms to the real successions and coexistences of the universe, and the belief of fools, which is mechanically produced by any accidental association that suggests the idea of a succession or coexistence to the mind."[199] Even in his discussions of logic and metaphysics, Mill was concerned to distinguish between the intellectual "clerisy," or "the wise," and the mass of men, or "the fools."

Rather than seeing a "fundamental cleavage" between Mill's views in the

195. James Mill, *Analysis of the Phenomena of the Human Mind*, 1:418.

196. In the *Examination*, Mill makes a similar point when he describes knowledge as something like "justified true belief": "The first requisite which, by universal admission, a belief must possess, to constitute it knowledge, is that it be true. The second is, that it be well grounded; for what we believe by accident, or on evidence not sufficient, we are not said to know" (*CW* 9:65n.). Note that this suggests that even his epistemology often exhibits a realist bent, not only his "philosophy of logic," as Geoffrey Scarre claims (see *Logic and Reality in the Philosophy of John Stuart Mill*, pp. 109ff.).

197. See Oskar Kubitz, *The Development of John Stuart Mill's "System of Logic,"* p. 185. See also John Stuart Mill, "Bain's Psychology," *CW* 11:341–74.

198. John Stuart Mill, *System of Logic*, *CW* 7:564; quoted in Oskar Kubitz, *The Development of John Stuart Mill's "System of Logic,"* p. 185.

199. James Mill, *Analysis of the Phenomena of the Human Mind*, 1:407.

Logic and in the *Examination*, I believe that his metaphysical commitment in both works is a realist one.[200] Too much of his work in both books concerns the relation between our thoughts and existing things for his position to have been the subjectivist one his remarks occasionally suggest.[201] But the question that still remains is whether the form of realism Mill supported remained the same over time. As I have noted, his mode of expression in the *Logic* often suggests a materialist position, that is, the "commonsense" realism view. He speaks of objects—such as gold—being the cause of our actual sensations, without reference to "permanent possibilities of sensation." Yet by the time of the *Examination*, his view was clearly an immaterialist one. Did his view change between the first edition of the *Logic* in 1843 and the publication of the *Examination* in 1865?

The answer seems to be that it did not. Mill's explanation of our acquisition of the belief in an external world is basically identical to that given by James Mill in his *Analysis of the Phenomena of the Human Mind*.[202] We know, from John Mill's *Autobiography*, that the reading group that used to meet at the Grotes' during the mid-to-late 1820s reconvened in order to read the *Analysis* when it was published. It is also known that the group discussed the question of our belief in the existence of matter, and the constitution of matter itself.[203] There is no reason to think that Mill disagreed with his father's view at that time; moreover, when he published an edition of the work with his own notes appended, he did not raise any criticisms of this part of his father's argument.

Why, then, did Mill not divulge this view of matter in the *Logic*? It may be that, as Mill claimed, he believed logic was "neutral" on the issue of metaphysics, and he wanted members of different metaphysical schools to be able

200. Because of an alleged inconsistency between the realism in the *Logic* and the idealism in the *Examination*, Geoffrey Scarre has characterized Mill's philosophy as containing a "fundamental cleavage" (*Logic and Reality in the Philosophy of John Stuart Mill*, p. 161). R. P. Anschutz accuses Mill of trying to satisfy the followers of Berkeley as well as the realists (*The Philosophy of J. S. Mill*, p. 178), while Alan Ryan claims that Mill simply could not make up his mind on this issue (*J. S. Mill*, p. 222).

201. Nevertheless, perhaps because of the ambiguity, many of Mill's readers interpreted his view as a subjective, idealist one. H. F. O'Hanlon strongly criticized Mill for his philosophy of "Pure Idealism" (*A Criticism of John Stuart Mill's Pure Idealism*). David Masson suggested that, at times, Mill's view is more purely idealist than even Berkeley's (*Recent British Philosophy*, pp. 64–66 and 294–95). John Grote also pointed to the ambiguity in Mill's writings, noting that at some points Mill claimed that "all physical fact or relation is in fact our feeling of such fact or relation" and at others that "it is conceivable that further physiological knowledge may resolve all our feeling . . . into a sort of refined physical fact" (*Exploratio Philosophica*, p. 162).

202. See James Mill, *Analysis of the Phenomena of the Human Mind*, 1st ed., 1:262–63.

203. See John M. Robson, introduction, *CW* 31:xxi, n50. One of the group's members, Henry Cole, left a diary in which he recorded the topics of the meetings he attended.

to agree with his positions on logic. But he also may have felt that the view was not likely to gain a warm reception; as Alexander Bain pointed out in his biography of James Mill, the immediate reaction to James Mill's book was not enthusiastic.[204] John Mill wrote to Bain while planning his work on Hamilton that "the great recommendation of this project is, that it will enable me to supply what was *prudently left deficient* in the *Logic*."[205] Hence Mill may have seen his omission of a detailed immaterialist stance in the *Logic* to have been "prudent."

By the 1860s, however, it may have seemed to him that the view would meet with a more positive reception. One reason for this has to do with developments in physics between the publication of the *Logic* in 1842 and the 1860s. In 1844, Michael Faraday had published a paper entitled "Speculation Touching Electrical Conduction and the Nature of Matter." In this work, he suggested that matter should not be defined as consisting of extended, impenetrable atoms (as on the Newtonian model), but instead as a plenum of "powers" filling space.[206] As John Tyndall later put it, Faraday "immaterialised matter into 'centers of force.'"[207] Faraday appealed to the view of Ruggiero Boscovich, who in his *De viribus vivis* of 1745 had put forward a similar view, claiming that atoms should be conceived as points from which forces of attraction and repulsion emanate.[208] Faraday explained, "The atoms of Boscovich appear to me to have a great advantage over the more usual notion. His atoms . . . are mere centers of forces or powers, not particles of matter, in which the powers themselves reside."[209] This view is echoed in

204. Alexander Bain noted that when the *Analysis* appeared, "philosophy was then at low water mark in this country" (*James Mill*, p. 330).

205. John Stuart Mill to Alexander Bain, December 1861, *CW* 15:752; emphasis added. See also Alexander Bain, *John Stuart Mill*, p. 118.

206. See Michael Faraday, "A Speculation Touching Electrical Conduction and the Nature of Matter," 1844. Reprinted in *Researches in Electricity* 2:284–93; quotation is on p. 290. See also P. M. Harman, *Energy, Force and Matter*, p. 77.

207. John Tyndall, *Faraday as a Discoverer*, p. 90.

208. Faraday was influenced by his mentor at the Royal Institution, Humphrey Davy, who appealed to Boscovich in his attempt to reconcile views of the German *Naturphilosophie* tradition he had been exposed to by Coleridge, with the facts of chemistry he was familiar with through his own work (see L. P. Williams, *The Origins of Field Theory*, pp. 68–70).

209. Michael Faraday, "A Speculation Touching Electrical Conduction and the Nature of Matter," 1844. Reprinted in *Researches in Electricity* 2:284–93; quotation is on p. 290. In his notes, Faraday wrote of his "final brooding impression that particles are only centers of force; that the force or forces constitute matter" (quoted in R. A. R. Tricker, *The Contributions of Faraday and Maxwell to Electrical Science*, p. 75). In a letter published in the *Philosophical Magazine* in 1850, Faraday referred to his "speculation" "respecting that view of the nature of matter which considers its ultimate atoms as centers of force, and not as so many little bodies surrounded by forces, the bodies being considered in the abstract as independent of the forces and capable of existing without them. In the latter view, these little particles have a definite form and a certain limited size; in

Mill's claim that matter is nothing but causal "powers," that is, "power[s] of exciting sensations,"[210] and not a "substratum" in which these powers reside. By the time he wrote the *Examination*, this Boscovichean position, extended into "field theory," had become more prevalent, especially in the writings of Maxwell, who was engaged in mathematizing Faraday's lines of force.[211] Mill was aware of these developments in physical theory. In a letter of 1863, he informed his correspondent that, "with regard to Matter, there has long been a growing tendency in thinkers to regard its particles as mere centers of force—even as local centers arbitrarily assumed to facilitate calculation and not implying the hypothesis of an absolute minimum."[212] The growing prominence of field theory in the 1860s may explain why Mill felt confident enough to express his phenomenalist immaterialism in the *Examination*.[213]

In the 1865 edition of the *Logic*, Mill added numerous phrases and footnotes making his immaterialism more explicit, at least as a possible metaphysical position. (He still maintained the claim that logic is neutral with respect to metaphysics.) But he did not change his mode of expression in other parts of the *Logic*, where he continued to speak of objects as if they were material things. The reason for this may be gleaned from this comment of Mill's in the *Examination:* "I affirm with confidence, that this conception

---

the former view such is not the case, for that which represents size may be considered as extending to any distance to which the lines of force of the particles extend: the particle indeed is supposed to exist only by these forces, and where they are it is" (Michael Faraday, "Thoughts on Ray-Vibrations"; reprinted in *Experimental Researches in Chemistry and Physics*, pp. 366–72; quotation is on p. 367). In his earlier work on electrostatics in the late 1830s, Faraday had already accepted the Boscovichean molecule (see L. P. Williams, *The Origins of Field Theory*, p. 86).

210. John Stuart Mill, *Examination of Sir William Hamilton's Philosophy, CW* 9:201.

211. John Herschel made reference to Boscovich and these developments in physics in his 1845 presidential address to the British Association for the Advancement of Science, in which he also praised Mill's *System of Logic* (p. xli).

212. See John Stuart Mill to J. Stuart Stuart-Glennie, July 23, 1863, *CW* 15:871.

213. Mill probably knew of the work of Faraday through Alexander Bain, who was surely aware of Faraday's 1844 paper, having been studying Faraday's work since at least 1840. Bain even wrote a paper entitled "On the Constitution of Matter" that was published in the *Westminster Review* in 1841. In this article he expressed a view of matter that is not quite the view put forward by Faraday in 1844, but shares some features with it: according to Bain, "The atoms of which bodies are composed are excessively small; they are not in contact with one another; every two adjoining atoms . . . are both attracted and repelled by one another" (see Alexander Bain, "On the Constitution of Matter," p. 83). Around the time he published the paper, Bain wrote his first letter to Mill at the urging of a mutual friend. When Mill responded, he praised Bain's article on matter (see Alexander Bain, *Autobiography*, p. 111). The manuscript of the letter is not extant; its existence and contents are known only via Bain's report (see *CW* 13:487). Mill may also have known something about Boscovich's work through the work of Joseph Priestley, the chemist, Unitarian minister, and early utilitarian, whose edition of David Hartley's *Observations on Man* was used by the reading group to which Mill belonged. In this work Priestley appended his own introductory essays, in one of which he discussed the view of matter expressed by "Father Boscovich" (see David Hartley, *Observations on Man*, p. xliii).

of Matter includes the whole meaning attached to it by the common world, apart from philosophical, and sometimes from theological, theories."[214] Unlike Berkeley, then, who believed one should "think with the learned, and speak with the vulgar," Mill believed he was both speaking and thinking with the vulgar.[215] That is, he thought that most people, though using language that suggests a commonsense realist view, in fact hold something closer to his own phenomenalist immaterialism. So Mill could continue to write in such a way as to suggest the existence of a material "substratum" underlying our possible and actual sensations, even while rejecting the claim that any such material realm exists.

One important consequence of Mill's immaterialist realist position bears on the issue of unobservable properties and entities. We have already seen that Mill's ultra-empiricism rejects the possibility of knowledge of unobservable entities and properties. Even his "deductive-difference" method cannot yield knowledge of unobservables; any postulated entities or properties remain "hypothetical" until they are "brought to the test of observation." However, Mill's metaphysical commitment leads to a stronger and more striking conclusion: he rejected the very *existence* of at least a certain class of unobservables. As we have seen, matter on Mill's view is nothing but the "permanent possibilities of sensation," or (and he seemed to think of these phrases as equivalent) "power[s] of exciting sensations." Unobservable entities or properties, which cannot possibly cause sensations, would seem therefore to have a problematic status for Mill. Entities or properties that may be unobserved, or even unobservable given our present location or abilities of our sense-enhancing instruments, can be said to exist, as long as it is possible that they could potentially exercise the power of causing sensations at a future time. But entities or properties that are "unobservable in principle"—such as the luminiferous ether, which was considered undetectable by any of our senses, no matter how augmented by instrumentation—cannot be said to exist on Mill's metaphysical view. Thus Mill not only rejected the possibility of having knowledge of these unobservables; he rejected the very possibility of their existence.

Whewell's Critique of the "Young Mill"

After the publication of *System of Logic* Jones wrote to Whewell, "Have you seen young Mill's book—. . . he wrestles stoutly with you—there is much

214. John Stuart Mill, *Examination of Sir William Hamilton's Philosophy*, CW 9:183.
215. George Berkeley, *A Treatise concerning the Principles of Human Knowledge*, p. 120.

that must interest us—He bestows a length of tedious pages on the first elements of dialectics which will assuredly frighten the public."[216] Whewell
replied by noting that "when he comes to induction he appears to me to write
like a man whose knowledge is new (indeed he confesses that he had much
of it from Herschel and me)—and not very well appreciated. . . . Of course I
shall not notice Mills's [sic] book til I come to a new edition, and probably no
otherwise than by altering, as I have said, the mode of presenting my doctrine."[217] Actually, Whewell did not directly respond to Mill even in the 1847
second edition of *The Philosophy of the Inductive Sciences.* Instead, two years
later he published a work devoted to answering Mill: *Of Induction, with Especial Reference to Mr. J. Stuart Mill's "System of Logic."*[218]

More than one-quarter of this work is devoted to a critique of Mill's view
of Kepler's discovery. According to Whewell, Mill illegitimately attempted
to set up a distinction between description and induction. He characterized
Mill as arguing in the following way: "When particular facts are bound
together by their relation in *space,* Mr. Mill calls the discovery of this connection *Description,* but when they are connected by other general relations,
as time, cause and the like, Mr. Mill terms the discovery of the connection
Induction."[219] Mill asserted that the discovery of Mars's orbit, in which the
particular facts were connected with the spatial property of an elliptical
curve, was merely a description, while the discovery that the planetary orbits
are connected by the causal property of being acted upon by the Sun's gravitational force was an induction.[220] Whewell claimed in his reply that there is
no obvious argument to warrant such a distinction. If inference is needed to
discover the property connecting particular facts, then it is wrong to call the
act a mere (noninferential) description, whether the property is a spatial one
such as an elliptical curve or a causal one such as gravitational force. Whewell noted "that the orbit of Mars is a Fact—a true description of the path—

216. Richard Jones to William Whewell, April 6, 1843, WP Add.Ms.c.52 f. 79.

217. William Whewell to Richard Jones, April 7, 1843, WP Add.Ms.c.51 f. 227. He told Jones
later that his 1844 essay "On the Fundamental Antithesis of Philosophy" "contains an answer to
J. Mill's main argument against me," but Mill's name is not mentioned in the essay (April 18, 1844,
WP Add.Ms.c.51 f. 235). One of the last paragraphs of the essay concerns Mill's critique of
Whewell's notion of necessary truth (see William Whewell, "On the Fundamental Antithesis of
Philosophy," p. 480).

218. In a letter to J. D. Forbes, Whewell explained, "I must also print an answer which I have long
been meditating to John Mill, though I am afraid in these times no one will read philosophy—even
controversial philosophy" (William Whewell to J. D. Forbes, October 10, 1849, WP O.15.47 f. 68).

219. William Whewell, *Of Induction,* p. 21.

220. John Stuart Mill, *System of Logic,* CW 7:299.

does not make it the less a case of Induction," because inference was needed in order to discover this true description.[221]

As we have seen, Mill's claim about Kepler arose from his desire to reject any internal or subjective element in knowledge. In *Of Induction*, Whewell recognized that the real point of contention between them here had to do with the *source* of a conception or colligating property used in inference. Recall that in Whewell's antithetical epistemology, the conceptions are modifications of Fundamental Ideas that are supplied by our minds in our contemplation of the world around us. Thus the conceptions used in colligating particular facts have an ideal, subjective source, yet correspond to the relations that exist in the physical world. Mill vehemently denied this claim. "Conceptions," he wrote, "do not develop themselves from within, but are impressed from without." That is, "the conception is not furnished *by* the mind till it has been furnished *to* the mind."[222] This is why he insisted, as we have seen above, that finding a colligating conception is merely a matter of describing what is observed outside the mind, rather than being an inference involving conceptions provided from within. In his response to Mill, Whewell claimed that Mill was ignoring how difficult it often is to find the appropriate conception with which to colligate a set of particular facts; indeed, as we have seen, Mill rather naively believed that Kepler's task in discovering the elliptical orbit was an almost trivially simple one.[223] In the *Logic*, Mill claimed that Whewell's "Colligation of the Facts by means of appropriate Conceptions, is but the ordinary process of finding by a comparison of phenomena, in what consists their agreement or resemblance."[224] Whewell rather archly responded by noting that "of course" discovering laws involves finding some general point in which all the particular facts agree, "but it

221. William Whewell, *Of Induction*, p. 23. Interestingly, Mill appeared to agree with Whewell about the inferential nature of observation. In his article on Berkeley's theory of vision, Mill claimed that "the information obtained through the eye consists of two things—sensations, and inferences from these sensations." He noted that sensations themselves "are merely colours variously arranged, and changes of colour, [and] that all else is inference, the work of the intellect, not of the eye" ("Bailey on Berkeley's Theory of Vision," *CW* 11:250). Mill was anxious to point out, however, that the inference is one of the association of ideas; we infer that a particular object or person is presently causing our sensations of color, and this inference is based on our past experience of having these same sensations while looking, for example, at a particular person (see p. 253). (Of course, given his view of matter described in the previous section, what Mill meant as the "cause" of our having certain visual sensations must merely be the presence of a certain cluster of actual and possible other sensations, not the presence of a material cause for the visual sensation.) But it is not the case that we contribute a conception from our minds in making this inference.

222. John Stuart Mill, *System of Logic*, CW 8:653, and p. 655.

223. William Whewell, *Of Induction*, p. 31.

224. John Stuart Mill, *System of Logic*, CW 8:648.

appears to me a most scanty, vague, and incomplete account" to suggest that the commonality is found merely by observation, with no inference at all.[225]

Moreover, Whewell used *Of Induction* to object to Mill's mischaracterization of his view as being a nonrealist one. Mill wrongly accused Whewell of believing that these conceptions, because they are supplied by the mind, are merely ideal, mental constructs that correspond to nothing in the world but simply organize our experience in useful ways.[226] Mill asserted, for instance, that Whewell held that the conception of an ellipse "did not exist in the facts themselves."[227] Whewell had argued against this understanding of his view in his response to Herschel's review of his works. Contrary to the way in which Herschel and Mill interpreted his view, Whewell believed that our conceptions do "exist in the facts," in the sense that they provide shared properties and relations that really exist between objects and events, even though it takes an "act of the intellect" to find them there. Of the conception of the ellipse, Whewell wrote, "Kepler found it in the facts, because it was there."[228] Drawing upon an analogy earlier employed by Bacon, Whewell claimed that Nature is a book, and her laws are written within it, but we cannot read these laws without knowing the language in which they are written. The laws exist "in the facts," but our acquisition of knowledge of the laws requires that we develop and use the rules of grammar that exist in our minds.[229] In chapter 1 we examined the theological justification for this claim. Mill was therefore wrong to claim that Whewell was a nonrealist. Of course, Whewell was not the same kind of realist as Mill; he clearly rejected the immaterialism of Mill and Berkeley.[230] And Mill would have disdained the theologically based realism of Whewell's epistemology, had he recognized it. Indeed, we have seen that Mill explicitly denied the suggestion that the "universe of thought and that of reality" have been made to coincide.

Whewell also objected to the radically empiricist tenor of Mill's philosophy. He recognized that Mill's inductivism rejected inference to theories referring to unobservable entities or properties. As we have seen, Whewell

225. See William Whewell, *Of Induction*, pp. 41–42.

226. Some modern commentators similarly interpret Whewell. See, for example, Gerd Buchdahl, "Deductivist versus Inductivist Approaches in the Philosophy of Science."

227. John Stuart Mill, *System of Logic*, CW 7:294.

228. William Whewell, *Of Induction*, p. 23.

229. Ibid., p. 34.

230. In an early letter from Cambridge to his grammar schoolmaster, Whewell suggested that his correspondent read Berkeley if he wanted to see how the "science of the Mind" could "triumph over common sense," and described this science as one in which "you may pass the bounds of space and time and travel on to all Eternity without coming to any conclusions, or rather may arrive at a dozen contradictory conclusions all equally certain" (William Whewell to George Morland, June 15, 1814, in Isaac Todhunter, *William Whewell, D.D.*, 2:5).

complained that "Mr. Mill rejects the hypothesis of a luminiferous ether, 'because it can neither be seen, heard, smelt, tasted, or touched.'"[231] Even though Mill deleted this passage from later editions of the *Logic*, he continued to reject the postulation of causes that are unobservable, or even unobserved. (Indeed, as we have seen, he rejected the existence of any entity that is unobservable in principle, because it could not have the power of exciting sensations.) In two later works, Whewell associated Mill with Comte, whom Whewell criticized for "rejecting the inquiry into causes."[232] Whewell's method, in contrast, did allow for the inquiry into unobservable causes; like Bacon, he believed this to be the ultimate aim of science. Thus Whewell noted that "to exclude such inquiries, would be to secure ourselves from the poison of error by abstaining from the banquet of truth."[233] The history of science shows that it is both possible and important to seek these kinds of causes.

Further, Whewell strongly criticized Mill for the methodology he developed based on this radically empiricist epistemology. In *Of Induction*, Whewell rejected the methods of experimental inquiry as extreme oversimplifications of scientific discovery.[234] Mill had noted that the method of difference can only be applied in controlled laboratory settings; he was correct that the method could be useful in such cases, as it is today, along with the other methods, in devising studies in medical research.[235] However, Mill had also called these methods "the *only* possible modes of experimental inquiry," and suggested that they are not difficult to apply, at least in the physical sciences.[236] Indeed, he claimed that he wanted to find rules of induction analogous to the rules of the syllogism, that is, something like a "discovery machine," in the very sense that Whewell rejected.[237]

But the real complaint Whewell had of Mill's methods concerned Mill's means of justifying them: "Who will carry these formulae through the history of the sciences, as they have really grown up; and shew us that these four methods have been operative in their formation?"[238] Mill himself had not justified his methods by showing that they have, in fact, been used to make

---

231. William Whewell, *Of Induction*, p. 63.

232. See William Whewell, "Comte and Positivism," p. 356, and *On the Philosophy of Discovery*, chap. 21.

233. William Whewell, *On the Philosophy of Discovery*, p. 233.

234. William Whewell, *Of Induction*, p. 44.

235. Interestingly, though, Mill himself claimed that his methods of experimental inquiry were *not* useful in medical science because of the intermixture of effects (see *CW* 7:451 and "On the Definition of Political Economy," *CW* 4:337–38).

236. John Stuart Mill, *System of Logic, CW* 7:406.

237. See ibid., 7:283, where Mill explained that he sought a method "which might be for induction itself what the rules of the syllogism are for the interpretation of induction."

238. William Whewell, *Of Induction*, p. 45.

successful discoveries. Instead of surveying the history of science to find
whether scientists have used his methods of experimental inquiry, Mill
focused his examples on a narrow range of poorly understood cases. He spent
much time on William Wells's researches on dew, which Herschel had discussed in the *Preliminary Discourse*. Whewell blamed Herschel for suggesting that one or two examples are sufficient to understand scientific discovery. As he wrote to Jones, "Tell Herschel he has something to answer for in
persuading people that they could so completely understand the process of
discovery from a single example."[239] Moreover, the Wells example was inappropriate to illustrate inductive methods, according to Whewell, because it
was not really an original discovery, but rather a deduction of particular phenomena from already established principles.[240]

Mill's other favorite example concerns the work of Justus von Liebig in
physiological chemistry, specifically his theories regarding the cause of
death.[241] But, as Whewell noted in *Of Induction*, there are two problems here.
First, Liebig's theories were not yet verified, so Mill could not know whether
they were real discoveries.[242] As Whewell complained to Jones, these theories
were so recent that "the most sagacious physiologists and chemists cannot
yet tell which of them will stand as real discoveries."[243] Moreover, the science
of physiology was a new and not well-established one. Whewell exclaimed,
"Nor can I think it judicious to take so large a proportion of our examples
from a region of science in which, of all parts of our material knowledge, the
conceptions both of ordinary persons and men of science themselves, are
most loose and obscure, and the genuine principles most contested."[244]
Whewell believed that it was only possible to understand the process of
reaching truths by examining fields of knowledge in which truth is uncontested. Thus he held that "the philosophy of science is to be extracted from
the portions of science which are universally allowed to be the most certainly
established. . . . The first step towards shewing how truth is to be discovered,
is to study some portion of it which is asserted to as beyond controversy." For

239. William Whewell to Richard Jones, April 7, 1843, WP Add.Ms.c.51 f. 227. Herschel's belief
in the power of one or two good examples to exemplify these rules is consistent with his own
methodological claim that one or two strong instances can be enough, in certain cases, to justify
an inductive generalization. Whewell's disagreement is similarly consistent with his own rejection
of hasty "anticipations of nature."
240. See William Whewell, *Of Induction*, p. 50.
241. Ibid., p. 47.
242. Ibid.
243. William Whewell to Richard Jones, April 7, 1843, WP Add.Ms.c.51 f. 227.
244. William Whewell, *Of Induction*, p. 48. See also letters to Richard Jones, August 5, 1834 (WP
Add.Ms.c.51 f. 174), and August 21, 1834 (in Isaac Todhunter, *William Whewell, D.D.*, 2:185–88).

this reason Whewell also objected to Mill's introduction of moral and political subjects into his discussion of induction in the *Logic* even though, as we will see in chapters 4 and 5, Whewell himself was quite concerned with these topics. (Indeed, his original plan for the *Philosophy* included the "hyperphysical branches" of science, including morality and political economy.)[245]

Scientific Method and History of Science

These last criticisms grew out of Whewell's view of the relation between philosophical discussion of scientific method and the history of science. It is well known that Whewell claimed to have developed his philosophy of science from his study of the history of science; he wrote his *History of the Inductive Sciences* (1837) before the *Philosophy* (1840), claiming that any philosophy of the sciences must be "founded upon their history." As he put it in the preface to the *History*, "It seemed to me that our study of the modes of discovering truth ought to be based upon a survey of the truths which have been discovered."[246] Some commentators have claimed that on the contrary, Whewell first developed an a priori philosophy of science and then shaped his *History of the Inductive Sciences* to conform to his own view.[247] To a limited extent this is no doubt true. As we have seen, from his days as an undergraduate at Trinity College Whewell considered his "vocation" the advancement of the inductive method in the sciences. So from the very beginning he had an inductive view of scientific method, which influenced his writing of the *History* in one important sense: by leading him to the view that learning about scientific method must be inductive and therefore historical. As he wrote to Jones in 1831, "I do not believe the principles of induction can be either taught or learned without many examples."[248]

Examples, then, are needed to fill out the details of this broadly inductive view, and they are to come from knowledge of both current science and the history of science. That Whewell expected the required instances to proceed from these two sources is indicated in numerous letters as well as in his early induction notebooks. His earliest attempt at a draft of a work on induction appears in a notebook dated 1830. In it Whewell claimed that in order to judge the methodology of Bacon, it was necessary "to shew how this method has been exhibited and exemplified since it was first delivered." In order to do

245. William Whewell, *Of Induction*, pp. 5–6.
246. William Whewell, *The History of the Inductive Sciences*, 1:viii.
247. See, for example, Marion Rush Stoll, *Whewell's Philosophy of Induction*, and E. W. Strong, "William Whewell and John Stuart Mill."
248. William Whewell to Richard Jones, February 25, 1831, WP Add.Ms.c.51 f.99.

so, it is necessary to discuss the history of science; this recognition may explain why Whewell put aside his draft and began working on the *History*.[249] In this notebook and in several that follow over the next three to four years, he recorded many notes on recent discoveries in science, as well as citations from contemporary scientific works in which scientists expressed a view of proper scientific method.[250] Yet there are also numerous entries describing the histories of various scientific fields, interwoven with Whewell's early thoughts on an inductive philosophy of science. In one of these notebooks alone, he took reading notes on Humphry Davy's *Elements of Chemical Philosophy*, William Gilbert's *De Magnete*, and David Brewster's book on Newton, as well as on works by Georges Cuvier, Nicholas Copernicus, Galileo Galilei, Leonardo da Vinci, and William Harvey; described recent discoveries in optics by Jean Baptiste Biot, Thomas Young, Fresnel, François Arago, and G. B. Airy; and discussed historical material, giving details about the work of Aristotle, Euclid, Plato, Alhazen (ibn al-Haytham), Newton, Roger Bacon, Brahe, Kepler, Christian Huyghens, and Joseph von Fraunhofer. Whewell then used this examination of the history of optics and current research in the field to outline "Steps of the Induction of the Theory of Light."[251] In a notebook dated 1831–32, a discussion of the use of conceptions in induction includes notes on the scientific work of Archimedes, Pascal, Aristotle, Descartes, Marin Mersenne, Galileo, and Evangelista Torricelli.[252] In another notebook, dated December 1833, Whewell discussed Herschel's *Preliminary Discourse*, pointing out one problem with his friend's work: namely, that "we do not here find the view of physical science to which we hope to be led:— that if its history and past progress be rightly studied we shall acquire confidence in truth of all kind."[253] Moreover, in several letters to Jones in 1834, Whewell described himself as eager to get to his philosophy of science but determined to finish the history first.[254]

It is important to note that while writing both the *History* and the *Philosophy*, Whewell attempted to ensure he had a real understanding of the sci-

249. See William Whewell, notebook, WP R.18.17. f.12; for quotation see p. xv.

250. For example, in an entry dated July 1831 there is a quotation from John Dalton followed by a comment by Whewell: "[This is] an excellent description of induction and a good proof of the difficulty in presenting things inductively" (WP R.18.17 f.15, p. 40). The induction notebook of 1830 contains an extensive discussion of recent discoveries in geology (WP R.18. 17 f.12, pp. xxiv–xxv and 1ff).

251. See William Whewell, notebook, WP R.18.17 f.13; the notebook is undated, but it is headed "Induction IV"; the other induction notebooks are dated from 1826 to 1832.

252. William Whewell, notebook, WP R.18.17 f.15.

253. William Whewell, notebook, WP R.18.17 f.8.

254. See William Whewell to Richard Jones, July 27, August 5, and August 6, 1834; WP Add.Ms.c.52 f.173, f.174, and f.175.

entific work he was describing. Contrary to Robert Brown's sneer about him ("Yes, I suppose that he has read the prefaces of very many books"),[255] he did more than simply read: he actively engaged in scientific research in several areas. For example, at the time Whewell announced himself a candidate for the Chair in Mineralogy at Trinity, he had already published a paper on crystallography, and two on three-dimensional geometry, which later set the foundation for mathematical crystallography.[256] But he did not have much empirical knowledge of mineralogy, so he duly set about studying it; as we saw earlier, Whewell studied with Friedrich Mohs in Berlin and Vienna.[257] While in Germany he was impressed by the natural classification system in mineralogy, and rejected the artificial system then in vogue in England. He published a monograph on mineralogy, and several more papers.[258] Whewell also performed experiments with his friend (and future Royal Astronomer) Airy, in order to determine the mean density of the earth. In June of 1826 both men went to Cornwall, where they spent time in the Dolcoath copper mine comparing the effect of gravity on pendulums at the surface and at a depth of twelve hundred feet. (Bacon had suggested this experiment in his *Novum Organum*.)[259]

Some years later, Whewell became interested in tidal research, based in part on his friend and former student John Lubbock's work on the topic. (It was also an area that had greatly interested Bacon, who wrote an essay on the topic, entitled "On the Ebb and Flow of the Sea.")[260] He helped Lubbock

255. Reported by Charles Darwin in his "Autobiography," p. 131. Darwin related this story in order to illustrate the point that Robert Brown, though possessing some positive qualities, "was rather given to sneering at anyone who wrote about what he did not fully understand: I remember praising Whewell's *History of the Inductive Sciences* to him, and he answered—."

256. See William Whewell, "On the Double Crystals of Flour Spa," "On the Angle Made by Two Planes, or Two Straight Lines Referred to Three Oblique Coordinates," and "A General Method of Calculating the Angles Made by Any Plane of Crystals, and the Laws according to Which they are Formed." See also Harvey W. Becher, "Voluntary Science in Nineteenth Century Cambridge University to the 1850s"; Michael Ruse, "William Whewell: Omniscientist," p. 99; and H. Deas, "Crystallography and Crystallographers in England in the Early 19th Century."

257. See letter to Hugh James Rose, August 15, 1825, WP R.2.99 f.125.

258. See William Whewell, *An Essay on Mineralogical Classification and Nomenclature*, and Harvey W. Becher, "Voluntary Science in Nineteenth-Century Cambridge University to the 1850s."

259. For a description of this experiment, see William Whewell, *Account of Experiments Made at Dolcoath Mine, in Cornwall, in 1826 and 1828*. See also Richard Yeo, *Defining Science*, p. 53. The experiment was less than a success. The entry on Whewell in the *Dictionary of National Biography* rather laconically explains that "accidents to the instruments employed were . . . fatal to the success of the experiments" (Leslie Stephen, "Whewell, William," p. 1366). The fuller truth is that they kept dropping the pendulums down the shaft (see William Whewell, *Account of the Experiments Made at Dolcoath Mine*, p. 8). Two years later they tried again, and published their findings.

260. In this essay, Bacon had postulated a northward progressing tide over the whole globe (see Francis Bacon, *Works*, 5:443–58). But he realized that there was not enough empirical data to support this supposition strongly, and he urged mariners to record tidal times along the coast of West

obtain grants from the newly formed British Association for the Advancement of Science for his research, and suggested the term "cotidal lines" to designate lines joining high-water times.[261] Whewell also pushed for a large-scale research project of tidal observations. Aided by Captain (later Admiral) Beaufort, Hydrographer of the Navy, and with the support of the Duke of Wellington, he managed to organize simultaneous observations of tides at 100 British coast guard stations for two weeks in June 1834. In June 1835 he organized three weeks of observations along the entire coast of northwestern Europe and eastern America, including 101 ports in 7 European countries, 28 in America from the mouth of the Mississippi to Nova Scotia, and 537 in the British Isles, including Ireland.[262] This project resulted in over 40,000 readings; Whewell himself reduced all the data.[263] Eventually he presented sixteen papers to the Royal Society and several summary reports to the BAAS on the topic of the tides between 1833 and 1850. In recognition of his work in tidal research, Whewell was awarded a gold medal by the Royal Society in 1837.[264]

Whewell, then, had first-hand knowledge of the methods of empirical research. But he was also up-to-date on the discoveries of other scientists. Indeed, he consulted numerous scientists about their own discoveries or those of others throughout the history of their respective fields, and sent proof sheets of the *History* to them for their approval. For instance, Whewell asked his friend Airy to look over his section on the history of astronomy and to suggest references to French works on the polarization of light.[265] He sent queries on physics to J. D. Forbes[266] and on anatomical science to Richard Owen. While revising the *History* for its second edition, Whewell asked Fara-

Africa, among other places. After Bacon, in the late seventeenth and eighteenth centuries, analysis of the tides became mainly mathematical, utilizing models which were overly idealized, such as that of Pierre-Simon Laplace, which postulated an ocean covering the whole globe. But by the nineteenth century even the scant empirical data that existed showed the inadequacy of such models.

261. In his 1832 report to the BAAS, John Lubbock referred to "a series of points which form the crest of the tide-wave, . . . which I have called, at the suggestion of Mr. Whewell, cotidal lines." Quoted in David Cartwright, *Tides*, p. 111. See also William Whewell, "Essay towards a First Approximation to a Map of Cotidal Lines."

262. Whewell noted that the observations were made June 8–28, and occurred in 28 places in America, 7 in Spain, 7 in Portugal, 16 in France, 5 in Belgium, 18 in the Netherlands, 24 in Denmark, 24 in Norway, 318 in England and Scotland, and 219 in Ireland. See William Whewell, "Researches on the Tides—6th series. On the Results of an Extensive System of Tide Observations Made on the coasts of Europe and America in June 1835." See also David Cartwright, *Tides*, pp. 112–14.

263. William Whewell, "Researches on the Tides—6th Series," p. 291.

264. William F. Cannon, "William Whewell," p. 183.

265. See letters to Whewell from G. B. Airy, October 11, 1856 (WP Add.Ms.a.200 f.114), and April 21, 1831 (WP Add.Ms.a.200 f.9).

266. See, e.g., William Whewell to James David Forbes, February 19, 1840, WP O.15.47 f.51a.

day to check the section on his researches.[267] (Faraday responded by assuring Whewell that there were no errors to be corrected.)

So, while it would be an overstatement to say that Whewell had no ideas about philosophy of science until after he completed all three volumes of the *History*, it would also be a vast understatement to suggest that the *History* was written to conform to a fully fleshed-out a priori methodological position. Knowledge of both current scientific practice and the history of science were important to Whewell in developing his philosophy of science.

On the other hand, Mill's relationship with science was rather less intimate. In his *Autobiography*, he noted the deficiency of his training in mathematics, and regretted his lack of schoolroom experience in conducting or even observing experiments. While staying with the family of Samuel Bentham (brother to Jeremy) in France in 1820, Mill attended lectures at the Faculté des Sciences in chemistry and zoology.[268] At this time his cousin George—who was to become the finest systematic botanist of nineteenth-century Britain—introduced him to the study of botany.[269] This early exposure developed into a hobby in later life; Mill avidly collected specimens, and published a number of short notes in several botanical magazines.[270] At the time of his death he left a mass of notes and observations, prompting one contemporary to suggest that Mill had planned to have them printed as the foundation of a flora of Avignon. This same writer recalled "meeting, a few years since, the (at that time) parliamentary logician, with his trousers turned up out of the mud, and armed with the tin insignia of his craft, busily occupied in the search after a marsh-loving rarity in a typical spongy wood on the clay to the north of London."[271] But this is the extent to which Mill had any first-hand experience of the physical sciences. Indeed, he admitted to being fairly ignorant of both their history and their current practice. For example, in a letter of 1834 he admitted that his knowledge of mathematical and experimental science was "extremely superficial."[272]

267. See William Whewell to Michael Faraday, August 7, 1846, WP O.15.49 f.56.

268. John Stuart Mill, *Autobiography*, *CW* 1:59.

269. George Bentham went on to publish *Catalogue des plantes indigenes des Pyrénees et de Bas-Languedoc* (1826), *Handbook of the British Flora* (1858), *Flora Hongkongensis* (1861), *Flora Australiensis* (1863–78), and (with Joseph Hooker) *Genera Plantarum* (1862–83). See Simon Curtis, "The Philosopher's Flowers."

270. These are collected in *CW* 31. See the introduction to that volume by John Robson for a useful discussion of Mill's botanical studies.

271. E. H. Fox Bourne et al., *John Stuart Mill*, p. 47.

272. John Stuart Mill to John Pringle Nichol, *CW* 13:211. Given this admission, his later criticism of Hamilton's "want of familiarity with the physical sciences" seems a bit unfair (see *Examination of Sir William Hamilton's Philosophy*, *CW* 9:496n).

Thus Whewell's dismissal of Mill's scientific abilities, in a letter to Herschel soon after the publication of *System of Logic*, was not unjust. "Though acute and able," he wrote, "he is ignorant of science."[273] Nevertheless, Mill appeared to agree with Whewell's claim about the necessity of inferring a philosophy of science from knowledge of its practice and history. For example, in the preface to *System of Logic*, he explained that "on the subject of Induction, the task to be performed was that of generalising the modes of investigating truth and estimating evidence, by which so many important and recondite laws of nature have, in the various sciences, been aggregated to the stock of human knowledge."[274] He thus implied that he inferred his theory of induction from the method by which truths have, in the past, been discovered. He made a similar claim at the start of book 3.[275] Mill also suggested the necessity of inferring a philosophy of science from the practice of science and its history when, in his *Autobiography*, he expressed a debt to Whewell and Herschel, explaining that their books provided him with something needed before he could complete his work: namely, a comprehensive view of "the generalities and processes of the sciences."[276] Indeed, Mill went so far as to claim that without Whewell's *History* a "portion of this work would probably not have been written."[277]

Such comments notwithstanding, Mill clearly did not infer his philosophy of science from knowledge of science and its history. He wrote *System of Logic* without knowledge of most of the scientific examples included in the published version. The majority of these were added to the completed manuscript, on Mill's request, by his friend Bain. Bain described his reaction to seeing the completed manuscript in the following way:

> The main defect of the work . . . was in the Experimental Examples. I soon saw, and he felt as much as I did, that these were too few and not unfrequently incorrect. It was on this point that I was able to render the greatest service. Circumstances had made me tolerably familiar with the Experimental Physics, Chemistry and Physiology of the day, and I set to work to gather examples from all available sources.[278]

273. William Whewell to John Herschel, April 8, 1843, in Isaac Todhunter, *William Whewell, D.D.*, 2:315.

274. John Stuart Mill, *System of Logic*, CW 7:cxii.

275. Ibid., 7:283.

276. John Stuart Mill, *Autobiography*, CW 1:215.

277. See John Stuart Mill, *System of Logic*, CW 7:cxiii.

278. Alexander Bain, *John Stuart Mill*, p. 66. Mill, in his *Autobiography*, admitted that Bain "went carefully through the manuscript before it was sent to press, and enriched it with a great

Thus Mill could not have inferred his philosophy of science from these examples. And if he had done so from the examples given by Herschel and Whewell, surely he would have included more of these in his book (besides Herschel's example of Wells's research on dew); he would not then have needed Bain to provide new ones. Even with Bain's help, Mill found relatively few instances of his philosophy being exemplified in science; moreover, as we have seen, at least some of these examples are, as Whewell pointed out, either incorrect or inappropriate.

Indeed, Whewell's criticisms of Mill in this regard did not focus on the question of whether Mill's philosophy of science actually was inferred from science and its history, but rather on the fact that his methods had not been applied to a large number of appropriate historical cases. He complained that "if Mr. Mill's four methods had been applied by him . . . to a large body of conspicuous and undoubted examples of discovery, well selected and well analysed, extending along the whole history of science, we should have been better able to estimate the value of these methods."[279] On Whewell's view, even though Mill did not infer his philosophy from science and its history, it should be possible, if his methods are valid, to find many examples of their use in actual scientific practice throughout the history of science.

This sheds some light on what Whewell held to be the important relation between science and philosophy of science: not whether a philosophy of science is, in fact, inferred from knowledge of past and present scientific practice, but rather, whether a philosophy of science is *inferable from* such knowledge. Any valid philosophy of science must be shown to be exemplified in actual scientific practice throughout the history of science. Thus, even though Bacon did not infer his philosophy of science from a study of the history of science—and indeed, as Whewell noted, he did not have much history of modern science behind him—his philosophy can still be found to be valid if it is shown to be exemplified in the science that has come since his time. (To some extent this is the project Whewell himself had taken on.)[280] Even if Whewell did not, in fact, develop his philosophy of science only after his study of the history and practice of science was completed, he showed us in his works—through numerous apt examples—that his philosophy has

number of additional examples and illustrations from science; many of which, as well as some detached remarks of his own in confirmation of my logical views, I inserted nearly in his own words" (*CW* 1:255n).

279. William Whewell, *On the Philosophy of Discovery*, pp. 264–65.

280. Whewell saw this as his project from the beginning; see the 1830 notebook (notebook entry dated June 28, 1830, WP R.18.17. f.12, p. xv).

been embodied in the practice of science throughout its history. Mill was unable to do so.

Perhaps because Mill was fairly ignorant about the past history of physical science, he was also wrong about its *future* history. A further criticism that Whewell made of Mill's view of induction concerns Mill's claim that the future of scientific progress will be promoted by deduction rather than induction. Mill suggested that most inductive discoveries have already been made, so that nearly all of the remaining work will be to deduce the consequences of these discoveries.[281] Moreover, he noted that astronomy and chemistry, as well as the social sciences, are mainly deductive, because artificial experimentation, necessary for carrying out Mill's methods, is impracticable.[282] He claimed that in all the sciences, the deductive part is "the most difficult and important portion" of the scientists' work.[283] He even set himself in opposition to Bacon by claiming that "a revolution is peaceably and progressively effecting itself in philosophy the reverse of that to which Bacon has attached his name" (that is, a revolution of deduction against induction).[284]

In criticizing Mill, Whewell noted that he fit into the category of the "deductive minds" that Whewell had described in his Bridgewater Treatise. That is, Mill was one of those people for whom "the explanation of remarkable phenomena by known laws of Nature has . . . a greater charm . . . than the discovery of the law themselves."[285] Deductive explanation of phenomena has demonstrative force, and it is easy for "clear intellects" to follow the train of argument. Yet even a clear mind may be frustrated by the "mysterious step" that occurs in induction, the step by which the facts are colligated with an explicated and appropriate conception; this step, as we have seen, is a rational one, but it is not demonstrative, and it cannot be determined by any algorithmic process. Whewell explained that this "admiration for deductive skill," and the desire for demonstrative processes of reasoning, is what led to the influence of Descartes and his followers on the Continent. Yet it is not true that the inductive work in the physical sciences is mostly completed. To think it is could have frightening results: "to leave us too well satisfied with what we know already, to chill our hope of scientific progress, and to prevent our making any further strenuous efforts to ascend, higher than we have yet

---

281. John Stuart Mill, *System of Logic*, CW 7:482; quoted in William Whewell, *Of Induction*, pp. 73–74.

282. See John Stuart Mill, *System of Logic*, CW 7:251.

283. Ibid., 7:407.

284. Ibid., 7:482.

285. William Whewell, *Of Induction*, p. 76.

done, the mountain-chain which limits human knowledge."[286] In the case of science, Whewell chided Mill for being too much of a conservative. Given Mill's political motivation for developing his epistemology—to defeat an alternative view he believed would lead to conservatism—there is some irony in this.

286. Ibid., p. 77.

CHAPTER THREE

Reforming Science

It is unreasonable to accuse Mr. Darwin (as has been done) of violating the rules of Induction. The rules of Induction are concerned with the conditions of Proof. Mr. Darwin has never pretended that his doctrine was proved.
—*John Stuart Mill*, System of Logic, *1862 edition*

Even if we had no Divine record to guide us, it would be most unphilosophical to attempt to trace back the history of man without taking into account the most remarkable facts in his nature: the facts of civilization, art, government, writing, speech—his traditions—his internal wants—his intellectual, moral, and religious constitution. If we will look backwards, we must look at all these things as evidences of the origin and end of man's being.
—*William Whewell, presidential address to the Geological Society, 1838*

The publication of Charles Darwin's *Origin of Species* in 1859 was one of the most significant scientific events of the age. Not surprisingly, both Mill and Whewell evaluated Darwin's theory in terms of their respective views of scientific reasoning. In his 1862 edition of *System of Logic*, Mill indicated that Darwin's method was consistent with his own precepts of science—at least, that it did not contradict those precepts. Yet by claiming that Darwin "never pretended that his doctrine was proved," Mill instead only showed himself to be misunderstanding what Darwin was up to. On the other hand, Whewell understood all too well, and rejected Darwin's theory (in a letter to his friend J. D. Forbes) as "unphilosophical"—as not satisfying the conditions Whewell, in his presidential address to the Geological Society twenty-one years earlier, had set for a scientific explanation of man's origins. Although Whewell never accepted the truth of Darwin's theory, he was, as we will see, an important influence upon its development. In particular, Darwin was inspired by Whewell's view of the proper way to confirm theories in science.

This chapter will detail the controversy between Mill and Whewell over the confirmation of scientific theories. Whewell claimed that there were three important tests of the truth of a theory: prediction, consilience, and coherence. Mill rejected these tests, even though his own "Deductive Method" requires that a theory make successful predictions in order to be verified. Mill's misgivings about Whewell's alleged intuitionism once again resulted in misunderstanding. Indeed, these misgivings led Mill to claim, somewhat oddly, that "Dr. Whewell . . . pass[ed] over altogether the question of proof."[1] Whewell's discussion of verification was noticed by Darwin, however; in his *Origin of Species* he supported his theory of evolution by natural selection in a way quite similar to Whewell's test of consilience. Although Darwin thus seemed to embrace Whewell's method for confirming theories, Whewell himself did not accept Darwin's theory as proven (though he at least, unlike Mill, realized that Darwin was attempting to prove it). I will explain the reasons for Whewell's rejection of Darwin's theory. Contrary to a tradition started in Whewell's own time, I will show that his rejection was not merely the result of blind theological prejudice. He had religious faith, but he also had faith in science. On Whewell's view, Darwin's theory clashed with faith of both types.

Whewell on Natural Classification

Understanding Whewell's confirmation tests requires attention to his view of natural classification, so I start from here. In his *Essay on Mineralogical Classification and Nomenclature* (1828), Whewell discussed the controversy between proponents of natural and artificial systems of classifying mineralogical specimens.[2] As we saw in chapter 2, while studying with Friedrich Mohs he converted to the natural classification system. In his monograph on mineralogy, Whewell argued in favor of this method, claiming that minerals should be grouped together not according to definitions of classes chosen

---

1. John Stuart Mill, *System of Logic*, CW 7:304.

2. Whewell did not use the term "natural kinds"; as Ian Hacking notes, John Venn introduced this term in his 1866 *Logic of Chance* (see Hacking, "A Tradition of Natural Kinds"). Whewell used instead the terms "natural classification" and "natural groups of kinds." Hacking credits Whewell with introducing the issue of kinds to nineteenth-century British philosophy in his discussion of natural and artificial systems of biological classification in his *Philosophy of the Inductive Sciences*. In fact, Whewell discussed the issue in his earlier work on mineralogy ("A Tradition of Natural Kinds," p. 111).

arbitrarily by the mineralogist, but rather according to the way they are really organized in nature. Whewell thus accepted the existence of what philosophers have called (since John Venn's introduction of the term in 1866) "natural kinds." Whewell explained later, in the *Novum Organon Renovatum*, that "there are classifications, not merely arbitrary, founded upon some *assumed* character, but natural, recognized by some *discovered* character."[3] Thus natural classifications are found empirically. Whewell even made this point in his *Inaugural Lecture* for the Great Exhibition of 1851. He explained that "by [a priori] assuming fixed and uniform principles of classification we can never obtain any but an artificial system, which will be found, in practice, to separate things naturally related, and to bring together objects quite unconnected with each other."[4] Thus, when seeking natural classifications, scientists "seek something, not of their own devising and creating;—not anything merely conventional and systematic; but something which they conceive to exist in the relations of the plants themselves;—something which is without the mind, not within;—in nature, not in art."[5] That is, on Whewell's view, scientists seek to "carve nature at its joints," in Plato's famous phrase, not merely to classify things into nominal or conventional groupings.

It is due to the fact that we have the Idea of Kinds that we recognize the existence of classes or kinds in nature. Whewell noted that the Idea of Kinds, no less than the Ideas of Space and Time, is necessary for us to have any general scientific knowledge of the world; to know any general laws, we must have the notion that there are classes of things significantly like each other. We recognize natural kinds by noticing observed resemblances between individual things.[6] For example, one of the properties shared by individual members of the kind "gold" is ductility.[7] Whewell pointed out, however, that not every superficial resemblance is relevant to classifying things into natural kinds.[8] This is because the superficial properties are not themselves definitive of natural groupings.

In addition to sharing some observable properties, certain individuals share underlying traits or essences. To take a simple example not used by Whewell, the superficial properties typically exhibited by members of the

3. William Whewell, *Novum Organon Renovatum*, p. 17. See also p. 220.
4. William Whewell, *Inaugural Lecture*, p. 17.
5. William Whewell, *The Philosophy of the Inductive Sciences*, 1:492. See also *Novum Organon Renovatum*, p. 230.
6. William Whewell, *On the Philosophy of Discovery*, p. 366.
7. Ibid.
8. William Whewell, *The Philosophy of the Inductive Sciences*, 1:486–87. See also pp. 513–14.

kind "water" are these: liquidity at room temperature, colorlessness, odor-lessness, tastelessness. Yet there is also an underlying trait, which we now understand to be the chemical structure H_2O. A liquid might be brown, rank-smelling, and foul-tasting, but still be water (as, for instance, in the presence of polluting factors). And there might be a clear, odorless, tasteless liquid that is not water, because it does not have the chemical composition H_2O. Al-though the shared superficial properties are not used in defining the natural-kind term, they are important initially in drawing our attention to the possi-bility that there may exist a natural-kind category of which the instances are members. As Whewell explained, "Besides the Natural System, by which we *form* our classes, it is necessary to have an *Artificial System*, by which we *recognize* them."[9] He pointed out that Linnaeus, who had proposed an arti-ficial system of classification, recognized such a system as merely a means to reach a natural system.[10] After we establish that there is a natural kind shar-ing one or more underlying traits, the superficial properties are still useful in helping us recognize possible members of the kind.

Whewell did indicate something about what he believed to be the essen-tial, underlying traits that define certain kinds of natural classes. In the case of minerals, it is their chemical composition. He claimed in the *History of the Inductive Sciences* that for plants and other organic beings, the essence is the "general structure and organization" of the being, especially those organs most important for the preservation of life.[11] In *The Philosophy of the Inductive Sciences*, he suggested that reproductive relations between indi-viduals are also important.[12] Moreover, following Locke's analysis in the *Essay on Human Understanding*, Whewell believed that the underlying structural traits are responsible for the production of the superficial properties.[13] Thus,

9. William Whewell, *History of the Inductive Sciences*, 3:204.

10. See ibid., 3:267–68.

11. Ibid., 3:281. In his 1844 essay, Darwin criticized Whewell (and Lamarck) for holding this view (see in Charles Darwin, *The Works of Charles Darwin*, 10:149–50). He repeated the criticism, without reference to Lamarck or Whewell (but rather citing Richard Owen) in *On the Origin of Species* (2nd ed., p. 336).

12. Whewell referred to Georges Cuvier's claim that a species can be defined as "the collection of individuals descended from one another, or from common parents," noting that "we consider it a proof of the impropriety of separating two species, if it be shown that they can by any course of propagation, culture, and treatment, the one pass into the other" (see *The Philosophy of the Induc-tive Sciences*, 1:505).

13. Locke defined natural kinds by the sharing of a "real essence," in the sense of sharing some underlying structure or trait causally responsible for the superficial properties of a thing (see John Locke, *An Essay concerning Human Understanding*, bk. 3, chap. vi, sec. 2, p. 439). Yet, unlike Whewell, Locke believed that real essences are unknowable to us; thus we can never have knowl-edge of real or natural kinds (bk. 3, vi, 9, pp. 444–45).

he explained, "The elementary composition of bodies, since it fixes their essence, must *determine* their properties."[14]

This discovery of the underlying structure is a task for empirical science. Like other aspects of science, two elements are required: "the exact acquaintance with *facts*, and the general and applicable *ideas* by which these facts are brought together."[15] Specifically, we need the Idea of Resemblance (or Likeness, as Whewell called it in the earlier editions of the *Philosophy*), and also the Idea of Natural Affinity. The Idea of Resemblance enables us to see how things are similar and different from other things; without this idea our individual sensations would remain unconnected and unremembered.[16] The Idea of Natural Affinity enables us to see that certain resemblances point to essences shared by members of natural kinds; such resemblances are "indications" of a "natural affinity" between things.[17] These are the properties we should focus on. Whewell did not, however, have much to say about how certain superficial resemblances can lead us to the affinities, or essences, of natural kinds, except to note that a great many resemblances, or a "mass of resemblances," indicates a natural affinity, that is, some underlying essential commonality.[18] Indeed, he seems to be rather pessimistic regarding our ability to discover these essential properties: "The nature of connexion of Kinds and Properties is a matter in which man's mind is all but wholly dark. . . . For in how few cases—if indeed in any one—can we know what is the *essence* of any Kind."[19]

This pessimism explains why Whewell claimed that we could not often give complete definitions for membership in natural kinds.[20] He explained that the demand for such definitions in natural history is due to a prejudice that arises from our using mathematics as the model of scientific reasoning.[21] The study of natural history is a corrective to this prejudice, because when we study kinds, we find that not one of them can be "rigorously defined, yet all of them are sufficiently definite."[22] Nevertheless, kinds *have* rigorous definitions, even though we do not know what they are. Indeed, these

---

14. William Whewell, *History of the Inductive Sciences*, 3:196–97; emphasis added. See also *The Philosophy of the Inductive Sciences*, 1:513.

15. William Whewell, *History of the Inductive Sciences*, 3:241.

16. William Whewell, *The Philosophy of the Inductive Sciences*, 1:469–70.

17. See ibid., 1:540.

18. See ibid., 2:368–69.

19. William Whewell, *On the Philosophy of Discovery*, p. 367.

20. See ibid., pp. 365–66, and *The Philosophy of the Inductive Sciences*, 1:475.

21. William Whewell, *The Philosophy of the Inductive Sciences*, 2:369.

22. Ibid., 2:370.

definitions give necessary and sufficient conditions for an individual to be a member of a given kind.[23] When we do not know the necessary and sufficient conditions, "we may indeed give a specific description of one of the kinds, and may call it a definition; but it is clear that such a definition does not contain the essence of the thing."[24]

Whewell's position that we currently lack definitions for natural-kind terms, and that we may in fact rarely come to know them, is consistent with his view that true definitions tend to come at the end, rather than at the beginning, of a science.[25] Yet he made it clear that although we may not be able to define completely the differences between disparate kinds, "there must manifestly *be* fixed and definite differences, in order that we may have any knowledge" about things.[26] Moreover, God knows the essences of Kinds: "Every Kind of thing, every genus and species of object, appears to Him in its essential character, and its properties follow as necessary consequences. He sees the essences of things . . . while we, slowly and painfully, by observation and experiment . . . make out a few of the properties of each Kind of thing."[27] Just as we can move closer to God's knowledge of the laws of nature by explicating our Ideas and applying them to our carefully observed experience of the world, so too can we move closer to knowledge of essences as our Idea of Natural Affinity is clarified, and as we gain more knowledge of individual things.

Whewell also claimed that there is a natural ordering of these groupings. He noted that there are not only natural kinds, but also a "Natural Order" of kinds. "Genera are groups of species associated in virtue of natural affinity, of general resemblance, of real propinquity," he explained.[28] They are not formed by "arbitrary marks."[29] A natural system is formed by arranging according to the natural affinities of things, ascending from the lower natural groups to the higher.[30]

In addition to the classes that are composed of individual objects or substances, Whewell also spoke of classes of events or processes. As examples of

23. I here disagree with Michael Ruse's claim that Whewell endorsed a "property cluster" or "polytypic" view of natural kinds. See "The Scientific Methodology of William Whewell," p. 251.

24. William Whewell, *The Philosophy of the Inductive Sciences*, 2:370.

25. See William Whewell, "On the Uses of Definitions." See also *The Philosophy of the Inductive Sciences*, 1:11–16. As we will see in chapter 5, Whewell severely criticized the Ricardian economists and their Oxford followers such as Richard Whately and Nassau Senior for putting definitions at the start of the science of political economy.

26. William Whewell, *On the Philosophy of Discovery*, p. 366.

27. Ibid., p. 367.

28. William Whewell, *The Philosophy of the Inductive Sciences*, 1:490.

29. Ibid., 1:492.

30. Ibid., 1:499.

natural classes, he included those of planetary motion and optical diffraction. The members of these classes are individual cases of planetary motion (such as the orbit of Mars) and individual instances of optical diffraction. Individual cases of optical diffraction are recognized as belonging to this kind by their observed properties, for example by the fact that they involve light appearing to bend around obstacles. But they also share a kind essence. Whewell was rather less explicit on the question of what constitutes the kind essence for classes of events or processes than he was about classes of minerals or animals. But he seemed to believe that individual cases of optical diffraction share a cause, and that this shared cause constitutes the kind essence. Thus, in the case of event or process kinds, the kind essence can be seen as being constituted by the cause or causes of that event or process. The kind essence "determines," as Whewell put it, the similarity of observed properties in each case of optical diffraction, just as the kind essence of a class of minerals determines the observed properties of members of that class. Moreover, that a group of events constitutes a natural event kind can be inferred even before it is known what the essence of the kind is, just as we can be alerted to the existence of an object kind by shared superficial properties even before we know what underlying trait(s) constitute the essence of the kind. For instance, gold was assumed to be a natural kind, even before its chemical composition was known. Whewell would also say (were he being explicit about this) that the class "planetary motion" was assumed to be a natural kind, and not just a conventional one, even before its essence was known.

This broader notion of natural kinds—that is, a notion including object kinds and event kinds—suggests a particular metaphysical view. To say that there are natural event kinds of concern to science is to suggest that there are actually existing causal powers working in nature by laws which we can discover; that is, that we can carve nature at its joints with respect to processes or events as well as to objects. It is precisely this suggestion that Mill denied.

Mill on Kinds

Mill's position on kinds has been misunderstood by many commentators, no doubt because of his own rather misleading statements. Mill claimed to recognize the existence of natural kinds, that is, groupings that are "realities in nature."[31] In fact, in the *Autobiography* he reported that he was able to get back to work on the *Logic* in 1838 after a short hiatus once he accepted the

---

31. See John Stuart Mill, *Autobiography*, CW 1:191, 229; see also *System of Logic*, CW 7:137–41.

existence of "real kinds."[32] What he wrote about these "real kinds," though, shows that his view differs from Whewell's, and demonstrates that Mill actually rejected the existence of natural kinds. This position is more consistent with his ultra-empiricism.

Mill explicitly discussed the issue of kinds in two sections of the *Logic:* in book 1, on classification and the "five predicables" of scholastic philosophy, and in book 4, on classification as "subsidiary to induction." In the first discussion he attacked the essentialism of the "schoolmen" and "metaphysicians" for holding that there is a distinction between essential and accidental properties, essential properties being "that without which the thing could neither be, nor be conceived to be. Thus, rationality was of the essence of man, because without rationality man could not be conceived to exist."[33] Mill argued instead that so-called essential propositions are merely verbal ones. That is, he claimed that objects have essential properties only insofar as the class is described by a connotative name which gives the properties as part of the connotation.[34] So, in the case of man, "*man* cannot be conceived without rationality. But though *man* cannot, a being may be conceived exactly like a man in all points except that one quality. . . . All therefore which is really true in the assertion . . . is only, that if he had not rationality, he would not be reputed a man. . . . Rationality, in short, is involved in the meaning of the word man: is one of the attributes connoted by the name. The essence of man, simply means the whole of the attributes connoted by the word."[35] His rejection of Aristotelian essences, then, follows from his rejection of all necessity other than verbal necessity. Moreover, Mill also criticized Locke's analysis of essences, claiming it is flawed because it admitted "real essences" as the supposed cause of superficial properties of things.[36] Mill denied that individuals can be grouped according to the underlying traits or

32. See John Stuart Mill, *Autobiography*, CW 1:229.

33. John Stuart Mill, *System of Logic*, CW 7:110.

34. Ibid., 7:112; see also John Skorupski, *John Stuart Mill*, p. 87.

35. Ibid., 7:110–11; see John Locke, *An Essay concerning Human Understanding*, bk. 3, chap. vi, sec. 4. See also John Skorupski, *John Stuart Mill*, pp. 78, 87. Moreover, Mill explained that "the distinction between the essence of a class, and the attributes or properties which are not of its essence . . . amounts to nothing more than the difference between those attributes of the class which are, and those which are not, involved in the signification of the class name" (*System of Logic*, CW 7:121).

36. John Stuart Mill, *System of Logic*, CW 7:115. Mill similarly criticized the Scholastic philosophers: "Their notion of the essence of a thing was a vague notion of a something . . . which makes it the Kind of thing that it is—which causes it to have all that variety of properties which distinguish its Kind. But when the matter came to be looked at more closely, nobody could discover what caused the thing to have all those properties, nor even that there was anything which caused it to have them" (*CW* 7:127).

structure causally responsible for the production of their shared observed properties. This rejection of Lockean essences is based on his rejection of real causes other than invariable antecedents. As we have seen, Mill rejected the postulation of any unobservable causes at all. We can know only "uniformity of coexistence" of observed properties. In a later passage in the *Logic*, he noted that the most that can be said about members of kinds is that "general propositions may be, and are, formed, which assert that whenever certain properties are found, certain others are found along with them."[37]

The claim that kinds are defined by observed or superficial properties is a claim *denying* that kinds are natural and asserting that they are only nominal. And yet, Mill clearly affirmed the existence of certain kinds that are "natural." His distinction, though, between kinds which are natural and kinds which are merely conventional shows that even the former are not real kinds in the sense Locke intended nor in the way we define them today. Mill claimed that two characteristics distinguish "natural" from "conventional" kinds: first, in the case of natural kinds there are "innumerable properties" that are shared, rather than only one. So, for instance, in the case of the "natural" class "water," there are many shared properties, even ones "yet undiscovered." On the other hand, in the case of the conventional kind "white things," there is only the one "finite and determinate" shared property, namely, being white.[38] Moreover, Mill claimed that the shared properties of kinds that are natural "do not depend on causes, but are ultimate properties."[39] Thus he explicitly denied Locke's position that members of a natural class share underlying traits causally responsible for the production of their shared superficial properties.

In book 4 of the *Logic*, Mill returned to the issue of kinds in a chapter entitled "On Classification, as Subsidiary to Induction." It was perhaps his interest in the science of botany that caused him to return to the topic of classification in book 4, and to recognize it as an important operation "subsidiary" to induction. As we saw in chapter 2, Mill was something of an amateur botanical collector. One of his obituary writers suggested that the sections on classification in the *Logic* would have been different if Mill "had not been a naturalist as well as a logician."[40] And it may well be true: in botany, even today, it is not clear that plant species are natural kinds in the sense of sharing any essential, underlying properties; hence Mill may have

37. Ibid., 7:579.
38. Ibid., 7:123.
39. Ibid., 7:581.
40. E. H. Fox Bourne et al., *John Stuart Mill*, p. 47.

been influenced in his general position on natural kinds by the specific case with which he was familiar.[41] He did not point to his experience, however, as an influence on his view. Rather, he claimed that his reading of Whewell's *Philosophy* and Auguste Comte's *Cours de Philosophie Positive* had the greatest impact on his position concerning kinds and classification.[42]

The view that Mill developed in this chapter is, however, very unlike Whewell's, even though he quoted copiously from Whewell's discussion. Mill claimed here that "scientific classifications" must be natural, not conventional.[43] Unlike in the earlier discussion, he linked the distinction between natural and conventional kinds to the issue of *causes*. It would be best, if possible, to group things by a shared cause of the observed properties, but this is not possible. "Causes are preferable, both as being the surest and most direct of marks. . . . But the property which is the cause of the chief peculiarities of a class, is unfortunately seldom fitted to serve also as the diagnostic of the class. Instead of the cause, we must generally select some of its more prominent effects."[44] But note that his use of the term "cause" here is consistent with that in book 3, where he had defined *cause* as "invariability of succession . . . found by observation." Thus by "cause" here Mill did not mean any underlying cause, but rather a preceding phenomenon. So, ideally, natural classes would be defined by the antecedent phenomenon preceding the "chief peculiarities" of the group. Yet the antecedent phenomenon is not always the most useful to use as the defining one, because it is not the most noticeable or most recognizable property. Natural classifications, then, must be formed based on the "more prominent effects,"[45] that is, those observed properties "which contribute most . . . to render the things like one another, and unlike other things."[46] Interestingly, Mill criticized Whewell for what Mill saw as his rejection of the possibility of defining natural classes by necessary and sufficient conditions. Mill argued that it *is* possible, and also important, to define natural classes in terms of necessary and sufficient prop-

41. On the situation in botany, see Michael Ruse, "Biological Species," p. 239.

42. In a footnote appearing only in the manuscript version, but not in the printed version, Mill wrote: "The theory of scientific classification, in its most general aspect, is now very well understood, owing chiefly to the labours of the distinguished naturalists to whom science is indebted for what are called Natural Arrangements or Classifications. . . . Mr. Whewell, in his *Philosophy of Inductive Sciences*, has systematized a portion of the more general logical principles which those classifications exemplify; but this has been still more completely done by M. Comte, whose view of the Philosophy of Classification . . . is the most complete with which I am acquainted" (*System of Logic, CW* 8:713n).

43. See ibid., 8:719–20.

44. Ibid., 8:714.

45. Ibid.

46. Ibid., 8:716.

erties or "characters" which are "rigorously common to all the objects included in the group." He claimed that only by giving such definitions is it
possible to form general assertions or laws.[47] Yet again, the "characters" that
must be part of the definition of a natural group are all observed properties,
and they are not considered indications of an underlying unobserved cause.

Not surprisingly, Mill's view is quite similar to that of Comte. In the third
volume of his *Cours de Philosophie Positive,* Comte maintained that the art
of classification is most fully developed in biological science.[48] By seeing how
biologists form natural groups, scientists in other areas can learn how to do
so properly. As Comte described it, in forming natural groups, "the process is
to class together those species which present, amidst a variety of differences,
such essential analogies as to make them more like each other than like any
others."[49] That is, natural groups are formed by observed resemblances; thus
Comte noted that it is necessary to replace attention to interior characteristics with attention to exterior ones.[50] In England, it was not Whewell who
proposed a similar view, as Mill suggested by linking Comte and Whewell as
his influences; rather, it was John Herschel. Like Comte, Herschel suggested
that both Artificial and Natural classifications are grouped according to
observed resemblances, with no indication that these point to essential properties. Artificial classes are grouped by "a single head of resemblance among
individuals otherwise very different" (similar to Mill's "white things"), while
natural groups are those that share "a great variety of analogies" and yet are
made up of "objects which yet differ in many remarkable particulars" (like
Mill's "natural classes," which share "innumerable properties").[51]

Mill, then, rejected Lockean natural kinds. This analysis of Mill's view of
kinds is consistent with his metaphysical and epistemological commitments, as examined in the previous chapter. We saw there that his ultra-
empiricism led him to reject unobservable essences—not only our ability to
have knowledge of them, like Locke, but even their existence, since such
essences are not "powers of exciting sensation." Thus Mill denied the "supposed necessity of ascending higher, into the essences and inherent constitution of things."[52] He also rejected any unobserved causes, insisting that
he was only interested in "physical" ones, defined in Stewart's sense of
constant conjunction. Mill is guilty, as Whewell accused him, of following

47. Ibid., 8:721.
48. See Auguste Comte, *Cours de Philosophie Positive,* p. 325.
49. Ibid., p. 341.
50. Ibid., p. 347.
51. John Herschel, *Preliminary Discourse on the Study of Natural Philosophy,* p. 142.
52. John Stuart Mill, *System of Logic, CW* 7:326.

Comte in "rejecting the inquiry into causes." He did not believe that we can have knowledge of the causal order of nature; and as we have seen, the belief in natural kinds is the belief in groupings of things based on the causal structure of the world.[53]

Mill, Whewell, and Prediction

We turn now to the debate between Mill and Whewell over confirmation. Whewell claimed that there were three "tests of hypotheses." These are first, that "our hypotheses ought to *fortel* phenomena which have not yet been observed; at least phenomena of the *same kind*" as those which the hypothesis originally colligated;[54] second, that they should "explain and determine cases of a *kind different* from those which were contemplated in the formation" of the hypotheses;[55] and third, that hypotheses should "become more coherent" over time.[56] These verification criteria are known respectively as prediction, consilience, and coherence. Although the definitions of *prediction* and *consilience* imply a connection between Whewell's notion of kinds and his views on confirmation, he did not explicitly discuss this relation. Because of this commentators have, for the most part, ignored it.[57] On my view, however, understanding this relation is crucial to comprehending the way in which Whewell believed scientific theories are confirmed.

Whewell gave a number of examples of cases in which theories made successful predictions, including epicycle theory and its prediction of eclipses of the sun and moon, as well as the locations of planets on given days and times; the laws of refraction, and their prediction of the effect of any new combination of transparent lenses; electromagnetic theory, and its prediction of motions which had never before been observed; and Newton's theory and its many successful predictions, including the return of Halley's comet.[58] Although Whewell's notion of natural kinds is less prominent in his discussions of prediction than in his discussions of consilience, it still plays an important role. Each of these examples of prediction concerns previously

---

53. For a similar view of what follows from belief in natural kinds, see Richard Boyd, "What Realism Implies and What It Does Not," p. 129.

54. William Whewell, *Novum Organon Renovatum*, p. 86.

55. Ibid., p. 88.

56. Ibid., p. 91.

57. The exceptions to this are John Herschel, "Whewell on Inductive Sciences"; Michael Ruse, "The Scientific Methodology of William Whewell"; and William Harper, "Consilience and Natural Kind Reasoning." However, these writers have not fully examined Whewell's notion of natural kinds.

58. The first three examples are from the *Novum Organon Renovatum*, pp. 86–87, while the fourth is from *On the Philosophy of Discovery*, p. 273.

unknown facts about the event kind originally colligated by the theory. For instance, epicycle theory predicted new facts about the event kind "planetary motion" (which was defined in such a way as to include not only the moon but also the sun).[59]

Whewell's characterization of prediction includes a temporal element: a "successful prediction" occurs when some fact unknown at the time that the theory was discovered is predicted by the theory, and is afterwards found to be true. Such predictions, when they turn out to be successful, are evidence in favor of the hypothesis. (However, contrary to Mill's assertion, Whewell did not argue that successful predictions were *conclusive* evidence for a theory; indeed, as examples of hypotheses that led to successful predictions, he included several that were then known to be false, such as the epicycle theory.)[60] Moreover, Whewell claimed even more strongly that successful predictions were better evidence than mere explanations of known facts. He maintained that "to predict unknown facts found afterwards to be true is . . . a confirmation of a theory which in impressiveness and value goes beyond any explanation of known facts."[61]

Whewell argued for his two claims about predictive success by invoking an analogy between constructing a true theory and breaking a code. "If I copy a long series of letters of which the last half-dozen are concealed, and if I guess these aright, as is found to be the case when they are afterwards uncovered, this must be because I have made out the import of the inscription."[62] Successful prediction of formerly unknown facts, more so than the colligation or explanation of known facts, Whewell thus argued, is proof that we have broken the code of Nature, that we have "detected Nature's secret."[63] Additionally, he compared our ability to colligate facts into hypotheses by his methodology to knowing the alphabet of the language of Nature. As in learning a new language, however, this is not enough; we must also know how to construct intelligible words and sentences using the alphabet, and thus learn how to use "the legislative phrases of Nature."[64] Whewell noted that it is easier to learn the alphabet than to use these legislative phrases; thus successful predictions of unknown facts serve as stronger evidence than colligations

59. In his discussions of prediction, Whewell did not distinguish, as some modern commentators do, between predictions of new instances of known phenomena (e.g., the prediction of a solar eclipse) and predictions of novel phenomena (e.g., the electromagnetic motions that, Whewell noted, had never before been observed).

60. John Stuart Mill, *System of Logic*, CW 7:501.

61. William Whewell, *History of the Inductive Sciences*, 2:464.

62. William Whewell, *On the Philosophy of Discovery*, p. 274.

63. William Whewell, *Novum Organon Renovatum*, p. 87.

64. Ibid.

of known facts. Our prediction of some new fact is analogous to the attempt of a language student to form a proper sentence in her new language. When our prediction is confirmed, as when the attempted sentence elicits an appropriate response from a native speaker, it is a sign that we have spoken correctly: Nature is "respond[ing] plainly and precisely to that which we utter, [and] we cannot but suppose that we have in great measure made ourselves masters of the meaning and structure of her language."[65]

Whewell's view of the power of prediction seemed to be borne out in 1846, when a successful prediction appeared to provide stunning evidence for the truth of Newton's theory. In that year, the planet Neptune was discovered after its existence, its position, and even its mass had been predicted mathematically. Perturbations in the orbit of Uranus expected on Newtonian theory had led some to conclude that there must be an unobserved planetary body external to Uranus's orbit exerting an additional gravitational force on the planet. Using Newton's theory, the French mathematician U. J. J. Le Verrier calculated mathematically the mass and orbit of this postulated planet.[66] Acting upon Le Verrier's calculations, astronomers at the Berlin Observatory found the planet less than one degree from its expected location. This success was considered further and quite strong evidence for Newton's theory of universal gravitation. Whewell praised the discovery as a triumph of astronomy, which he termed the "Queen of the Sciences." He argued that predictive success is strong confirmation of a theory, because the agreement of the prediction with what is found to be true is "nothing strange, if the theory be true, but quite unaccountable, if it be not."[67] If Newtonian theory were not true, Whewell was suggesting, the fact that from the theory we could correctly predict the existence, location, and mass of Neptune would be bewildering, and indeed miraculous—equivalent to the feat of a nonspeaker of Russian forming an intelligible and meaningful question in that language which elicited a proper response from a native speaker.

Mill, on the other hand, claimed in the *Logic* that this kind of predictive success is *not* evidence for the theory's truth. "Such predictions and their fulfillment are, indeed, well calculated to impress the uninformed, whose faith in science rests solely on similar coincidences between its prophesies and what comes to pass," he declared. "But it is strange that any considerable stress should be laid upon such a coincidence by persons of scientific attain-

65. Ibid.
66. At roughly the same time, an Englishman, James Couch Adams, also made these calculations, which very nearly caused an international priority war between the British and the French (see Robert Smith, "The Cambridge Network in Action").
67. William Whewell, *On the Philosophy of Discovery*, pp. 273–74.

ments."[68] Instead, he argued, a theory's predictive success may indicate only that the theory has the same observable consequences as the true theory.[69] Thus, in response to Whewell's claim that correctly predicting the last few letters of a document is proof that you understand the code in which it is written, Mill pointed out that "even an erroneous interpretation which accorded with all the visible parts of the inscription would accord also with the small remainder; as would be the case . . . if the inscription had been designedly so contrived as to admit of a double sense."[70]

Mill's criticism of Whewell here is peculiar, because he and Whewell actually agreed on the importance of predictive success for verification (with one difference, as we will see). Mill too asserted that predictive success is evidentially important—indeed, in the case of most causal laws in science, he claimed that successful prediction was the *required* method of confirmation. As we saw in chapter 2, in cases of the "composition of causes," the proper method for determining laws is what Mill called the deductive method. The first step of this method is supposed to be a direct induction to individual causes using the methods of eliminative induction or a deduction from previous direct inductions. The second stage is that of "determining from the laws of the causes, what effect any given combination of those causes *will* produce."[71] This step he calls "ratiocination," because it is "a process of calculation."[72] The third stage is that of "verification by specific experience," by which it is confirmed that the first step of the deductive method was not merely "conjectural." Mill explained that "to warrant reliance on the general conclusions arrived at by deduction, these conclusions must be found, on careful comparison, to accord with the results of direct experience."[73] The second and third step, then, consist in making predictions from the initial inductions, and then comparing these predictions with what is, in fact, the case. (Mill noted that this need not be the actual order of discovery, merely "their correct logical order.")[74] Only after the initial inductive step, Mill claimed, can a hypothesis be verified by drawing consequences from it and then checking whether these consequences do in fact obtain. The "hypothetical method," he pointed out, "suppresses the first of the three steps," and

68. John Stuart Mill, *System of Logic*, CW 7:500.
69. For a discussion of modern versions of this argument, see Jarrett Leplin, *A Novel Defense of Scientific Realism*, pp. 152–63. See also Bas van Fraassen, *The Scientific Image*, and Arthur Fine, "The Natural Ontological Attitude."
70. John Stuart Mill, *System of Logic*, CW 7:502.
71. Ibid., 7:458; emphasis added.
72. Ibid., 7:458–59.
73. Ibid., 7:460.
74. Ibid., 7:491.

because of this is illegitimate.[75] This is how he characterized Whewell's view; thus Whewell's appeal to the second and third steps—to prediction and veri-fication—was considered by Mill insufficient for proof.

Mill's analysis of Whewell's position is grounded in his mistaken belief that Whewell allowed hypotheses with no inductive support to be confirmed by predictive success. Mill's criticism of prediction in the *Logic* occurs in the context of his rejection of hypotheses assumed merely in order to account for the phenomena, such as (on his view) Descartes' vortices or the luminiferous ether of the wave theorists. In criticizing Whewell at this junc-ture of his discussion, he implied that Whewell allowed predictive success to count as confirmation of these kinds of hypotheses. But as we have seen, this was not Whewell's position. Successful prediction is evidence of the truth of theories obtained by induction. Mill is led to claim erroneously that "Dr. Whewell . . . pass[ed] over altogether the question of proof."[76] He discounted Whewell's arguments about confirmation because he believed that they were meant to be used in testing the truth of mere "hypotheses."

Like Whewell, Mill believed that prediction is a test of a theory's truth, when that theory has been arrived at inductively. Indeed, it is a necessary test of the truth of a law that has been discovered by the deductive method, and thus is a stronger form of evidence than simply explaining the facts already known, which cannot function as confirmation in these cases. Mill seems, then, to have agreed with Whewell not only that successful predic-tion is evidence for the truth of an inductively discovered hypothesis, but also that prediction is stronger evidence than is the explanation of facts already known.

There is, however, one important way in which their views do diverge. Whewell's claim about the superiority of predictive success involved a *tem-poral* element; that is, he believed that some fact that was not known at the time a hypothesis was discovered, and was successfully predicted by the theory afterwards, is stronger evidence for the truth of that hypothesis than is its ability to explain or colligate facts already known at the time of discov-ery. (Such cases are referred to in recent discussions as "new evidence.") On Mill's view, however, the temporal element is not important. For instance, as an example of the successful application of ratiocination in the deductive method, he cited the case of Newton demonstrating that his law of universal gravitation deductively entailed Kepler's laws.[77] That Kepler's laws were

75. Ibid., 7:492.
76. Ibid., 7:304.
77. Ibid., 7:461.

already known to be true did not lessen their value as "predictive" evidence for Newton's law.

Whewell did not offer an argument for his claim about the superiority of "new evidence" over "old evidence," that is, of successful predictions over explanations of known facts. He merely stated that "if we can predict new facts which we have not seen, as well as explain those which we have seen, it must be because our explanation is not a mere formula of observed facts, but a truth of a deeper kind."[78] The intuition behind this claim, which has been made by numerous modern commentators, is that if a theorist knows what fact must be explained by his theory, it is simple for him to "cook" the theory in order to somehow explain the known fact; whereas, on the other hand, if a theory predicts a novel (unknown) fact, it is not due to the ingenuity of the theorist, but to the truth of the theory. However, this intuition has been contested in recent years.[79] Interestingly, the discovery of Neptune, Whewell's paradigmatic case of a novel prediction, may be considered not only as a prediction of a new, unexpected member of the object kind "planet," but equally well as an explanation of a known fact about the event kind "planetary motion," namely the perturbations of the planet Uranus. The problem for Newton's theory was described in this way by G. B. Airy in his report to the British Association for the Advancement of Science in 1832: "I need not mention that there are other subjects (the theory of Uranus, for instance) in which the existence of difficulties is known, but in which we have no clue to their explanation."[80] Thus whether the postulation of an unseen planet is considered an explanation of known facts or the prediction of a new fact depends on the way it is described, not on any objective features of the evidence itself or of its relation to the theory. So it would be odd, to say the least, if the way of *describing* the evidence made a difference to its evidential value. It should be noted that Whewell only supported the thesis about the superiority of new evidence in very strong cases of predictive success, such as the discovery of Neptune; he explained of such success that "it is a confirmation which has only occurred a few times in the history of science, and in the case only of the most refined and complete theories, such as those of Astronomy and Optics."[81]

78. William Whewell, *Of Induction*, p. 60.
79. See Stephen Brush, "Prediction and Theory-Evaluation," and my "Is Evidence Historical?"
80. G. B. Airy, "Report on the Progress of Astronomy during the Present Century," p. 189.
81. William Whewell, *History of the Inductive Sciences*, 2:464. Herschel's view of prediction differed from that of both Mill and Whewell. Even though he argued that "bold hypotheses" without full inductive support could be verified by successful prediction, Herschel did not claim additional evidential benefit over and above the value of explanation of known facts. (He alleged this only for the criterion of consilience, as we will see below.) He seemed to have thought that any such

Consilience, Coherence, and the Unification of Kinds

Whewell also discussed two further criteria for confirmation, consilience and coherence. His notion of natural kinds plays a central role in his characterization of these criteria. Though Mill, somewhat strangely, never directly discussed these confirmation tests, his rejection of Whewell's notion of natural kinds, and indeed his general denial of our ability to have knowledge of real causes, indicates that he would have rejected consilience and coherence as methods of verification.

Whewell explained that "the evidence in favour of our induction is of a much higher and more forcible character when it enables us to explain and determine cases of a *kind different* from those which were contemplated in the formation of our hypothesis. The instances in which this have occurred, indeed, impress us with a conviction that the truth of our hypothesis is certain."[82] Whewell called this type of evidence a "jumping together," or "consilience," of inductions. To understand what he meant by this, it may be helpful to detail the "jumping together" that occurred in the case of Newton's law of universal gravitation, Whewell's exemplary case of consilience. In book 3 of the *Mathematical Principles of Natural Philosophy*, Newton listed a number of "propositions." These propositions are empirical laws that were inferred from certain "phenomena" (which are described in the preceding section of book 3). The first such proposition, inferred from phenomena of "satellite motion," is that "the forces by which the circumjovial planets are continually drawn off from rectilinear motions, and retained in their proper orbits, tend to Jupiter's centre; and are inversely as the squares of the distances of the places of those planets from that centre." The result of another, separate induction from the phenomena of "planetary motion" is that "the forces by which the primary planets are continually drawn off from rectilinear motions, and retained in their proper orbits, tend to the sun; and are inversely as the squares of the distances of the places of those planets from

---

additional advantage of prediction over explanation was mostly psychological rather than logical; that is, such predictions cause us to increase our belief in the truth of a hypothesis, more than increasing the probability that it is true. He also suggested that new evidence has a promotional benefit in popularizing science. In a letter he sent to the *Guardian* after the discovery of Neptune, Herschel noted that the successful prediction of the new planet was such that he could not "conceive of anything better calculated to impress the general mind with a respect for the accumulated facts, laws and methods, as they exist at present, and the reality and efficiency of the forms into which they have been moulded. . . . We need some reminder of this kind in England, where a want of faith in the higher theories is still to a certain degree our besetting weakness." See John Herschel, "Letter to the *Guardian*," October 25, 1846; quoted in Robert Smith, "The Cambridge Network in Action," pp. 418–22.

82. William Whewell, *Novum Organon Renovatum*, pp. 87–88.

the sun's centre." Newton saw that these laws, as well as other results of a number of different inductions, coincided in postulating the existence of an inverse-square attractive force as the cause of various phenomena.[83] According to Whewell, Newton realized that these inductions "leap to the same point," that is, to the same law.[84] Newton was then able to bring together inductively (or "colligate") these laws, and facts of other kinds (for example, the event kind "falling bodies"), into a new, more general law, namely the universal gravitation law: "All bodies attract each other with a force of gravity which is inverse as the squares of the distances." By seeing that an inverse-square attractive force provided a cause for different event kinds—for satellite motion, planetary motion, and falling bodies—Newton was able to perform a more general induction, to his universal law.

In terms of natural kinds, what Newton found was that these different event kinds share an essential property, namely the same cause. He discovered that what makes "the orbit of Mars" a member of the class "planetary motion" is that it is caused to have the properties it does by an inverse-square attractive force of gravity between Mars and the other bodies in the universe. He also found that other event kinds share this kind essence. What Newton did, in effect, was to subsume these individual event kinds into a more general kind composed of sub-kinds that share a kind essence. Consilience of event or process kinds results in *causal unification*. More specifically, it results in unification of natural-kind categories based on a shared cause. Phenomena that constitute different event kinds, such as "planetary motion," "satellite motion," and "falling bodies," were found by Newton to be members of a unified, more general kind, "phenomena caused to occur by an inverse-square attractive force of gravity" (or, "gravitational phenomena"). In such cases, according to Whewell, we learn that we have found a *vera causa*, or a "true cause," one that really exists in nature, and whose effects are members of the same natural kind.[85] Moreover, by finding a cause shared by phe-

83. As Ernan McMullin notes, the notion of the "attractive force" was problematic for Newton; he was never satisfied in his search for an "agent cause" of gravitational behavior, a "cause of gravity," as he put it. But he did succeed in causally unifying disparate phenomena by the concept of this attractive force (see McMullin, "The Impact of Newton's *Principia* on the Philosophy of Science," p. 296). It is this aspect of Newton's theory that Whewell pointed to as the exemplar of consilience.

84. William Whewell, *Novum Organon Renovatum*, p. 88.

85. See William Whewell, *On the Philosophy of Discovery*, p. 191. Michael Ruse agrees that Whewell's notion of the *vera causa* lies at the center of his criterion of consilience ("The Scientific Methodology of William Whewell," p. 232). See also Robert Butts, "Consilience of Inductions and the Problem of Conceptual Change in Science," p. 81, and Michael Ruse, "Darwin's Debt to Philosophy," pp. 161–62.

nomena in different sub-kinds, we are able to colligate all the facts about these kinds into a more general causal law. Whewell claimed that "when the theory, by the concurrences of two indications . . . has included a new range of phenomena, we have, in fact, a new induction of a more general kind, to which the inductions formerly obtained are subordinate, as particular cases to a general population."[86] He noted that consilience is the means by which we effect the successive generalization that constitutes the advancement of science.[87]

Note that consilience is importantly different from the type of reasoning known as "inference to the best explanation," even when the "best explanation" is defined as a causal hypothesis.[88] The causal law expressing the essence of different event or process kinds is not one postulated merely because it explains or accounts for these different classes of facts. Rather, in the case of each class, the law has emerged from a process of inductive reasoning following Whewell's methodology. Thus the causal law is not imposed from above as a means of tying together different event kinds. Instead, it wells up from beneath in each separate case. What is important in the case of consilience, then, is not merely that a single causal mechanism or law can explain or account for different event kinds (as the inference to the best explanation allows), but rather that separate lines of induction lead from each event kind to the same causal mechanism or law, that there is a convergence of distinct lines of argument. At each step, there is inductive warrant for the causal law besides its explanatory utility.

Whewell discussed a further, related test of a theory's truth: coherence.[89] Coherence occurs when we are able to extend our hypothesis to colligate a different event kind without ad hoc modification of the hypothesis, that is, without suppositions that are added merely for the purpose of saving the phenomena, for which there is no independent evidence. When Newton extended his theory regarding an inverse-square attractive force, which colligated facts of planetary motion and lunar motion, to the natural class "tidal activity," he did not need to add any new suppositions to the theory in order to

86. William Whewell, *Novum Organon Renovatum*, p. 96.

87. See William Whewell, *The Philosophy of the Inductive Sciences*, 2:74.

88. Numerous commentators on Whewell have interpreted his view as a form of "inference to the best explanation" or as the related view of abductive or retroductive inference, including Charles Peirce (who coined the term "abduction"). See Peirce, "Lecture on the Theories of Whewell, Mill, and Comte"; Larry Laudan, "William Whewell on the Consilience of Inductions"; Menachem Fisch, *William Whewell, Philosopher of Science*, p. 168n13; and Ernan McMullin, *The Inference That Makes Science*.

89. William Whewell, *Novum Organon Renovatum*, p. 91.

colligate correctly the facts of this event kind (facts about particular tides).[90] The situation was rather different for phlogiston theory (which explained combustion, before the discovery of oxygen, by the presence of an "essence" in the burning body called "phlogiston"). On Whewell's view, phlogiston theory colligated facts about "chemical combination." But when the theory was extended to colligate facts about the weight of bodies, it was unable to do so without an ad hoc and implausible modification (the assumption that phlogiston has "negative weight").[91] The emission or particle theory of light was claimed by its adherents to be capable of colligating the facts of different kinds (indeed, the same facts that the wave theorists claimed they could colligate). However, according to Whewell and other proponents of the wave theory, the particle theory could do so only by the admission of suppositions that were ad hoc or implausible. For example, in order to account for the facts of the velocity of light (in particular, the uniformity of velocity), the particle theorists needed to suppose that light particles are emitted with different velocities based on the mass of the luminous body, but that, as Humphrey Lloyd put it, caricaturing the particle theory, "among these velocities is but one which is adapted to our organs of vision."[92] (Of course, the wave theorists were themselves accused of postulating an implausible or ad hoc "luminiferous ether"; it was argued by Whewell and others that there was independent evidence for the ether's existence, and hence that it was not an ad hoc supposition.) In noncoherent cases such as these, the modifications have no independent evidence; they are added to the theory solely because they are

90. Facts about the tides were not known systematically until Whewell engaged in his tidology research program. Part of the importance of his project, Whewell believed, was to strengthen the claim that Newton's law could colligate these facts without modification (see his "On the Empirical Laws of the Tides in the Port of London," pp. 15–17).

91. See William Whewell, *Novum Organon Renovatum*, pp. 92–93. Whewell explained there that "the doctrine of Phlogiston brought together many facts . . .—combustion, acidification, and others,—and very naturally prevailed for a while. But the balance came to be used in chemical operations, and the facts of weight as well as the facts of combination were to be accounted for." He argued that the facts colligated by phlogiston theory were all facts of a single event kind, "chemical combination." A new event kind, "weight of bodies," needed to be colligated by any theory colligating chemical phenomena. ("Weight of bodies" was known to be an event kind by virtue of sharing a common cause, namely gravitational force.) However, he was wrong to claim that phlogiston theory only colligated facts of one kind. "Acidification" and "combustion" are each event kinds, no less than are "polarization" and "double refraction," which, as we saw above, Whewell agreed are different classes of phenomena colligated by the wave theory. Note that I argue below that the wave theory was not fully consilient; one reason is because it lacked a single causal law which could colligate all the facts of optical phenomena. Similarly, phlogiston theory lacked such a causal law. Thus Whewell's assertion that phlogiston theory was never consilient is correct, though not for the reason he suggested.

92. Humphrey Lloyd, "Report on the Progress and Present State of Physical Optics," pp. 300–301. See also p. 296.

needed in order to colligate a new or problematic set of facts. As this discussion suggests, coherence is closely related to consilience. The distinction between them is as follows. In cases of consilience, two independent colligations of different event kinds yield the same conclusion. In cases of coherence, the result of one or more colligations of one or more event kinds is extended, without ad hoc modification, to another event kind. This latter process generally occurs over time as theoretical systems are developed.[93]

In addition to Newton's law of universal gravitation, Whewell sometimes cited the wave theory of light as an example of a consilient theory. However, he was somewhat ambivalent over the extent to which the wave theory of light was consilient. (This may be surprising, because Whewell is generally cited as one of the wave theory's most keen supporters in the nineteenth century.)[94] For example, in his Bridgewater Treatise, he noted that for proponents of the wave theory (presumably including himself), "the theory of undulations is conceived to be established in *nearly* the same manner, and *almost as certainly*, as the doctrine of universal gravitation."[95] Whewell's relative tentativeness about the wave theory can be seen by comparing his "Inductive Tables" of astronomy and optics. In introducing these tables, he admitted that the table of astronomy shows the convergence toward simplicity (and thus consilience) to a "greater degree" than does that of the wave theory of optics.[96] And indeed, in the case of optics, the table is somewhat incomplete: there are numerous entries reading "imperfect colligations," and while the base of the table for astronomy includes both the name of the theory ("The Theory of Universal Gravitation") and a causal law ("All bodies attract each other with a force of *Gravity* which is inversely as the squares of the distances"), in the case of optics the base includes only the name of the theory ("The Undulatory Theory of Light") and no specific law. Later, in his 1858

93. William Whewell, *Novum Organon Renovatum*, pp. 90–91. Malcolm R. Forster, "Unification, Explanation, and the Composition of Causes in Newtonian Mechanics," and William Harper, "Consilience and Natural Kind Reasoning," both claim that consilience *results in* coherence. Their claim arises from the fact that they equate coherence with simplicity; thus they quote Whewell's statement that "consilience of inductions gives rise to convergence towards simplicity and unity" (Harper, p. 129; Forster, p. 79). Actually, consilience and coherence are similar means to obtain the same result, namely simplification in the sense of (causal) unification.

94. Geoffrey Cantor, for example, claims that Whewell was committed to the truth of the wave theory as early as 1831 (see *Optics after Newton*, pp. 197–99), while Jed Buchwald dates Whewell's embrace of the theory to 1832 (*The Rise of the Wave Theory of Light*, p. 296). While Whewell certainly believed that the wave theory was probably true, he did not believe it had been conclusively confirmed the way Newton's theory had been.

95. William Whewell, *Astronomy and General Physics, Considered with Reference to Natural Theology*, p. 105; emphasis added.

96. William Whewell, *The Philosophy of the Inductive Sciences*, 2:78.

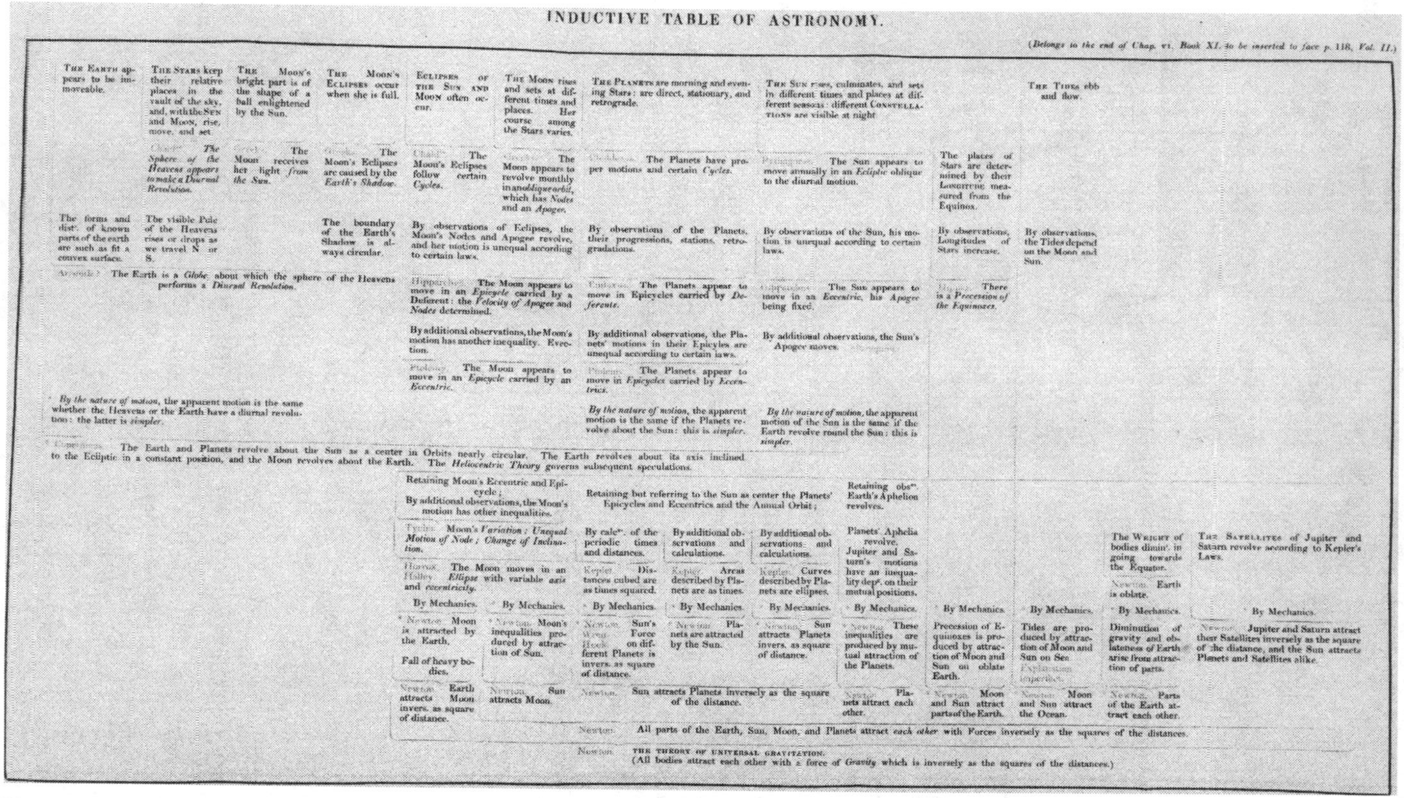

INDUCTIVE TABLE OF ASTRONOMY.

(Belongs to the end of Chap. vi. Book XI. to be inserted to face p. 116. Vol. II.)

FIGURE 4. Inductive Table of astronomy. From William Whewell's *Philosophy of the Inductive Sciences.* Author's collection.

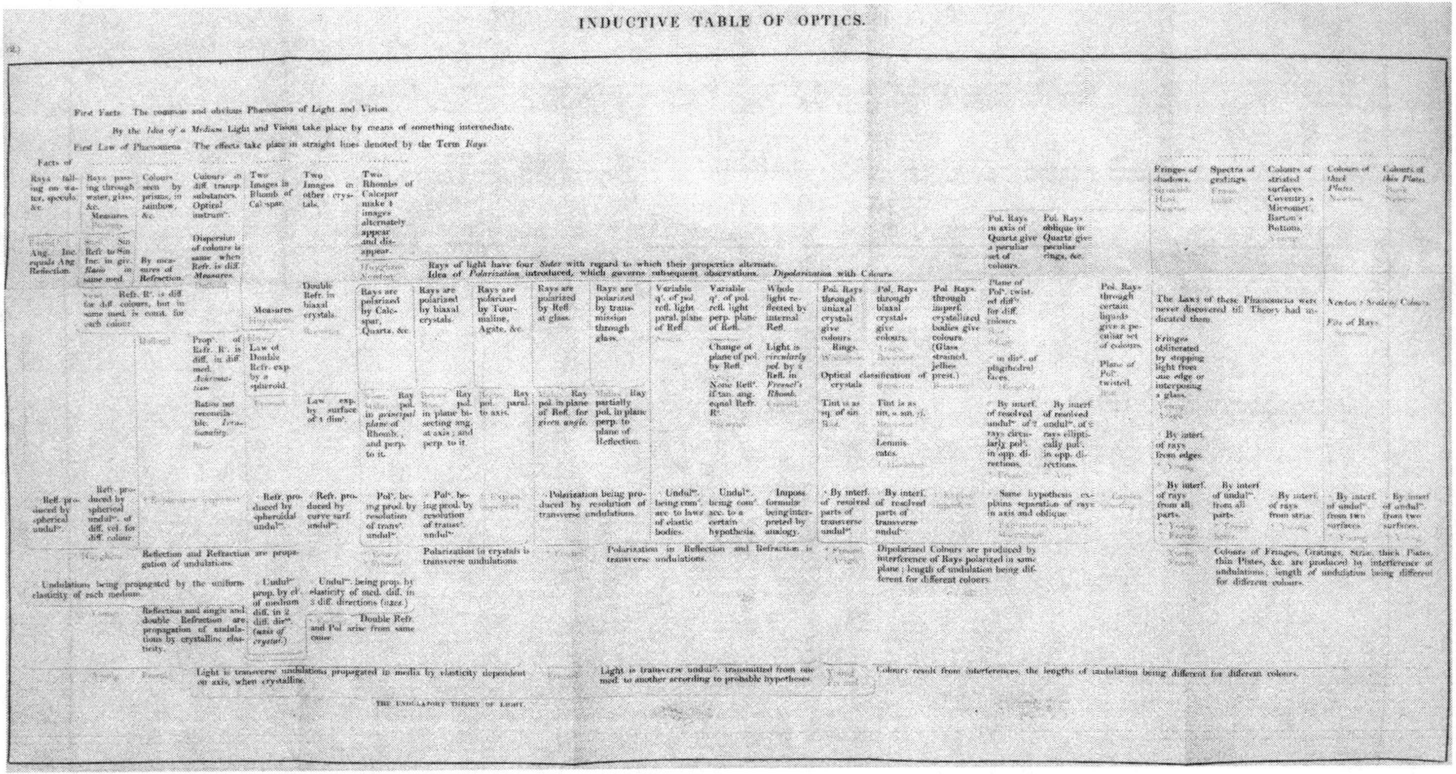

FIGURE 5. Inductive Table of optics. From William Whewell's *Philosophy of the Inductive Sciences*. Author's collection.

Novum Organon Renovatum, Whewell noted that "Kepler discovered that the planets describe ellipses, before Newton explained why they select this particular curve, and describe it in a particular manner [that is, because they act in accordance with his law of universal gravitation]. The laws of reflection, refraction, dispersion, and other properties of light have long been known; the causes of these laws are at present *under discussion*."[97]

Why, in 1858, was a proponent of the wave theory claiming that the causes of the laws were still "under discussion"? Surely the cause of each of these phenomena, according to a wave theorist, was the propagation of transverse vibrations through the ether. Whewell was perhaps referring to the fact that there was, as yet, no single specific causal law (and certainly, no mathematical law) uniting the various event kinds, as there was in the case of Newton's inverse-square law. This is why Whewell was not prepared to claim unambivalently that the wave theory of light was consilient. What he wanted was perhaps something like the equations for the electromagnetic field provided by James Clerk Maxwell in his Third Memoir of 1865. There was another problem for the wave theory as well: namely, that it seemed to be (in the complaint Whewell used against Descartes' vortices) "inconsistent mechanically": the ether seemed both highly elastic (in order to carry rapid vibrations at high velocities) and absolutely solid (because only solids were known to transmit transverse waves). Moreover, the wave theorists needed to suppress the longitudinal component of waves transmitted through an elastic solid. They still needed to work out the details of the inferred ether and the waves supposedly being carried through it. Until they did so, Whewell could not consider the theory to have the same level of consilience as the universal gravitation law.

We have seen that consilience was, for Whewell, an important confirmation test of causal laws regarding events or processes. But he also claimed that the criterion of consilience was useful in confirming empirical laws regarding objects. Such laws are important in the classification sciences such as mineralogy and botany, which seek natural classifications or taxonomies of objects. When two independent methods of classification yield the same groupings of objects, and hence the same empirical laws about the properties of these objects (such as "All metals are crystalline when solid"), Whewell maintained, this is proof that we have found the real—or natural—classification. This is analogous to his claim that, in the case of event kinds, consilience occurs when the respective colligations of two different kinds yield the same causal law. Whewell used this criterion to argue that a natural sys-

97. William Whewell, *Novum Organon Renovatum*, p. 118; emphasis added.

tem of classification of minerals had been discovered: "An arrangement by external characters which gives us classes possessing a common chemical character;—a chemical order which brings together like and separates unlike minerals;—such classifications have the evidence of their truth in their agreement with one another."[98]

Consilience, then, tells us that we have found a law that colligates members of a natural kind—either members of a natural object kind or a natural event kind.[99] In the case of object kinds, consilience is, Whewell suggested, evidence for the truth of a noncausal empirical law such as "All metals are crystalline when solid," and thus is evidence that we have grouped certain individuals into a kind ("metals") that is natural and hence supportive of the law. In the case of event kinds, consilience is evidence for having found a causal law, such as the inverse-square law discovered by Newton. This is at the same time evidence for the naturalness of the kind "gravitational phenomena."

We are now in a position to see why a common argument against Whewell's criterion of consilience is flawed. A number of commentators have claimed that consilience is fundamentally a relativistic or contextual criterion. According to this view, whether some theory is consilient depends on our knowledge-state with respect to the classes involved—specifically, whether we know that certain classes of facts are members of the same general type of thing.[100] One critic notes that for Descartes, planetary and terrestrial motions were phenomena of the same type, because both were the result of vortices; thus, relative to the Cartesian system, Newton's theory was *not* consilient.[101] However, as we have seen, the criterion for consilience is that a theory connects two (or more) different natural kinds into a more general natural kind *by virtue of inductively inferring* the same cause for each natural sub-kind. The issue, then, is not the relativistic one of whether we previously *knew* or *suspected* that these sub-kinds all belong to a more gen-

98. William Whewell, *The Philosophy of the Inductive Sciences*, 1:540–41. Whewell made the same claim in his *An Essay on Mineralogical Classification and Nomenclature*, p. iii.

99. Herschel recognized this in his review of Whewell's *History* and *Philosophy* ("Whewell on Inductive Sciences," p. 227). See also Michael Ruse, "The Scientific Methodology of William Whewell," p. 240.

100. See Larry Laudan, "William Whewell on the Consilience of Inductions"; Malcolm Forster, "Unification, Explanation, and the Composition of Causes in Newtonian Mechanics"; William Harper, "Consilience and Natural Kind Reasoning"; and Margaret Morrison, "Unification, Realism and Inference."

101. Larry Laudan, "William Whewell on the Consilience of Inductions," p. 374. For similar claims see Robert Butts, "Consilience of Inductions and the Problem of Conceptual Change in Science," pp. 74–75; Menachem Fisch, "Whewell's Consilience of Inductions"; and Margaret Morrison, "Unification, Realism and Inference."

eral kind, but rather whether they do all fit into the same more general kind by virtue of sharing the same cause, and whether we have reached this conclusion by the proper sort of inductive reasoning (thus conferring on the conclusion inductive warrant), and not merely by postulating a common cause hypothetically. Whewell famously believed that Descartes did not reach his conclusion by proper reasoning. On the other hand, Newton's theory was consilient because it satisfied both conditions.

This erroneous criticism of Whewell's criterion of consilience may arise from a confusion of his view with that of Herschel. Herschel's brief discussion of consilience in the *Preliminary Discourse on the Study of Natural Philosophy* focuses on the subjective, psychological aspects of consilience, and does not explicitly include the requirement of causal unification under more general natural kinds.[102] Herschel explained that

> the surest and best characteristic of a well-founded and extensive induction . . . is when verifications of it spring up, as it were, spontaneously, into notice, from quarters where they might be least expected, or even among instances of that very kind which were at first considered hostile to them. Evidence of this kind is irresistible, and compels assent with a weight which scarcely any other possesses.[103]

Thus Herschel's view of consilience is prone to the relativistic criticism. Whewell also did at times stress the psychological aspect of consilience, noting the impact of "the unexpected coincidence of results."[104] Nevertheless, he emphasized the logical aspect of consilience, which does not depend on the psychological. Whewell's consilience concerns not only the psychological element of surprise, but also the logical element of causal unification of different event or process kinds into more general kinds, whose members share a common cause.

Somewhat oddly, Mill did not directly address the issue of consilience.[105]

---

102. See John Herschel, *Preliminary Discourse on the Study of Natural Philosophy*, p. 170.
103. Ibid., p. 170.
104. William Whewell, *The Philosophy of the Inductive Sciences*, 2:67.
105. Mill used the term "consilience" in several passages of the *Logic*, but in a looser sense. For example, in discussing the science of "ethology" in book 6 of the *Logic*, he remarked that there must be "a consilience of *a priori* reasoning and specific experience" in verifying the laws of the science. In the first two editions, when he made this statement Mill credited Whewell with introducing the term (*CW* 8:874). In discussing verification more generally in the social sciences, Mill wrote, "The ground of confidence in any concrete deductive science is not the *a priori* reasoning, but the consilience between its results and those of observation *a posteriori*" (ibid., 8:896–97). In the third edition of 1852, Mill substituted the word "accordance" for "consilience." In his *Of*

Indeed, in the *Logic* he appears to have conflated Whewell's arguments in favor of prediction with those supporting consilience, as for instance when he accused Whewell of claiming that predictive success is an infallible sign of a theory's truth; in fact, Whewell claimed this only for the criterion of consilience.[106] In the only passage in which Mill made reference to something like consilience, he rejected it: "Any law of nature is deemed to have gained in point of certainty, by being found to explain some complex case which had not previously been thought of in connection with it: and this indeed is a consideration which it is the habit of scientific inquirers to attach rather too much value than too little."[107] But even here Mill appears to have concentrated only on the psychological impact of consilience, and not on the logical one. Thus Mill, like recent commentators, seems to have confused Whewell's view and Herschel's.

The main reason it is apparent that Mill would reject consilience as a confirmation criterion is that he rejected the notion of natural kinds on which, as we have seen, Whewell's notion of consilience is based. Mill rejected the claim that we can group individual objects together according to any underlying, unobservable structure or trait causally responsible for the production of observed properties. Moreover, he rejected the notion that types of phenomena can be known to be united in nature by virtue of sharing a common, unobservable cause. This is part of his general denial that we can have knowledge of the causal structure of nature. He thus denied the very basis of Whewell's criterion of consilience.

The Value of Consilience

Whewell claimed that consilience is "a criterion of reality, which has never yet been produced in favour of falsehood." He also noted that "there are no instances, in which a doctrine recommended in this manner has afterwards been discovered to be false."[108] He can thus be seen as making an inductive argument in favor of the confirmation value of consilience: namely, the argument that since consilient theories in the past have all turned out to be true, we can infer that (probably) all consilient theories are true, and thus

---

Induction of 1849, Whewell had pointed out that Mill was using the term in a way different from his own usage.

106. See John Stuart Mill, *System of Logic*, CW 7:501.

107. Ibid., 7:462.

108. William Whewell, *Novum Organon Renovatum*, p. 90, and *On the Philosophy of Discovery*, p. 192.

that our current consilient theories are (probably) true.[109] Whewell has been much ridiculed for this inductive argument by Mill and his twentieth-century followers.[110] As his detractors today like to point out, the one positive instance strongly adduced by Whewell for his inductive argument has been shown to be a counterinstance: Newton's theory, which seemed irrefutable in Whewell's time, has since been shown to be (strictly speaking) false. Newton's theory employs conceptions that, it turns out, are not true of the world, such as absolute space and absolute time. The cause postulated by the theory, the inverse-square attractive force of gravity propagated through space and time, is also nonexistent (it is not a *vera causa*). Thus, although Newton's theory was highly consilient, more so than any other theory of Whewell's time—it brought together different event kinds, by having inductive warrant for the inference that they shared a common cause—it turned out to be false. Whewell's inductive argument, then, initially seems to fail to establish any confirmatory value for consilience.

However, even though Newton's theory has been proved false, there is something about it that has not been disconfirmed, even today. Newton's theory still seems to have been correct in bringing together event kinds that belong to a more general kind, showing something true about the natural-kind structure of the physical world, although it was wrong about what the shared kind essence of the natural kind was. Newton demonstrated that the phenomena of free fall, pendulum motion, lunar acceleration, satellite motion, and planetary motion share a cause, and in this way constitute a general event kind. This insight has not been proved false, although the particular shared cause he postulated has been rejected. Einstein's general theory of relativity proposed instead that the cause of these phenomena (and others) is the curved structure of space-time. It may well be that this proposed cause will be replaced by another; but so far it seems unlikely that a later theory will separate these different phenomena and attribute to each of them different causal structures or mechanisms. These event kinds do seem to be sub-kinds of a more general kind that is connected in nature. Thus the one truly consilient theory endorsed by Whewell is still believed to have gotten things right, at least in terms of the natural-kind structure of the physical world.[111]

109. Whewell also makes an additional argument for consilience's value: he claims that consilience is like the coinciding testimony of two witnesses. In my "Consilience, Confirmation and Realism," I show how this argument can be understood as a form of "common-cause argument." But I here focus on the argument upon which Whewell most relies.

110. See, for example, Bas Van Fraassen, "Empiricism in Philosophy of Science," p. 267.

111. See my "Consilience, Confirmation, and Realism," where I show how Whewell's criterion of consilience can be used to argue for a natural-kind realism.

Of course, that Newton's theory was wrong about the particular cause he postulated does show that Whewell went too far in claiming that consilience is conclusive evidence for a theory, in the sense of proving that a theory will *never* be shown to be false. And it is certainly possible that even the natural-kind structure Newton's theory proposed might turn out to be wrong. It would have been more accurate—as well as more consistent with his general view of the progress of science—for Whewell to have claimed that hypotheses satisfying the condition of consilience are our best theories, but are still subject to correction. He suggested this kind of revision when he asserted, referring to Newton's Rules of Reasoning in book 3 of the *Mathematical Principles of Natural Philosophy*, that "the really valuable part of the Fourth Rule is that which implies that a *constant verification*, and, if necessary, rectification, of truths discovered by induction, should go on in the scientific world. Even when the law is, or appears to be, most certainly exact and universal, it should be constantly exhibited to us afresh in the form of experience and observation."[112]

Consilience and Darwin's *Origin of Species*

As we have seen in the earlier chapters of this book, the debate between Whewell and Mill did not take place in an intellectual vacuum: both men were influenced by other thinkers, as well as by each other. At the same time, their works made an impression on many of those who read and discussed them. Whewell's work in particular had an impact upon numerous men of science. I have already noted that in addition to inventing the word *scientist*, he provided terminological assistance to Michael Faraday, Charles Lyell, Forbes, and others. Yet his terminological influence was emblematic of a broader impact upon the science of the age. Whewell's work had this effect because he was respected as an expert in science—so much so that Lyell asked Whewell to succeed him as president of the Geological Society, even though Whewell had conducted no research in the field.

One particularly notable instance of the influence of Whewell's writings on the science of the day relates to the topic of this chapter, the debate between Mill and Whewell over the confirmation of theories. In his *Origin of Species*, Darwin seemed to appeal to Whewell's notion of consilience in justifying his theory of evolution by natural selection.[113] Thus ironically, although

---

112. William Whewell, *On the Philosophy of Discovery*, p. 196; emphasis added.
113. See D. R. Oldroyd, "How Did Darwin Arrive at His Theory?"; Michael Ruse, "Darwin's Debt to Philosophy" and *The Darwinian Revolution*; and Paul Thagard, "Darwin and Whewell."

Whewell never accepted evolutionary theory, he was an important influence upon Darwin's argument for it.

Whewell's view of the organic origin question—the "mystery of mysteries," as his friend Herschel had put it—was well known at the time.[114] Species, according to Whewell, were natural kinds with essential properties, properties that were fixed throughout time. As he wrote emphatically in the *History of the Inductive Sciences*, "Species have a real existence in nature, and a transmutation from one to another does not exist."[115] Moreover, the adaptation of species to their environment—such as the "fit" between the growing cycle of plant life and the seasons—proved the intelligent design of the universe. Whewell was one of the contributors to the series of Bridgewater Treatises, funded by bequest of the Earl of Bridgewater in order to show how various disciplines gave evidence for this intelligent design. In his *Astronomy and General Physics, Considered with Reference to Natural Theology* (1833)— one of the most successful of the series—Whewell argued that the lawful structure of the physical world implied that there must be a creator of the laws of nature. As he put it, "Laws imply a Lawgiver." (Later, Darwin rather cheekily used a quotation from this discussion on the frontispiece of the *Origin:* "But with regard to the material world, we can at least go so far as this— we can perceive that events are brought about not by insulated interpositions of Divine power, exerted in each particular case, but by the establishment of general laws.")[116] When Robert Chambers's transmutationist *Vestiges of the Natural History of Creation* appeared in 1844, Whewell responded by publishing *Indications of the Creator* (1845), a compilation of extracts from his earlier works showing evidence of design in the natural world.

In his *History*, Whewell argued further that in studying organic structures it is necessary to make reference to the end or goal for which they were designed. He invoked Kant in arguing against Étienne Geoffroy Saint-Hilaire, who had argued that the structure and function of an animal should be studied only by attending to analogies between it and other organized beings.

114. John Herschel to Charles Lyell, February 20, 1836. Darwin-Lyell Collection, American Philosophical Society. Published in Walter F. Cannon, "The Impact of Uniformitarianism," pp. 304–11.

115. William Whewell, *History of the Inductive Sciences*, 3:479; this comment remained unchanged from the 1837 first edition. On the other hand, even in this work Whewell had seemed to allow that the origin of species was susceptible to a natural (i.e., nontheological) explanation, which prompted Charles Lyell to write, "Whewell in his excellent treatise on the Inductive Sciences, appears to me to go nearly so far as to contemplate the possibility at least of the introduction of fresh species being governed by general laws" (see May 24, 1837, in Katherine M. Lyell, ed., *Life, Letters and Journals of Sir C. Lyell*, 2:12).

116. William Whewell, *Astronomy and General Physics, Considered with Reference to Natural Theology*, p. 267.

Whewell countered that it is necessary to assume the existence of an end, or final cause, as our guide in the study of animal organization.[117] We cannot, he argued, understand the function of a structure without knowing the end or purpose it serves in the functioning of the whole organism. He thus endorsed teleological explanations in biology. In the third edition of this work, published in 1857, Whewell modified his view in accordance with Richard Owen's theory of archetypes (discussed in chapter 1). In this final edition of his *History* he endorsed instead Owen's "unity of plan" explanations of the structures of organic beings. Not all structures have a purpose in the functioning of the whole organism.[118] But all can be understood by being seen as instantiations of an archetype structure, designed by the Creator for some purpose. Thus Whewell explained that "final Causes . . . appear . . . not merely as contrivances for evident purposes, but as modifications of a given general Plan for special given ends."[119]

Darwin was aware of Whewell's views on species, the design question, and scientific methodology, both through his reading and by his personal acquaintance with the older man. He had met Whewell through his friendship with the botanist John Henslow while at Cambridge between 1828 and 1831. Whewell is known to have attended Henslow's botany lectures with Darwin, and it may be that during discussions of these lectures he recommended Herschel's *Preliminary Discourse* to his colleague.[120] (Darwin later claimed that reading Herschel's book "stirred up in me a burning zeal to add even the most humble contribution to the noble structure of natural science.")[121] Whewell continued to take an interest in Darwin's scientific and professional advancement after Darwin returned to London from his *Beagle* expedition. While president of the Geological Society, Whewell pressed Dar-

---

117. See William Whewell, *History of the Inductive Sciences*, 3:377–81.

118. Étienne Geoffroy Saint-Hilaire's "unity of type" explanations, unlike Richard Owen's, were evolutionary in that he proposed that historical species give rise to modern forms.

119. William Whewell, *History of the Inductive Sciences*, 3:560.

120. See Adrian Desmond and James Moore, *Darwin*, p. 81, and Michael Ruse, *The Darwinian Revolution*, p. 62.

121. Charles Darwin, *Autobiography*, p. 67. After reading the book, Darwin wrote to W. D. Fox, "If you have not read Herschel in Lardner's Cyclo.—read it directly" (February 15, 1831, in *The Correspondence of Charles Darwin*, 1:118). See also letter to Herschel, sent with a copy of the *Origin*: "Scarcely anything in my life made so deep an impression on me [as reading the *Preliminary Discourse*]: it made me wish to try to add my mite to the accumulated store of natural knowledge" (November 11, 1859; in *The Correspondence of Charles Darwin*, 7:371). Darwin met Herschel in 1836 during the *Beagle*'s call at the Cape of Good Hope (see Howard Gruber and Paul Barrett, *Darwin on Man*, p. 28n). Herschel later asked Darwin to write the Geological section of the *Manual of Scientific Inquiry* that Herschel edited and published in 1849. For more on the influence of Herschel on Darwin, see Michael Ruse, *The Darwinian Revolution* and "Darwin's Debt to Philosophy."

win to accept the post of secretary. Darwin turned down the invitation, but Whewell persisted, and eventually Darwin agreed.[122]

Darwin respected Whewell's intelligence and breadth, calling him one of the "best conversers on grave subjects to whom I have ever listened."[123] He also admired Whewell's scientific work, and criticized Thomas Carlyle for not doing so: in his *Autobiography*, Darwin wrote of Carlyle, "He laughed to scorn the idea that a mathematician, such as Whewell, could judge, as I maintained he could, of Goethe's views on light."[124] Darwin made use of Whewell's work on the tides, asking him to send his papers, and later remarked that Whewell "will always rank among the great investigators of the theory of tides."[125] In his "Journal of Researches" he thanked Whewell for explaining the undulations that would probably result on the shore during an earthquake.[126] And although he disagreed with Whewell's view of natural classification in living things, he referred to Whewell's view in respectful terms in his 1844 essay.[127]

Darwin is known to have appreciated Whewell's *History of the Inductive Sciences;* in his diary he noted that he went through the *History* carefully for the second time in autumn of 1838.[128] The annotations in his copy of the *History* show that he read with special care the section on teleological explanations in physiology. Darwin also read Whewell's Bridgewater Treatise: he made numerous implicit and explicit references to this work in his notebooks,[129] showing that his concern to explain adaptation was framed in terms

---

122. See Charles Darwin to William Whewell, February 16, 1839, WP Add.Ms.c.88 f. 4.

123. Charles Darwin, *Autobiography*, p. 106.

124. Ibid., p. 138.

125. Darwin's opinion of Whewell's importance as a tidal investigator was made to Leslie Stephen (see his "Whewell, William," *Dictionary of National Biography*, 1885–90, 20:1372). See also letter to John Henslow, March 1834, in *The Correspondence of Charles Darwin*, 1:371.

126. Charles Darwin, in *The Works of Charles Darwin*, 3:296n.

127. Charles Darwin, 1844 essay, in ibid., 10:149–50.

128. According to Edward Manier, Whewell is the sixteenth most frequently cited author in Darwin's notebooks and early manuscripts, between John Henslow (#15) and Georges Cuvier (#17). See *The Young Darwin and His Cultural Circle*, pp. 18–19.

129. For instance, in the C notebook Darwin wrote, "Whewell in comment few will dispute says civilization is hereditary; i.e. instincts of wisdom virtue? . . . but more especially the powers of reasoning" (C notebook, p. 72; quoted in Howard Gruber and Paul Barrett, *Darwin on Man*, p. 262; see also William Whewell, *Astronomy and General Physics, Considered with Reference to Natural Theology*, pp. 254–68). Later in this notebook Darwin wrote, "Musalman's [*sic*] of the Peninsula are, generally speaking, a much fairer race than the Hindu's, in the same tracts; and that in their appearance and manners they are as opposite as day and night; yet we know how remote the periods at which they both left the lands of their forefathers. . . . quote Whewells [*sic*] Bridgewater Treatise (p. 26) about plants from Cape of Good Hope continuing for some time to flower at their own periods" (C notebook, p. 91; quoted in Gruber and Barrett, *Darwin on Man*, p. 266). Less flatteringly, Darwin noted elsewhere, "Mayo . . . quote Whewell as profound, because he says

of not only his reading of William Paley at Cambridge but also his reading of Whewell.[130]

Darwin was paying close attention to Whewell's methodological views as well. In a letter to Whewell, he praised his *History*, noting that "to see so clearly the steps by which all the great discoveries have been come to is a capital lesson to every one, even to the humblest follower of science and I hope I have profited by it."[131] Later, after reading Herschel's joint review of the *History* and the *Philosophy*, he recorded in a notebook the exclamation "*I must study* Whewell on Philosophy of Science."[132] There is no direct evidence that Darwin ever followed up on this plan; his library did not contain a copy of the book, and he did not refer to any particular passages of it. But he was made aware of many of the details of Whewell's methodology, including his confirmation criterion of consilience, by reading Herschel's review.

As we have seen, Darwin was interested in Whewell's work in science as well as methodology in the late 1830s through the '50s, that is, while working out his evolutionary theory. Later, when writing the *Origin*, and soon after completing it, he frequently showed his concern with whether or not the book would appear to have been based on a correct, "scientific," methodology. He was anxious that his work, unlike the anonymously published and much-reviled *Vestiges of the Natural History of Creation*, be considered scientifically (that is, methodologically) sound. He was particularly distressed by a scathing review of *Vestiges* written by Whewell's friend, the geologist Adam Sedgwick. Sedgwick had focused his wrath on the author's evident lack of understanding of proper (inductive) method: "The author is intensely hypothetical, and builds his castles in the air, misconceiving the principles of science, or misunderstanding the facts with which it has to deal; or, what is worse still, distorting them to serve his purpose. He does all this, apparently, without having any just conception of the methods by which men, after the toil of many generations, have ascended, step by step, to the higher elevations of physical knowledge—without any even glimmering conception of what men mean when they tell us of Inductive Science and its sober

---

length of days adapted to duration of sleep of man!!! Whole universe so adapted!!! And not man to planets.—instance of arrogance!!!" (notebook D, p. 49; quoted in ibid., p. 347).

130. Susan F. Cannon claims that Darwin's work was very much formulated in response to Whewell's Bridgewater Treatise and his 1838 presidential address to the Geological Society ("The Whewell-Darwin Controversy," p. 380). Cannon's analysis is flawed, however, by her mistaken reading of Whewell as a follower of the German Romantics.

131. Charles Darwin to William Whewell, April 16, 1839, in *The Correspondence of Charles Darwin*, 2:186.

132. Charles Darwin, from list entitled "Books to Be Read," Darwin Collection, University Library, Cambridge. Quoted in Michael Ruse, "Darwin's Debt to Philosophy," p. 166.

truths."[133] Because of such ignorance, Sedgwick even suggested that the book was written with "the science gleaned at a lady's boarding school."[134] After reading Sedgwick's review, Darwin wrote to Lyell, "It is a grand piece of argument against mutability of species, and I read it with fear and trembling."[135] His continuing worries about the methodology of his book were expressed to Asa Gray in 1859: "What you hint at generally is very, very, true: that my work will be grievously hypothetical, and large parts by no means worthy of being called induction, my commonest error being probably induction from too few facts."[136]

And indeed, Darwin's worry was prophetic: many of the reviews of his book raised the criticism that Darwin's theory was not sufficiently "inductive." For example, Sedgwick complained that Darwin had "departed from the true inductive track."[137] Owen made the same criticism in his review.[138] Although he had expected such criticisms, they nevertheless rankled Darwin; he expressed this frustration to one of his more sympathetic reviewers, noting that "until your review appeared I began to think that perhaps I did not understand at all how to reason scientifically."[139] Hostile reviewers were demanding "direct proof"—by which they seemed to mean something like the blind collection of facts and simple inductive generalization of the "naïve inductive" method, the type of view of Francis Bacon promulgated by T. B. Macaulay in his 1837 article. Of one of the purveyors of such a claim, Darwin wrote, "On his standard of proof, *natural* science would never progress."[140] He countered that his theory contained much "indirect evidence."[141] By "indirect evidence," Darwin appeared to suggest something like Whewell's notion of consilience.

133. Adam Sedgwick, "Review of *Vestiges of the Natural History of Creation*," p. 2. Sedgwick added similar criticisms of the argument in *Vestiges* to his introduction to the fifth edition of *A Discourse on the Studies of the University of Cambridge* (1850).

134. Adam Sedgwick, "Review of *Vestiges of the Natural History of Creation*," p. 27.

135. Charles Darwin to Charles Lyell, October 8, 1845, in *The Correspondence of Charles Darwin*, 3:258.

136. Charles Darwin to Asa Gray, November 29, 1857, in ibid., 6:492.

137. Adam Sedgwick, "Objections to Mr. Darwin's Theory of the Origin of Species," *Spectator*, March 24, 1860, pp. 285–86 and April 7, 1860, pp. 334–35.

138. Richard Owen, "Darwin on the Origin of Species," *Edinburgh Review, or Critical Journal* 111 (1860): 487–532.

139. Charles Darwin to Henry Fawcett, July 20, 1861, in *The Correspondence of Charles Darwin*, 9:212.

140. Charles Darwin to Charles Lyell, referring to William Hopkins, June 1, 1860, in ibid., 8:233. In a letter to T. H. Huxley on December 5, 1860, Darwin, exasperated, exclaimed that "all the talk about Baconian induction is cant and rubbish" (in ibid., 8:514). Nevertheless, in his *Autobiography* Darwin claims that he had "worked on true Baconian principles, and without any theory collected facts on a whole-sale scale" (*Autobiography*, p. 119).

141. F. J. Pictet, in his review of the *Origin*, claimed that he would not accept Darwin's theory until he saw with his own eyes the evolution of a new organ! (Pictet later changed his mind, and

In his introduction to the first edition of the *Origin*, Darwin noted the importance of providing an alternative causal explanation—besides design—for the modification of species to their environment. He made certain to distinguish his own view from that of the author of the *Vestiges*, complaining that one problem with the view propounded there was that the author never really explained the cause of coadaptations of organic beings to each other and to their respective environments.[142] Darwin believed that his theory of descent with modification by natural selection provided a causal explanation for such adaptation, as well as for many other kinds of phenomena. Throughout the book he pointed to numerous phenomena that his theory explained causally but that special creation or Lamarkian evolutionary theories could not.[143] In his concluding chapter, Darwin pointedly criticized the "eminent" authors who support special creation, claiming, "To my mind it accords better with what we know of the laws impressed on matter by the Creator, that the production and extinction of the past and present inhabitants should have been due to secondary causes, like those determining the birth and death of the individual."[144] That is, Darwin's theory of evolution by natural selection was superior to that of special creation because it "accorded with" what we know already about the laws of nature, and because it assigned natural causes to the history of species that were similar to the natural causes in the history of individuals.

After it became clear that many of the attacks launched against his theory were being framed in terms of his supposedly faulty methodology, Darwin told T. H. Huxley, "I have got fairly sick of hostile reviews. Nevertheless, they have been of use in showing me when to expatiate a little and to introduce a few new discussions."[145] Perhaps for this reason, in the second edition of the *Origin* he more clearly appealed to the consilience of the theory, writing, "I cannot believe that a false theory would explain, as it seems to me that the theory of evolution does explain, the several large classes of facts above specified."[146] In

---

accepted Darwin's theory on the basis of its "indirect evidence.") (See David Hull, *Darwin and His Critics*, pp. 51 and 146.) Referring scathingly to this kind of view, Darwin wrote to Henry Fawcett in 1861: "About thirty years ago there was much talk that geologists ought only to observe and not to theorize; and I well remember someone saying that at this rate a man might as well go into a gravel-pit and count the pebbles and describe the colours. How odd it is that anyone should not see that all observation must be for or against some view if it is to be of any service!" (September 18, 1861, in *The Correspondence of Charles Darwin*, 9:269).

142. Charles Darwin, *On the Origin of Species*, 1st ed., pp. 3–4.

143. See, for example, ibid., pp. 343–44.

144. Ibid., p. 488.

145. Charles Darwin to T. H. Huxley, December 2, 1860, in *The Correspondence of Charles Darwin*, 8:507.

146. Charles Darwin, *On the Origin of Species*, 2nd ed., p. 388.

the sixth edition, published in 1872, Darwin more emphatically exclaimed: "It can hardly be supposed that a false theory would explain, in so satisfactory a manner as does the theory of natural selection, the several large classes of facts above specified. . . . It is a method used in judging of the common events of life, and has often been used by the greatest natural philosophers. The undulatory theory of light has thus been arrived at."[147]

Darwin's method of justifying his theory in the *Origin* is similar to Whewell's notion of consilience in several ways. First, Darwin's theory explains not just many facts, but many *kinds* of facts. That is, the theory provides an explanation for facts in the realms of classification, geography, paleontology, embryology, comparative anatomy, and geology. Indeed, before writing the *Origin*, Darwin spent much time observing, experimenting, and collecting data on the different kinds of phenomena and phenomenal laws that he wanted his theory to explain.[148] In the sixth edition, he added a chapter entitled "Miscellaneous Objections to the Theory of Natural Selection," in which he outlined more explicitly the classes of facts that can be explained by his theory but not by the theory of special creation.[149] It is possible to see these "classes of facts" as being similar to Whewell's event kinds. For example, the adaptation of organisms to their environment can be seen as a kind, whose members (the polar bear's white fur, the woodpecker's thin beak) share a single cause, which the special creationists claimed was God's intentional design; Darwin claimed it was evolution by natural selection. Another way in which Darwin's strategy is similar to consilience is that he wanted to include in his theory the laws or theories accepted by the experts in various fields, much as Whewell believed that a consilient theory includes (by causally unifying) the already known phenomenal laws of various kinds.[150] Thus Darwin stressed

147. Ibid., 6th ed., in *The Works of Charles Darwin*, 16:438–39. In a letter to Asa Gray in 1859, Darwin already expressed his confidence about the theory in these terms: while admitting that there were still some difficulties with the theory, Darwin noted, "I cannot possibly believe that a false theory would explain so many classes of facts, as I think it certainly does explain" (November 11, 1859, in *The Correspondence of Charles Darwin*, 7:369).

148. Dov Ospovat notes that Darwin collected data and undertook experiments on embryology (his measurements of young pigeons), geographical distribution (especially the transports of seeds), paleontology and extinction (contemplation of data on aberrant forms), and classification (experiments on the lawn at Down and compilation of data on variation in large genera). He also read voluminously the work of others. See Ospovat, *The Development of Darwin's Theory*, p. 170.

149. As one of his reviewers noted, Darwin's theory "provides particularly admirable explanations for the unity of organic composition, the representative or rudimentary organs, and the species plus genera which form natural series. It corresponds equally well with many paleontological facts. It accords well with the specific resemblances which exist between two consecutive faunas, with the parallelism which one observes sometimes between the sequences of palaeontological and embryological development, etc." (F. J. Pictet, "*On the Origin of Species* by Charles Darwin"; translated and reprinted in David Hull, *Darwin and His Critics*, p. 146.)

150. See Dov Ospovat, *The Development of Darwin's Theory*, p. 114 and pp. 148–49.

that his theory explained phenomenal laws such as the law of embryonic resemblance.[151]

Moreover, Darwin believed his theory to be superior not only because it provided an explanation for many different kinds of facts, but because it provided a *causal* explanation, one framed in terms of natural causes similar to those appealed to in science. For instance, his theory provided a cause for the observed cases of homological structures. Darwin claimed that neither the Lamarckian evolutionary view nor the doctrine of special creation gave a natural cause for this phenomenon. But, as he noted, his theory did provide an explanation: because homologous structures descended from a common ancestor, and because the changes that eventually resulted in the branching off of different species happened gradually, the main pattern of structure remained the same.[152] Hence Darwin's theory provided the causal unification of Whewell's criterion of consilience as applied to classes of phenomena, by showing that all these different kinds of phenomena share a common cause, namely the alteration of original organisms by gradual modification, through the mechanism of natural selection. It seems, then, that we have in Darwin an example of an eminent scientist of Whewell's own time who embraced a key part of his methodology.[153]

It remains to be asked why Darwin appealed to consilience as evidence for his theory. As we have seen, while working on the *Origin* he was certainly aware of Whewell's great influence and reputation among the scientific community, and was engaged with his thoughts about methodology and science. He knew that Whewell's theological commitments, as well as his expressed views of species, would make it unlikely that Whewell would support Darwin's theory. It was a shrewd pragmatic move to incorporate the methodological pronouncements of a powerful potential critic. (Along similar lines, as we have seen, Darwin used a comment from Whewell's Bridgewater Treatise as one of the epigraphs to the *Origin*.) Yet it does not seem likely that Darwin was being hypocritical here; he probably believed that consilience was the best way to confirm a theory.[154] We may ask, however, if he did intend his theory to be justified by appeal to Whewell's criterion of consilience, why did he mention neither consilience nor Whewell by name? It may be that Darwin

---

151. See ibid., p. 165.

152. Charles Darwin, *On the Origin of Species*, 1st ed., pp. 434–35.

153. Of course, Darwin rejected Whewell's claim that species were natural kinds sharing an (unchanging) essence. Yet unlike Mill, he believed that we could have knowledge of an unobserved causal order of nature; there was therefore no theoretical bar to his acceptance of a confirmation criterion based on the causal unification of diverse kinds of phenomena.

154. Here I agree with Michael Ruse, *The Darwinian Revolution*, p. 63.

feared this would provoke Whewell to a harsh public response, similar to his response against Chambers's *Vestiges of the Natural History of Creation.*

While Whewell appears to have influenced Darwin, Mill's work did not appear to have any impact upon him at all. There are no references to Mill's work in Darwin's notebooks dating before the publication of the *Origin,* so it is not clear that he ever read *System of Logic.* If he had read it, surely Darwin would have seen that his theory could not be verified by the kind of direct observations Mill required.[155] After publishing the *Origin,* however, he was happy enough to accept Mill as a methodological ally; when he saw Henry Fawcett's review of the *Origin,* Darwin wrote to Huxley, "I have read few first pages of the MacMillan's article, and it pleases me that he quotes Mills [*sic*] Logic and declares that I have philosophised in the right spirit."[156] Darwin was presumably also pleased to learn, from Fawcett, that Mill personally approved of his method; in a letter to Darwin on July 16, 1861, Fawcett reported, "I was spending an evening last week with my friend Mr. John Stuart Mill, and I am sure you will be pleased to hear from such an authority that he considers that your reasoning throughout is in the most exact accordance with the strict principles of logic. He also says the method of investigation you have followed is the only one proper to such a subject."[157] Darwin may have been less pleased to read the footnote added in 1862 to the *System of Logic,* cited at the start of this chapter, in which Mill indicated that "Mr. Darwin has never pretended that his doctrine was proved."[158] But of course Darwin *did* claim that his theory was supported by evidence, though not conclusive evidence. Interestingly, perhaps because Mill rejected the notion of consilience as evidence of a theory, he misread Darwin's intentions; recall that Mill had also accused Whewell of not being concerned with proof. Mill believed that Darwin, like Whewell, used the "hypothetical method," which could not lead to proof until the missing first, inductive step was provided.[159]

---

155. As Darwin wrote to J. D. Hooker of F. W. Hutton's positive review, "He is one of the very few who see that the change of species cannot be directly proved, and that the doctrine must sink or swim as it groups and explains phenomena. It is really curious how few judge it in this way, which is clearly the right way" (April 23, 1861, in *The Correspondence of Charles Darwin,* 9:99).

156. Charles Darwin to T. H. Huxley, December 5, 1860, in ibid., 8:514.

157. Henry Fawcett to Charles Darwin, July 16, 1861, in ibid., 9:204.

158. John Stuart Mill, *System of Logic,* CW 7:498–99n.

159. Janet Browne has claimed that Mill "sanctioned" Darwin's work in the 1862 edition of the *System of Logic,* and showed that "the natural history sciences . . . could be brought in to an acceptably rigorous philosophical framework" (*Charles Darwin,* 2:186). Yet a closer look indicates otherwise. Mill's discussion of Darwin appeared in a footnote to his section on "the indispensableness of scientific hypotheses." Hypotheses are valuable for being "necessary steps in the progress of something more certain," because they give "inducement" to trying one experiment rather

Darwin was pleasantly surprised by Whewell's reaction to his work. (Of course, he had had rather low expectations.) Upon receiving his copy of the *Origin*, Whewell wrote,

> My dear Mr. Darwin, I have to thank you for a copy of your book on the "Origin of Species." You will easily believe that it has interested me very much, and probably you will not be surprised to be told that I cannot, *yet at least*, become a convert to your doctrines. But there is so much of thought and fact in what you have written that it is not to be contradicted without a careful selection of the ground and manner of the dissent.[160]

Darwin was so pleased with this response that he sent Whewell's letter to Lyell, writing, "Possibly you might like to see enclosed note from Whewell, merely as showing that he is not horrified with us."[161] Certainly, Whewell's response to Darwin is in strong contrast with those of Sedgwick and Owen (who had, like Whewell, been friendly to Darwin beforehand), and with Whewell's own reaction against the *Vestiges*.[162]

This is not to suggest, of course, that Whewell became a convert to evolutionism. He continued to reject not only Darwin's specific theory of evolution, but any view that included the origin of mankind as part of a natural process. Indeed, when he published the seventh edition of his Bridgewater Treatise in 1864, Whewell presented several arguments in the new preface against the "recent" claim that "the structure of animals has become what it is by the operation of external circumstances and internal appetencies" rather than by intelligent design.[163] He criticized this kind of view for "assert[ing] the world to be the work of chance." Raising the old arsenal of pro-

than another or because they serve as a first attempt to trace order in the complicated set of appearances (*System of Logic*, CW 7:496). Although such hypotheses are often false, they have what modern philosophers of science would call "heuristic value." Of Darwin's hypothesis, Mill asked, "is it not a wonderful feat of scientific knowledge and ingenuity to have rendered so bold a suggestion, which the first impulse of every one was to reject at once, admissible and discussable, even as a conjecture?" (ibid., 7:499n) This is hardly a very strong endorsement of Darwin's scientific work.

160. William Whewell to Charles Darwin, January 2, 1860, in *The Correspondence of Charles Darwin*, 8:6; emphasis added.

161. Charles Darwin to Charles Lyell, January 4, 1860, in ibid., 8:15.

162. Compare the letter to Darwin from Adam Sedgwick, November 24, 1859, in ibid., 8:396–98. See also Sedgwick's review ("Objections to Mr. Darwin's Theory of the Origin of Species") and Richard Owen's review ("Darwin on the Origin of Species"). Although he wrote a nasty critique, Sedgwick remained friendly with Darwin afterwards (see Michael Ruse, *The Darwinian Revolution*, p. 23).

163. William Whewell, *Astronomy and General Physics, Considered with Reference to Natural Theology*, 7th ed., p. xvi.

design arguments and examples, he claimed that only intelligent design could account for the intricacy and perfect adaptation of structures found in nature, such as the eye.[164]

In correspondence with other clergymen, Whewell similarly expressed critical views of Darwin's theory. For example, when an Aberdeen cleric with whom Whewell was not acquainted wrote to him in 1864, asking his opinion of Darwin's work, he replied, "It still appears to me that in tracing the history of the world backwards, so far as the palaetiological sciences enable us to do so, all the lines of connexion stop short of a beginning explicable by natural causes; and the absence of any conceivable natural beginning leaves room for, and requires, a supernatural origin." In this letter, Whewell referred to Darwin's views as "speculations" and "a vast array of hypotheses," and noted that "still there is an inexplicable gap at the beginning of his series." He rejected the claim that Darwin's theory was confirmed, asserting that "most of his hypotheses are quite unproved by fact."[165] Because of the nature of his criticisms of Darwin's theory, some have argued that Whewell's rejection of it was merely a reflection of his theological prejudices.[166] However, this is an overly simplistic position. Whewell seemed to allow that his theological views might have to accommodate an evolutionary account of man's origins, if such an account were ever scientifically confirmed.

In a letter to his friend Forbes—written a month before his preface to the seventh edition of the Bridgewater Treatise—Whewell confessed,

> I have myself taken no share in the discussions on the antiquity of man; but I will not conceal from you that the course of speculation on this point has somewhat troubled me. I cannot see without some regrets the clear definite line, which used to mark the commencement of the human period of the earth's history, made obscure and doubtful. . . . It is true that a reconciliation of the scientific with the religious views is still possible, but it is not so clear and striking as it was. But it is still a weakness to regret this; and no doubt another generation will find some way of look-

164. Ibid., p. xx. Whewell had made this point in the first edition as well (see p. 228).

165. On the other hand, Whewell refused the suggestion that he publish a work critical of Darwin's book, as he had when the *Vestiges* was published, because "a person who ventures into the controversies which are at present agitated ought to have a great deal of specific knowledge, which I do not possess." Whewell to Rev. D. Brown of Aberdeen, October 26, 1863, in Isaac Todhunter, *William Whewell, D.D.*, 2:433–35.

166. See, for instance, R. M. Young, *Darwin's Metaphor*, and Adrian Desmond and James Moore, *Darwin*.

ing at the matter which will satisfy religious men. I should be glad to see my way to this view, and am hoping to do so soon.[167]

In his *History of the Inductive Sciences*, Whewell had asserted that man's origin must stand outside science and law; on this question, he believed, geology "says nothing, but she points upwards."[168] There could be no "scientific view" of the origin of man. By 1864, however, Whewell seemed to accept that there was a scientific view, and that it was an evolutionary one.[169] Publicly, as we have seen, he continued to reject this position. But in this letter to Forbes he indicated that he took quite seriously the possibility that the evolutionary view might turn out to be scientifically sound. If so, it would need to be reconciled with the religious point of view.

Whewell had always believed that "all truths must be consistent with other truths": that truths of natural science and those of theology do not conflict, even if we do not have full insight into how they coincide.[170] Moreover, he held that apparent conflict between science and scripture did not necessarily mean that science was to blame. In his *Philosophy*, he had noted that "when a scientific theory, irreconcilable with [the Bible's] ancient interpretation, is clearly proved, we must give up the interpretation, and seek some new mode of understanding the passage in question, by means of which it may be consistent with what we know."[171] Earlier, in his review of Lyell's *Principles of Geology*, Whewell explained, "We do not conceive that those who endeavour to fasten their physical theories on the words of scripture are likely to serve the cause either of religion or science."[172] Indeed, in 1864 he refused an invitation to sign a "Declaration of students of the natural and

167. William Whewell to J. D. Forbes, January 4, 1864, in Isaac Todhunter, *William Whewell, D.D.*, 2:435–36. The preface to the seventh edition of Whewell's Bridgewater Treatise is dated "Trinity Lodge, Feb. 13, 1864" (p. xxii). In a letter to Forbes of July 24, 1860, Whewell had expressed both sides of his ambivalence. On the one hand, he claimed, "I reckon Darwin's book to be an utterly unphilosophical one." On the other, in discussing the famous British Association for the Advancement of Science meeting in Oxford between T. H. Huxley and Bishop "Soapy Sam" Wilberforce, he criticized the latter, noting that "perhaps the Bishop was not prudent to venture into a field where no eloquence can supercede the needs of precise knowledge" (WP O.15.47 f. 83).

168. William Whewell, *History of the Inductive Sciences*, 3:488.

169. By 1865 or even earlier, most British biologists were evolutionists, while hardly any were in 1859. See Michael Ruse, *The Darwinian Revolution*, p. 229, and Alvar Ellegård, *Darwin and the General Reader*.

170. Whewell expressed this view privately in a letter to Hugh James Rose in 1826 (December 12; WP R.2.99 f.27) and publicly in his *History of the Inductive Sciences*, 3:487; this claim also appeared in the first edition.

171. William Whewell, *The Philosophy of the Inductive Sciences*, 1st ed., 2:148.

172. William Whewell, "Lyell's *Principles of Geology*, Volume 2," p. 117.

physical sciences," which claimed to support a "harmonious alliance between Physical Science and Revealed Religion," but which was seen by many
scientists as attempting to put theological restraints on scientific inquiry. (Of
Whewell's old friends and acquaintances, only Sedgwick and David Brewster
signed; Augustus DeMorgan and Herschel were quite vocal in their public
denunciations of the document.)[173] That same year Whewell wrote the letter
to Forbes. He was then seventy years old, and he was strongly committed to
his theological views; yet it was not part of these views that well-confirmed
scientific theories conflicting with our understanding of scripture must be
rejected. That Darwin's theory seemed to conflict with his theological views
was not a sufficient reason for Whewell to reject it categorically. Thus, in a
private letter (where such speculations could do no harm), he could even suggest the possibility of a reconciliation of the scientific and religious views.[174]

   This reconciliation was not yet necessary, however, for Whewell did not
believe that Darwin's theory was well confirmed. Even though Darwin used
the criterion of consilience to some degree, his theory was not in fact completely consilient. Whewell had at least four reasons for doubting the consilient status of this theory. First, while evolution by natural selection did
explain different kinds of phenomena, it is not so clear that Darwin reached
his theory by numerous inductions from each of these facts.[175] Recall that for
Whewell, it was the convergence of separate lines of induction that was important: the reading of the cause independently from each class of facts. This was
necessary to ensure that in each case, the assertion of the cause as a cause for
a particular event kind has high inductive warrant (is more likely than not).
Otherwise, on Whewell's view, we cannot be certain that we have a *vera
causa*. Darwin did not emphasize that separate processes of induction led him

---

173. See W. H. Brock and R. M. Macleod, "The Scientists' Declaration," p. 41.

174. Whewell may have had another reason for his willingness to "wait and see" about Darwin's theory: namely, that it could be viewed as actually uniting theology and science, by implying
that unity of type and design were not incompatible. Asa Gray later made this point, writing in
Nature that Darwin had performed a great service to natural science by "bringing back to it Teleology; so that, instead of Morphology vs. Teleology, we shall have Morphology wedded to Teleology" ("Charles Robert Darwin," p. 81). Darwin replied to Gray, "What you say about Teleology
pleases me especially," and claimed that "I do not think any one else has ever noticed the point"
(Charles Darwin to Asa Gray, June 5, 1874; in *Life and Letters of Charles Darwin*, 2:367). Perhaps
Whewell did notice. See Dov Ospovat, *The Development of Darwin's Theory*, pp. 148–50.

175. Thus, although one reviewer framed Darwin's discovery in terms of Whewell's discoverers' induction, Whewell did not agree. W. B. Carpenter, in his review of the *Origin*, noted, "The history of every science shows that the great epochs of its progress are those not so much of new discoveries of *facts*, as those of new *ideas* which have served for the colligation of facts previously
known into general principles, and which have thenceforward given a new direction to inquiry. It
is in this point of view that we attach the highest importance to Mr. Darwin's work" ("Darwin on
the Origin of Species," reprinted in David Hull, *Darwin and His Critics*, pp. 113–14).

to his mechanism of natural selection. Indeed, Darwin believed that any hypothesis, no matter how it was first obtained, could be proved to be probably true by passing the test of consilience. In a letter of 1860 he explained, "It seems to me fair in Philosophy to invent *any* hypothesis and if it explains many real phenomena it comes in time to be admitted as real."[176] Thus it may well have seemed to Whewell that Darwin, like Herschel, was using consilience as a means to justify not a real induction, but a bold hypothesis.

Moreover, recall Whewell's ambivalence about the unificatory status of the wave theory of light. Although this theory exhibited much consilience, he believed it had not attained the truly consilient status of Newton's law of universal gravitation, in part because it was unable to give a specific causal law that unified all optical phenomena. Darwin's theory suffered from the same problem: he was not able to give the causal law by which variations in organisms are produced, nor the causal mechanism by which they are inherited by offspring.[177] He himself admitted that "our ignorance of the laws of variation is profound."[178] Without the causal laws of genetics, Darwin's theory could have had, at best, a status for Whewell similar to that of the wave theory of light. And third, Darwin's theory had not yet stood the test of time in order to become coherent, the way the wave theory had done, even without a causal law.

Whewell had a final reason for believing that Darwin's theory was not fully consilient. In his 1838 presidential address to the Geological Society, he had laid out the challenge for any naturalistic account of man's origin. "Even if we had no Divine record to guide us," Whewell argued, "it would be most unphilosophical to attempt to trace back the history of man without taking into account the most remarkable facts in his nature."[179] He claimed that accounting adequately for the origin of species was impossible without accounting for man's origin as an intellectual and moral being. (Although Darwin made only one brief mention of humans in the *Origin*, it was obvious to his readers what the implications of his theory were for man's origins.) When he referred to Darwin's *Origin* as a "most unphilosophical book," he was indicating that Darwin had not explained how an evolutionary view could

176. Charles Darwin to Charles James Fox Bunbury, February 9, 1860, in *The Correspondence of Charles Darwin*, 8:76.

177. In his 1861 edition of the *Physical Geography of the Globe*, Herschel explicitly criticized Darwin on these grounds, noting that "we can no more accept the principle of arbitrary and casual variation of natural selection as a sufficient condition, *per se*, of the past and present organic world, than we can receive the Laputan method of composing books (pushed *à outrance*) as a sufficient account of Shakespeare and the Principia" (quoted in David Hull, *Darwin and His Critics*, p. 61).

178. Charles Darwin, *On the Origin of Species*, 1st ed., p. 167.

179. William Whewell, "Presidential Address," p. 642.

account for the appearance of a rational and moral creature. As he had written in his *Of the Plurality of Worlds* six years before the appearance of the
Origin, "The introduction of reason and intelligence upon the Earth is no
part nor consequence of the series of animal forms. It is a fact of an entirely
new kind."[180] Whewell believed that Darwin's theory did not colligate the
most important fact about the object kind "man," and so could not be fully
consilient.[181]

In a notebook entry, Darwin referred to Whewell's presidential address,
complaining that "Whewell thinks . . . gradation between Man and animal,
small point in tracing history of Man.—granted—but if all other animals have
been so formed, then man may be a miracle, but induction leads to other
views."[182] His claim about induction is not quite right, however. Only if we
take it as a given that man and animals are the same type of being can we project to humans inductive conclusions that we draw from our experience of animals. But this was precisely what Whewell denied. By the 1840s, Whewell
had developed the view that man's nature as an intellectual and moral being
was inherently progressive. Interestingly, then, he held that man's nature was
evolutionary—yet it was so in a special sense: man's nature came to approximate more and more closely the Ideas in the Divine Mind. Animals did not
progress in this way. Whewell drew an absolute distinction between man and
beast. This distinction was based not only on the developmental aspect of
man—unlike animals, man has the "capacity of progress," both in his intellectual and moral faculties[183]—but also on the fact of man's rationality. Animals have only instinct, while man has reason. In his *Indications of the Creator*, Whewell's rejection of Chambers's "speculations" was based on this
special nature of man. On similar grounds he denied the analogy Darwin drew
between man and other organic beings. The special nature of man required
that there be special causes for man's appearance in the natural world.

Thus Whewell's rejection of Darwin's theory was also related to another
of Whewell's methodological positions, his denial of Herschel's characteriza-

180. William Whewell, *Of the Plurality of Worlds*, p. 164.
181. Darwin himself recognized the need to explain how man's reason and morality could arise
by evolution by natural selection; he did so in his *Descent of Man*, published five years after Whewell's death. I am not suggesting that Whewell, had he still been alive then, would have changed
his mind about Darwin's theory. In fact, he seems to have prejudged the issue, believing that man's
intellectual faculty could not arise from other organic beings (see, e.g., Whewell, *Indications of the
Creator*, pp. 43–44). I merely note that during his lifetime, Whewell had methodological grounds
for questioning the probable truth of Darwin's theory.
182. Notebook C, p. 55; quoted in Howard Gruber and Paul Barrett, *Darwin on Man*,
pp. 256–57.
183. William Whewell, *Of the Plurality of Worlds*, p. 81. See also *The History of Scientific
Ideas*, 2:276.

tion of the *vera causa* principle. In his *Preliminary Discourse,* Herschel had
noted that scientists seek *verae causae,* or true causes, explaining that these
are "causes recognized as having a real existence in nature, and not being
mere hypotheses or figments of the mind."[184] He required that *verae causae*
be analogous to causes that are already known to have produced similar
effects in other cases.[185] Whewell, however, rejected this interpretation of the
vera causa principle, believing it to be overly restrictive. He agreed with Herschel that causes are not to be "arbitrarily assumed," and that we must have
"good inductive grounds" for believing in their existence. Thus he explained
in his *Philosophy of the Inductive Sciences* that *verae causae* are "those
which are justly and rigorously inferred."[186] However, he disagreed with Herschel's condition that there must be an analogous connection to known
causes. Were we to limit ourselves in this way, Whewell argued, science could
never progress. This type of interpretation "forbids us to look for a cause,
except among the causes with which we are already familiar. But if we follow
this rule, how shall we ever become acquainted with any new cause?"[187]

On these grounds Whewell criticized Lyell (and uniformitarian geology
in general) for using Herschel's *vera causa* principle to rule out the possibility
of catastrophic causes of geological change, that is, for not allowing that
causes of unknown kinds and intensity may have been at work in the past.[188]
On the other hand, in his *Preliminary Discourse,* Herschel referred to Lyell's
theory of climatic change as a prime example of a *vera causa.*[189] Darwin, who

184. John Herschel, *Preliminary Discourse on the Study of Natural Philosophy,* pp. 144–45.
In his *Mathematical Principles of Natural Philosophy,* Newton had expressed four "Rules of Reasoning in Philosophy." The first of these methodological precepts read, in part, "We are to admit no
more causes of natural things than such as are both true and sufficient to explain their appearance."
By requiring causes that are "true causes," Newton was clearly rejecting an instrumentalist view
of theories, in which causes may be postulated as long as they "save the appearances," regardless of
whether they are actually true. Further, the rule has seemed to some commentators to suggest that
there must be empirical evidence for the existence of a cause independently of its explanatory
power in a particular case. For instance, Newton's eighteenth-century supporter Thomas Reid
interpreted Newton's methodological injunction in this way in his *Essays on the Intellectual Powers of Man.* The concept of the *vera causa* is discussed in Vincent Kavaloski, "The Vera Causa Principle."

185. As Herschel explains, *verae causae,* when not directly observable in the case at hand, "are
not to be arbitrarily assumed; they must be such as we have good inductive grounds to believe do
exist in nature, and do perform a part in phenomena analogous to those we would render an account
of" (*Preliminary Discourse on the Study of Natural Philosophy,* p. 197).

186. William Whewell, *The Philosophy of the Inductive Sciences,* 1st ed., 2:189. See also D. B.
Wilson, "Herschel and Whewell's Versions of Newtonianism."

187. William Whewell, *The Philosophy of the Inductive Sciences,* 1st ed., 2:442.

188. See William Whewell, "Lyell's *Principles of Geology* volume 1," and "Lyell's *Principles of
Geology,* Volume 2."

189. William Whewell, *On the Philosophy of Discovery,* p. 191. Michael Ruse has argued that
Herschel eventually came to accept Whewell's less-empirical notion of the *vera causa.* He bases

accepted Herschel's characterization of the *vera causa* principle, believed that there was a strong enough analogy between man and other species to warrant the assertion of the same cause at work in their origins, and argued that his mechanism of natural selection was a *vera causa*. Whewell rejected this. (So, for that matter, did Herschel, to the dismay of Darwin.)[190]

The Popularization of Induction

Notwithstanding the many differences between their views of science, Whewell still saw Mill as engaged to a certain extent in the same task as the one he shared with Richard Jones and Herschel. He opened *Of Induction* by invoking Bacon and reminding his readers that more recently, the topic of induction had been treated in books published by Herschel, Mill, and himself. He then noted that these books "have circulated to such an extent as to shew that discussions concerning Induction may hope for a certain degree of popularity."[191] So even though he and Mill differed to such a degree on the details of the inductive philosophy (and even, as we have seen, on its future prospects), Whewell saw their books as sharing in the project of bringing the public to see the importance of induction in science.[192] Thus, when the *Logic* appeared, Whewell wrote to Herschel, "Jones will tell you of a new book by young Mill about the philosophy of science, suggested in a great degree by your book on the same subject and by mine. There is in new books of this kind a satisfaction in which both you and I may have a share. I mean that notions and expressions, which were new and strange when we began to write, are now familiarly referred to as part of the uncontested truth of the matter."[193]

---

this claim on Herschel's acceptance of the wave theory of light, which postulated unobservable causes of optical phenomena—the motion of unobservable light waves in an unobservable ether (see *The Darwinian Revolution*, p. 59). However, Herschel's view of *verae causae* does not rule out theoretical or unobservable causes, as long as they are analogous to other known causes. For instance, the motion of waves in a medium was known empirically to have certain effects in other cases (such as water waves), and very often wave theorists of light did describe optical effects in analogy with these cases, strengthening their case for inferring the action of undulations in an unseen medium, the ether. See Herschel's review of Whewell ("Whewell on Inductive Sciences," p. 234), and also Vincent Kavaloski, "The Vera Causa Principle," p. 59. It seems likely that Darwin's admiration for Herschel's notion of the *vera causa* is what enabled him to accept Lyellian geology in the 1830s. See Michael Ruse, *The Darwinian Revolution*, p. 61.

190. Like Whewell, Herschel rejected the analogy between man and other species.

191. William Whewell, *Of Induction*, p. 2.

192. Similarly, Herschel praised the recent works on logic by Whewell and Mill in his address to the British Association for the Advancement of Science in 1845. He explained that "the common pursuit of truth is of itself a brotherhood," implying that he, Mill, and Whewell were brothers in their endeavor (see John Herschel, "Address," pp. xl–xli).

193. William Whewell to John Herschel, April 8, 1843, in Isaac Todhunter, *William Whewell, D.D.*, 2:315.

Mill, on the other hand, did not see the debate this way. To some degree he respected Whewell as an educational reformer and as an academic philosopher. In 1844 he wrote to Comte that Whewell "merits all our respect, for he has introduced important improvements into the instruction at Cambridge and attracted interest in the great philosophical questions [of our day]. He found the spirit of philosophy dormant; he is one of those who did their utmost to awaken it."[194] Yet, as Alexander Bain recounted, when presented with the opportunity to meet his famous opponent on friendly ground, Mill refused, apparently viewing Whewell as a personal enemy.[195] As we have seen, Mill believed there was more at stake than merely a debate over inductive method in science. Strong moral and political motives were at work in the development of his epistemology and methodology, and in his decision to debate Whewell. In the next two chapters we will explore further these considerations.

194. John Stuart Mill to Auguste Comte, October 5, 1844, in Oscar Haac, *The Correspondence of John Stuart Mill and Auguste Comte*, p. 260.

195. Alexander Bain tells us that "it is no wonder, as he told me once, that he avoided meeting Whewell in person, although he had had opportunities of being introduced to him (I have no doubt, through his old friend James Garth Marshall, of Leeds, whose sister Whewell married)" (Alexander Bain, *John Stuart Mill*, p. 94).

CHAPTER FOUR

Reforming Culture: Morality and Politics

The contest between the morality which appeals to an external standard, and that which grounds itself on internal conviction, is the contest of progressive morality against stationary—of reason and argument against the deification of mere opinion and habit. The doctrine that the existing order of things is the natural order, and that, being natural, all innovation on it is criminal, is as vicious in morals, as it is now at last admitted to be in physics, and in society and government.
—*John Stuart Mill, "Dr. Whewell's Moral Philosophy," 1852*

There is a Review of my Lectures, looking in my Morality, in the Westminster Review. It is plainly by John Mill and I am rather amused to hear what is the amount of what can now be said of the best of Bentham's school in favour of their master. He offers to put the results of our controversy on an issue which I am quite willing to accept. Whether the pleasures of animals—pigs, geese, lions, or the like for instance—are of the same *moral* value as those of man. He says *yes*, I say *no*.
—*William Whewell to Richard Jones, 1852*

As we have seen, Mill's avowed "mission" in *System of Logic* was to drive intuitionism or a priorism from its "stronghold" in the physical sciences. His motivation for this undertaking was his belief that the a priori philosophy could be invoked to justify the existing status quo in society, and thus that it was dangerously backward-looking, in opposition to the types of reform he endorsed. Thus scientific knowledge was seen as a model for knowledge in moral and political realms. If intuitionism could be shown to be superfluous in science, then its appeal in the other areas would be diminished. Not only Whewell's epistemology but also his moral philosophy came

REFORMING CULTURE: MORALITY AND POLITICS

under attack in Mill's battle against intuitionism. In his review of Whewell's *Elements of Morality, Including Polity*, Mill wrote, "We do not say the intention, but certainly the tendency, of his efforts, is to shape the whole of philosophy, *physical as well as moral*, into a form adapted to serve as a support and justification to any opinions which happen to be established."[1] More specifically, he accused Whewell of justifying the practices of slavery, cruelty to animals, and forced marriages.[2] This is why the defeat of the a priori school, on Mill's view, was "no mere matter of abstract speculation," but rather was "full of practical consequences."[3] And this is why the debate with Whewell could not remain an arid intellectual controversy, but rather became, in Mill's mind, a struggle literally between the forces of good and evil: between those who wished to reform society's inequities and those who desired to justify the status quo.[4] Thus there arose a certain amount of personal animosity toward Whewell on Mill's part. While Whewell considered the two of them engaged in the same task to some extent, for Mill the conflict was personal as well as professional.

Whewell too saw scientific knowledge as an exemplar of the kind of knowledge that could be gained in other realms, such as morality and politics. Yet, unlike Mill, he was also independently motivated to understand the methods and history of science, and applied himself for many years in order to do so. This is not to suggest that Whewell was no partisan in the moral and political debates of the day. He was a strong and vocal opponent of utilitarianism, believing utilitarian moral philosophy to be inadequate philosophically, appalling morally, and distressingly irreligious. Moreover, part of his motivation for taking on the Professorship in Moral Philosophy in 1838 was to counter the prevailing utilitarian view at Cambridge. His major works in ethics were written expressly as replacements for William Paley's utilitarian tract as the required reading for undergraduates at the university.[5] Further, Whewell was opposed to numerous political positions advocated by the utilitarian radicals, including their policies in political economy, as we will see in the following chapter. So he disagreed with the utilitarians on both moral

---

1. John Stuart Mill, "Dr. Whewell's Moral Philosophy," *CW* 10:168; emphasis added.
2. See ibid., 10:200.
3. John Stuart Mill, *Autobiography*, *CW* 1:269–70.
4. For a similar view, see John Robson, *The Improvement of Mankind*, p. 125, and Stefan Collini, introduction, *CW* 21:xiii.
5. Ernest Albee noted that the book's acceptance at Cambridge, soon after it was published, showed that the University represented "the prevailing tendency of the time and country" (*A History of English Utilitarianism*, p. 168). He explained that "utilitarianism had been . . . distinctly in the air for more than a generation" before Bentham published his *Principles of Morals and Legislation* (ibid., p. 167).

and political grounds. Although Whewell did not shape his view of science expressly to counter utilitarianism, he was certainly not loath to employ his epistemology against moral and political positions he found alarming.

In this chapter I will examine the political views of Mill and Whewell. We will see that although Mill considered Whewell a consistent supporter of the status quo, Whewell's political position was actually more complex and, indeed, more similar to Mill's view, than he suspected. Moreover, although they disagreed vehemently and publicly over moral philosophy, I will show that both men shared rather similar views about the role of moral philosophy in reshaping society. Both rejected the simplistic moral system of Jeremy Bentham, which grounded man's motivations for acting virtuously in physical pleasure and pain. Instead, they constructed moral philosophies that emphasized the necessity of creating morally excellent characters which would naturally find happiness in acting virtuously. Creating this kind of character required the proper type of education. Thus both Mill and Whewell believed that it was essential to widen the scope of this kind of education, in order to reform society.

Mill's Political Evolution

In his youth, Mill was considered one of the "Philosophical Radicals," the utilitarian followers of Bentham and James Mill in the 1820s and 1830s, including George Grote, John Arthur Roebuck, Charles Buller, Joseph Hume, and Sir William Molesworth.[6] Their explicit purpose was, as a reviewer of J. S. Mill's *Autobiography* put it, "the reform of the world."[7] The problem, as they saw it, was the existence of aristocratic privilege; thus Mill defined "philosophical radicalism" as "enmity to the Aristocratical principle."[8] They regarded the British government as an aristocracy, in which the Tories and Whigs competed for the privilege of protecting the interests of the aristocratic few against the democratic many.[9] Bentham, in his *Plan of Parliamentary Reform* (1817), attacked the "moderate reform" being championed by the Whigs, saying that "Tories and Whigs drive the same road to despotism."[10] (Later in

6. See Joseph Hamburger, *Intellectuals in Politics*, p. 1.
7. Henry Reeve, "*Autobiography* of John Stuart Mill," *Edinburgh Review* 139 (1874), p. 99; quoted in ibid., p. 29; see also John Stuart Mill, *Autobiography*, CW 1:137.
8. John Stuart Mill, "Fonblanque's England under Seven Administrations," CW 6:353.
9. See Stefan Collini et al., *That Noble Science of Politics*, p. 105.
10. Jeremy Bentham, *Plan of Parliamentary Reform*, in *The Works of Jeremy Bentham*, 3:436–622.

the century, his followers would have mixed feelings about what they saw as small liberal victories—Catholic Emancipation, partial reforms of the electoral system—believing them to be instances of winning the battle while losing the war.)[11] Indeed, the Philosophical Radicals preferred that there be only two parties, the Tory party of the aristocracy, and the Radical party of the people. As Mill put it, "the problem will then be reduced to its simplest terms: who is for the aristocracy and who for the people, will be the plain question."[12]

The Philosophical Radicals gained some members in the House of Commons in the elections of December 1832, due to proreform sentiment: after the passage of the Reform Bill of that year, the electorate did not distinguish between Whigs and Radicals, regarding as a "reform" candidate anyone who supported the passage of the bill. Yet the Philosophical Radicals were not very effective in Parliament, being unwilling to compromise and work on less radical reforms with the Whigs or even with other radical groups such as the Owenites.[13] In his *Autobiography*, Mill expressed the disappointment he felt at this, noting of the Philosophical Radicals in Parliament that "on the whole they did very little to promote any opinions."[14] They did not, that is, successfully apply the ideas of utilitarianism in order to "reform the world."

By the mid-1830s, Mill's own publicly expressed political views had begun to diverge from those of the Benthamite Philosophical Radicals. He began to understand that the Benthamite approach to politics—based on its assumption that men were always, and everywhere, alike—was flawed. The way men were at present, Mill came to believe, need not be their nature for all time. Moreover, the type of government appropriate to a society varied according to the stage of civilization reached by that society. Thus, as we saw in the introduction, Mill rejected the Benthamite claim that democracy was always the best form of government. In this way he historicized politics, believing that historical context determined which political system was appropriate for a particular society. Mill changed his position on the secret ballot based on such considerations: although in the 1830s he had supported it, by the 1850s he did not. In the 1830s he believed that the aristocratic domination of the political process necessitated the secret ballot, while by the 1850s the growth in political influence of the middle classes had solved that

11. See Joseph Hamburger, *Intellectuals in Politics*, pp. 31–32.
12. John Stuart Mill, "Parliamentary Proceedings of the Session," *CW* 6:300.
13. See Joseph Hamburger, *Intellectuals in Politics*, p. 114.
14. John Stuart Mill, *Autobiography*, *CW* 1:203.

problem and raised a new one—the selfishness of the voters. This new problem called for a new solution, namely an open ballot, in which the voters' selfish impulses would be brought out into the open and in that way defused.[15]

Another shift away from philosophical radicalism came when Mill read Alexis de Tocqueville. Like Tocqueville, whom he admired greatly, Mill worried about the "tyranny of the majority" over the individual.[16] He and Tocqueville, in fact, identified a new challenge to freedom: rather than worrying only about the oppression of society by tyrants and aristocrats, they were concerned as well about the domination of the individual by society. In one of his essays on Tocqueville, Mill expressed his fear that democracy would impose on the individual a single set of mass values, thereby bankrupting the national culture. For this reason, although Mill was sympathetic to many of the aims of the Owenites, and later became optimistic about the benefits of a modified socialist economic system, he worried about the loss of individuality that seemed to be endemic to these positions. At the same time, however, he began to endorse the need for an intellectual clerisy. What concerned him was not merely the tyranny of the majority over the individual, but the tyranny of the uncultivated masses over the future direction of society. In order to form "the best public opinion," he argued, "there should exist somewhere a great social support for opinions and sentiments different from those of the mass."[17]

When Mill published his review of Tocqueville's work in 1835, he was publicly rebuked in a pamphlet written by his former friend Roebuck, who noted that Mill "believes, if I mistake not, in the advantages to be derived from an Aristocracy of Intellect."[18] As Roebuck rightly saw, Mill did have hopes that an intellectual elite could reform the world, in part by being moral, political, and intellectual guides to the masses.[19] Mill had already expressed this hope in his articles concerning "the spirit of the age" (1831), in which he

15. See Bruce L. Kinzer, "J. S. Mill and the Secret Ballot," and John Stuart Mill, *Thoughts on Parliamentary Reform, CW* 19:331–38, and *Considerations on Representative Government, CW* 19, chap. 10. Reformers continued to support the ballot until it finally triumphed under William Gladstone in 1872.

16. Tocqueville returned Mill's admiration, writing, "At present I think that an enlightened man of good sense and good will would be a [Philosophical] Radical in England" (Alexis de Tocqueville, *Journeys to England and Ireland* [1835]; quoted in Joseph Hamburger, *Intellectuals in Politics*, p. 297).

17. John Stuart Mill, "De Tocqueville on Democracy in America (II)," *CW* 18:198.

18. John Roebuck, "Democracy in America," p. 3. According to Roebuck, their friendship had ended abruptly in 1833, after Roebuck advised Mill to terminate his relationship with the married Harriet Taylor. See Nicholas Capaldi, *John Stuart Mill*, p. 111.

19. See T. H. Heyck, *The Transformation of Intellectual Life in Victorian Britain*, pp. 193–94, and John Stuart Mill, "Coleridge," *CW* 10.

noted that "reason itself will teach most men that they must, in the last resort, fall back upon the authority of still more cultivated minds, as the ultimate sanction of the convictions of their reason itself."[20] He believed it was "natural" that the mass of people would choose those possessing the most highly cultivated minds to lead them. In *On Liberty* he explained that "the initiation of all wise and noble things, comes and must come from individuals; generally at first from one individual. The honour and glory of the average man is that he is capable of following that initiative."[21] Thus Mill disdained purely democratic systems, at least until the minds of most citizens had become highly cultivated. As he explained, "no government by a democracy . . . ever did or could rise above mediocrity."[22] On these grounds Mill denied the possibility of self-government to certain societies, such as those in the colonies of Britain. The British needed to serve as a "government of leading strings" that could, perhaps in the distant future, help "as a means of gradually training the people to walk alone."[23]

In order truly to reform society, Mill came to believe, there must be a "change of character . . . in the uncultivated herd."[24] Individuals, not merely political institutions, needed to be reformed. By 1843 his friend John Sterling observed that "Mill was brought up in the belief that Politics and Social Institutions were everything, but he has been gradually delivered from this outwardness, and feels now clearly that individual reform must be the groundwork of social progress."[25] In the Early Draft of his *Autobiography*, written in the 1850s, Mill noted that many of the institutional reforms he had advocated had been effected. "But these changes," he realized, "have been attended with much less benefit to human well being than I should formerly have anticipated, because they have produced very little improvement in that on which depends all real amelioration in the lot of mankind, their intellectual and moral state."[26] Thus humankind must be improved before society

20. John Stuart Mill, "Spirit of the Age II," January 23, 1831, *CW* 22:244.

21. John Stuart Mill, *On Liberty, CW* 18:269.

22. Ibid. See also "Spirit of the Age III, pt. 1," February 6, 1831, in *CW* 22:252.

23. John Stuart Mill, *Considerations on Representative Government, CW* 19:175–76. Although Mill worked in the East India Company for over thirty years, he never traveled to India and may never actually have met an Indian (see Trevor Lloyd, "John Stuart Mill and the East India Company," pp. 44–79). The strong current of paternalism in nineteenth-century liberal thinkers such as Bentham, the Mills, and Macaulay is discussed in Uday Singh Mehta, *Liberalism and Empire.*

24. John Stuart Mill, *Early Draft of Autobiography, CW* 1:112, 238.

25. Reported by Caroline Fox, March 9, 1843, in her *Memories of Old Friends*, 2:8–9. In his review of Mill's posthumously published *Autobiography*, John Morley referred to this shift in Mill's emphasis to "a fundamental re-constitution of accepted modes of thought" as a "crisis of middle age." In "Mr. Mill's Autobiography," pp. 15, 16, and 19.

26. John Stuart Mill, *Early Draft of Autobiography, CW* 1:244.

could be improved.[27] The role of the intellectual class—and eventually, of government—was to direct this transformation of the intellectual and moral state of the masses. Like Plato, Mill believed that "the business of rulers is to make the people whom they govern wise and virtuous."[28] Although the development of his political views troubled his old Radical associates (Francis Place claimed that "if James Mill could have anticipated that his son John Stuart should preach so abominable a heresy [as plural voting] . . . he would have cracked his skull"),[29] Mill himself believed that his political views became more, rather than less, radical. While he had once been satisfied with "superficial improvement," he now saw himself as one who sought truly radical change—indeed, change that would completely "regenerate society"[30] by bringing about a "complete renovation of the human mind."[31]

The regeneration of society, then, required a radical transformation of individuals. Mill believed that this renovation could be brought about by a certain kind of education and sustained by a particular type of liberty. This education was to be intellectual, political, cultural, and moral. In his *Thoughts on Parliamentary Reform* (1859) and his *Considerations on Representative Government* (1861), Mill stressed the importance of a literate electorate; indeed, he argued for a minimum education test as a condition of a universal franchise. "If it is asserted that all persons ought to be equal in every description of right recognized by society," he announced, anticipating the criticisms of his former Radical associates, "I answer, not until all are equal in worth as human beings. . . . A person who cannot read, is not as good, for the purpose of human life, as one who can."[32] Suffrage gives a power not only over oneself,

27. As Mill wrote to Gustave D'Eichthal, "You imagine that you can accomplish the perfection of mankind by teaching them St. Simonianism, whereas it appears to me that their adoption of St. Simonianism, if that doctrine be true, will be the natural result and effect of a high state of moral and intellectual culture previously received: that it should not be presented to the minds of any who have not already attained a high degree of improvement, since if presented to any others it will either be rejected by them, or received only as Christianity is at present received by the majority, that is, in such a manner as to be perfectly inefficacious" (February 10, 1830, *CW* 12:49).

28. John Stuart Mill, "Grote's *Plato*," *CW* 11:435.

29. Francis Place, reported in a letter from John Bowring to Edwin Chadwick, December 23, 1868, Chadwick Papers, University College, London; quoted in Joseph Hamburger, *Intellectuals in Politics*, p. 274.

30. See John Stuart Mill, *Autobiography*, *CW* 1:236, and "Grote's *Plato*," *CW* 11:387. The need to concentrate on the character of individuals became apparent as early as his mental crisis, during which he realized the need to place less emphasis on "the ordering of outward circumstances" and greater stress on "the internal culture of the individual" (see *Autobiography*, *CW* 1:146). But Mill claimed that it was under the influence of Harriet Taylor that he realized the importance of this insight, and began to put it into practice (see also Joseph Hamburger, *John Stuart Mill on Liberty and Control*, pp. 23–30).

31. See John Stuart Mill, "Grote's *History of Greece* [V]," *CW* 25:1162.

32. See John Stuart Mill, *Thoughts on Parliamentary Reform*, *CW* 19:323.

but over others, he argued, and the better educated are more qualified to have this power.[33] Rejecting the Radicals' old call for "one man one vote," Mill supported "plural voting": in his *Thoughts on Parliamentary Reform* he proposed that all adult men and women in Britain should be enfranchised (after passing a simple literacy and mathematics test),[34] but that those with a superior education should receive extra votes, up to six or more.[35] He later agreed with Thomas Hare's plan for minority representation, because it allowed the educated elite a disproportionate voice.[36]

Mill believed that fullest participation in government should be limited to those who have achieved the necessary level of education and cultivation to ensure that they would desire the best government, namely the one that served the interest of the general good. At the same time, he envisioned an educative role for participation in government. In *Subjection of Women*, for example, Mill discussed the cultivating effect of participation in public affairs, maintaining this as one reason for lifting the restrictions against women playing a role in government.[37] In his review of Grote's *History of Greece*, he praised ancient Athens for its openness to all citizens, claiming that this "formed a course of political education."[38] Although the masses were not yet ready for full participation and thus for democratic governance, Mill believed that they could be made more prepared by limited forms of participation, such as serving on juries and participating in voluntary associations.[39] He expressed this dual vision of the different levels of participation in his review

33. Ibid. See also John Stuart Mill, *Considerations on Representative Government, CW* 19:473–79.

34. See John Stuart Mill, *Thoughts on Parliamentary Reform, CW* 19:327.

35. Ibid., 19:325. Mill thus disagreed with Walter Bagehot, who wanted the franchise to be extended, but weighted toward property rather than education.

36. See Paul N. B. Kern, "Universal Suffrage without Democracy."

37. See John Stuart Mill, *Subjection of Women, CW* 21:293. Stefan Collini has noted that "one could without strain regard his whole notion of political activity itself as an extended and strenuous adult-education course" (introduction, *CW* 21:xlviii).

38. John Stuart Mill, "Grote's *History of Greece* [II]," *CW* 11:324.

39. See Graeme Duncan, "John Stuart Mill and Democracy," pp. 74–75. See also Alan Ryan, "Two Concepts of Politics and Democracy," and Richard Ashcraft, "Class Conflict and Constitutionalism in J. S. Mill's Thought." The "emancipatory function" of political participation has been most recently discussed in Nadia Urbinati, *Mill on Democracy*, p. 4. Urbinati, however, downplays the educative purpose of political participation, and the *need* for such an education, according to Mill. For instance, in her analysis, democratic citizenship means treating people as adults who can understand and publicly defend their choices (see p. 10). But she de-emphasizes Mill's claim that some adults (e.g., those in India) *cannot* be treated as adults, and that even in "civilized" nations such as England everyone needs to be educated morally as well as intellectually before they are fully capable of being treated this way. As a result, Urbinati ignores the irreducibly historical nature of Mill's political thought, and its equal emphasis on the Coleridgian as well as the Benthamite elements.

of Grote's *Plato*, where Mill praised Plato for recognizing that "the work of government is Skilled Employment," but criticized him for denying any role at all to the unskilled.[40]

The cultivation required by members of society before they are ready for democratic government is not only intellectual and political, but also cultural and, especially, moral. Drawing on his experience in overcoming his break-down years earlier, Mill emphasized that citizens must be taught to appreciate art, literature, and music. Further, they must attend to history, culture, and even nature. As he noted in "Utilitarianism," published the same year as *Representative Government*, "A cultivated mind—I do not mean that of a philosopher, but any mind to which the fountains of knowledge have been opened, and which has been taught, in any tolerable degree, to exercise its faculties—finds sources of inexhaustible interest in all that surrounds it; in the objects of nature, the achievements of art, the imagination of poetry, the incidents of history, the ways of mankind past and present, and their prospects for the future."[41] Moreover, as he elaborated in this essay and as I will discuss later in this chapter, Mill claimed that a cultivated mind was one that had learned to found his own happiness on the good of the whole, and whose possessor no longer suffered from a "miserable individuality." (This was also a concern of Tocqueville's.) So moral education was required as well.

The other requirement for transforming the human mind, according to Mill, was a certain kind of liberty. He stressed the importance of freedom of action and freedom of opinion. In *On Liberty*, he supported the right of individuals to choose "different experiments of living,"[42] and opened the book, on the frontispiece, with a quote from Alexander von Humboldt: "The grand, leading principle, towards which every argument unfolded in these pages directly converges, is the absolute and essential importance of human development in its richest diversity." In order for the human mind to be transformed in such a way as to allow for the regeneration of society, it was necessary that men be able to convert themselves to individuals who can think and make decisions for themselves. Recall that in *System of Logic*, Mill had noted that we are not constrained by our characters; that, in fact, we have the ability to change them. In the later work *On Liberty*, this claim reemerged as the position that we *must* desire to form our own characters, that is, to make our own choices about how to live and what to believe. In part this had an educative purpose, to "exercise" our faculties of perception, judgment, and

---

40. Thus Plato, like most "profound thinkers," erred in only seeing one half of the truth. See John Stuart Mill, "Grote's *Plato*," *CW* 11:436.

41. John Stuart Mill, "Utilitarianism," *CW* 10:216.

42. See John Stuart Mill, *On Liberty*, *CW* 18:261.

moral preference.[43] If we merely follow custom we are not fully human, but "ape-like" or even like a "steam engine."

But the liberty to diverge from custom was necessary, Mill judged, not only because it provided individuals with the exercise of their rational and judgmental faculties. The liberty to engage in different "experiments in living" was also crucial in order to counter the "*despotism* of custom."[44] Mill's worries about the mediocrity of mass culture led him to endorse the individual's liberty to rise above conformity to custom.[45] In part, his exhortation for individual experiments in living was meant as a call to the intellectual class to provide examples of nonmediocrity on which those without the ability to conceive of new forms of living could model their lives.[46] Beckoning back to his belief in the value of many-sidedness, that is, his notion that truth is often partially contained in several incompatible beliefs, Mill suggested that only by the promulgation of many incompatible kinds of lifestyles could the best of these be determined. However, the ultimate purpose of this liberty of lifestyle is not to reject custom for all time, but to establish custom on a better basis. He admitted in *On Liberty* that "it is important to give the freest scope possible to uncustomary things, in order that it may in time appear which of these are fit to be *converted into custom*."[47] The role of the intellectual class, or the individual genius, is to "set the example of a more enlightened conduct and better taste and sense in human life."[48] Mill's claim about the type of liberty required to reform society was therefore historicized. Given the current tyranny of a custom of mediocrity, liberty to diverge from

43. Ibid., 18:262.

44. Ibid., 18:272; emphasis added.

45. Mill's description of Plato would suit himself; in his review of George Grote's *Plato*, Mill claimed that Plato's mission was not to fight against Sophistry, but "the Commonplace" (*CW* 11:403).

46. See, for example, John Stuart Mill, *On Liberty, CW* 18:267–68. As he noted in his review of George Grote's *History of Greece*, without the freedom to express individuality, there can be no genius, and society is forced into mediocrity. Comparing ancient and modern times, Mill wrote, "The difference ... is most closely connected with the wonderful display of individual genius which made Athens illustrious, and with the comparative mediocrity of modern times. Originality is not always genius, but genius is always originality; and a society which looks jealously and distrustfully on original people—which imposes a common level of opinion, feeling, and conduct, on all its individual members—may have the satisfaction of thinking itself very moral and respectable, but it must do without genius" (*CW* 11:320–21). Mill had expressed a similar view in his 1832 essay, "Genius" (*CW* 1:329–39). His correspondence with Harriet Taylor in the years before they were married gives the strong impression that they considered themselves geniuses whose originality in personal relations was looked down upon by "respectable," but "common," people.

47. John Stuart Mill, *On Liberty, CW* 18:269. Though Mill did add that this was not the only reason for freedom of action, and that it was meant to apply to more than just the class of "mentally superior people" (ibid., 18:270).

48. Ibid., 18:267.

custom was necessary in order to reform society on a more enlightened basis. As he noted,

> There has been a time when the element of spontaneity and individuality was in excess, and the social principle had a hard struggle with it. The difficulty was then to induce men of strong bodies and minds to pay obedience to any rules which required them to control their impulses. . . . But society has now fairly got the better of individuality; and the danger which threatens human nature is not the excess, but the deficiency, of personal impulses and preferences.[49]

Eventually, the goal Mill foresaw was not the continued multiplicity of opinions and lifestyles; rather, he suggested that it was only "while mankind are imperfect [that] there should be different . . . experiments of living."[50]

Mill made the same point about freedom of opinion. In *On Liberty*, he noted that freedom of opinion and expression are goods "*until* mankind are much more capable than at present of recognizing all sides of the truth."[51] Indeed, Mill pointed out that "as mankind improve, the number of doctrines which are no longer disputed or doubted will be constantly on the increase, and the well-being of mankind may almost be measured by the number and gravity of the truths which have reached the point of being uncontested." He stressed that this "narrowing of the bounds of diversity of opinion" is not only desirable, but "necessary." Such a convergence of opinion was so important that Mill claimed it outweighed the value to be had by the constant clashing of opposing views.[52]

It is now possible to see more starkly the roots of Mill's concerns about intuitionism. He believed that it stood in the way of moral and social reform, because it allowed currently held customs and opinions to be raised to the

---

49. Ibid., 18:264.

50. Ibid., 18:260–61. His view remained the same as it had been in the earlier essay on Coleridge, in which Mill had noted that it was "in the present imperfect state of mental and social science" that it was important to have "antagonist modes of thought" ("Coleridge," *CW* 10:122).

51. John Stuart Mill, *On Liberty, CW* 18:269.

52. Ibid., 18:250–51. Paul Feyerabend has claimed that his famous call for methodological pluralism (sometimes, rather, "anarchism") was merely his application of Mill's views in *On Liberty*. This point has been examined by Elisabeth A. Lloyd, "Feyerabend, Mill, and Pluralism," and Kent W. Staley, "Logic, Liberty, and Anarchy." As this last paragraph demonstrates, Feyerabend (and these authors) overstate the similarities between the views of Feyerabend and Mill. For Mill the diversity of opinion is not an end, but merely a means to reach objective truth. He clearly did not want intuitionism to remain a live option, for example, in the way that Feyerabend claimed should be the case for astrology. Feyerabend, unlike Mill, denied that it would be a good thing for doctrines one day to be "no longer disputed or doubted."

status of "necessary truths," which are inviolable. In *On Liberty*, he remarked that custom gives the illusion of self-evidency. "People are accustomed to believe, and have been encouraged in the belief by some who aspire
to the character of philosophers, that their feelings on subjects of this nature
are better than reasons and render reasons unnecessary."[53] The view that reasons are not required to justify beliefs was odious to Mill as contributing to
the defense of the status quo, especially since the current status quo was a
mediocre one and the beliefs being justified were *false*. As he explained, "The
cessation on one question after another, of serious controversy, is one of the
necessary incidents of the consolidation of opinion; a consolidation as salutary in the case of true opinions, as it is dangerous and noxious, when the
opinions are erroneous."[54] Recall that Mill claimed in his *Autobiography*
that intuitionism was the chief danger *"in these times."* Once custom had
been properly established on less mediocre grounds, and opinion founded
upon truth, intuitionism would no longer be so dangerous.

As we have seen, Mill believed that Whewell's philosophy was the paradigmatic example of an antireformist view justifying the current status quo.
Because of this he criticized both Whewell's philosophy of science and his
moral philosophy. Indeed, Mill's own philosophy of science was devised in order to counter what he saw as Whewell's type of "intuitionism," because of the
alleged consequences of this view for politics. However, as we will now see,
Whewell's views were not as unambiguously conservative as Mill believed.
In fact, in some respects both men shared the same political temperament.

Whewell's Politics

In modern scholarship, Whewell has been called everything from a "Liberal
Anglican" to a staunch "Tory."[55] Yet an accurate description of his position is
neither as clear-cut, nor as extreme, as these characterizations suggest. To
gain a more nuanced view of Whewell's politics, it is useful first to distinguish between his positions on political issues within the university and his
opinions about political issues in the wider society. Let us begin with the former of these.

53. John Stuart Mill, *On Liberty*, CW 18:220.
54. Ibid., 18:250.
55. To take just two examples, Richard Brent characterizes Whewell as one of the "Liberal
Anglicans" at Cambridge (more specifically, a "peelite don of liberal inclinations"), while Boyd
Hilton describes Whewell and Jones as Tories (see Brent, *Liberal Anglican Politics*, p. 196 passim,
and Hilton, *The Age of Atonement*, p. 52).

In 1800, Cambridge University was considered a "branch of the Church of England."[56] Degrees were granted only to members of the established church, chapel attendance was mandatory, and all students were required to study some theology. After the repeal of the Test and Corporations Act of 1828, Dissenters could matriculate, but they could not obtain the B.A. without declaring themselves members of the Church of England; nor could they obtain the M.A., B.D., or doctorate in law, physics, or divinity without subscribing to the three articles of the Thirty-sixth Canon.[57] This requirement essentially precluded non-Anglicans from receiving Oxbridge degrees. Yet a growing, middle-class population of Dissenters began to be more outspoken about their rights to participate in the national educational system. At Cambridge, controversy over this issue was sparked in 1834, when Thomas Turton, Regius Professor of Divinity, published a pamphlet claiming that the admission of Dissenters to degrees would subvert the aims of education, and would lead to "a flood of dangerous speculation and cold skepticism and reckless infidelity overspreading the land."[58] Whewell's friend and assistant tutor Connop Thirlwall responded with an open letter countering Turton's claim that the role of the university was to instill religious values and beliefs. On the contrary, its role should be distinct from the role of the church. Moreover, even if the university should instill religious values and beliefs, forcing unwilling students to attend chapel services was not an efficacious means of doing so. Thus Thirlwall argued that compulsory chapel attendance should be eliminated. After the publication of this letter, he was called upon to resign his assistant tutorship by Christopher Wordsworth, then Master of Trinity (and younger brother to the poet), ostensibly for disagreeing publicly with the university's policy on compulsory chapel attendance, but more likely because Wordsworth himself was opposed to the admission of Dissenters to degrees.[59] (This controversy did not impede Thirlwall's career within the church; he was later appointed to the bishopric of St. David's.)[60]

Whewell entered the fray with two pamphlets. Earlier, he had supported the repeal of the Test and Corporations Act of 1828 and the Catholic Emancipation Act of 1829. (In a letter to Sir Robert Peel on this issue in 1845, Henry Goulbourn described Whewell as someone "who has always been of very lib-

56. Martha Garland, *Cambridge before Darwin*, p. 1.
57. As Richard Brent puts it, "The repeal of the Test and Corporation Acts thus presented the paradox of admitting Dissenters to the political nation, while continuing their exclusion from the national provision for higher education" (*Liberal Anglican Politics*, p. 185).
58. See Thomas Turton, *Thoughts on the Admission of Persons without Regard to their Religious Opinions to Certain Degrees in the Universities of England*.
59. See Martha Garland, *Cambridge before Darwin*, p. 16.
60. See ibid., p. 27.

eral opinions in such matters.")[61] In the first pamphlet, published before Thirlwall was dismissed, he objected to Thirlwall's claim that worship upheld by penalties is detrimental to religion. In fact, Whewell countered, this is not so. Students must be obligated to attend worship services. Without directly addressing the volatile issue of admission of Dissenters to degrees, Whewell claimed that *some* distinction between churchmen and Dissenters ought to be made, perhaps in the granting of fellowships. For, as he concluded, the university is one of the main supports of the church, and the church is "the very heart of our social body." In his second pamphlet, Whewell called the dismissal of Thirlwall "a matter of deep and serious regret,"[62] and claimed that they were not so far apart in their opinions as it had originally seemed. Whewell here explicitly agreed with the justness of admitting Dissenters to degrees, but noted that this should not result in a weakening of the Church of England, nor of the special relationship between it and the university. Oddly, however, he claimed that allowing Dissenters to take degrees should not result in changes to the rules obligating attendance at religious services. Although Whewell admitted that these rules should "be administered in a tolerant and liberal spirit," suggesting perhaps that Dissenters might be excused from following them, he went on to claim, "I understand this as implying that unconditional conformity to general rules, which has always been rendered by the Dissenters who have hitherto been among us."[63] Thus, although he seemed to want to distance himself from the overt prejudice and intolerance of Turton, Whewell continued to support a provision with which most Dissenters could not honestly comply (though there seem to have been a number of Dissenters who attended the chapel services, given Whewell's comment on the matter.) Later, as Master of Trinity, he continued to support compulsory orders and celibacy for Fellows of the College.

On other university issues, Whewell was not opposed to change, but generally believed that reforms should come slowly. As he grew older, he became (as is often the case) more conservative, and more protective of the power of

61. Henry Goulbourn to Robert Peel, January 1845, Peel Papers, British Museum, Add.Ms.40445, ff. 5–6; quoted in Norman Gash, *Reaction and Reconstruction in English Politics 1832–1852*, p. 95n. Whewell expressed his disagreement with those opposing Catholic Emancipation as early as two letters to his aunt in 1821 and 1822 (in Janet Stair Douglas, *The Life and Selections from the Correspondence of William Whewell, D.D.*, pp. 63, 73–74).

62. Whewell wrote to Adam Sedgwick, "The Master's request to him to resign the tuition I entirely disapprove of, and expressed my opinion against it to the Master as strongly as I could" (May 27, 1834, in Janet Stair Douglas, *The Life and Selections from the Correspondence of William Whewell, D.D.*, pp. 163–64). On this controversy see also Isaac Todhunter, *William Whewell, D.D.*, 1:91–92, and D. A. Winstanley, *Early Victorian Cambridge*, p. 77.

63. See William Whewell, *Additional Remarks on Some Parts of Mr. Thirlwall's Two Letters on the Admission of Dissenters to Academical Degrees*, pp. 17–19.

his college. When Whewell first became Master of Trinity, he submitted a proposal to promote a closer correspondence between the mathematical triposes and the lectures of the math professors, which would have increased the role of the university and its professors at the expense of the function of the colleges and their tutors. This proposal was defeated—showing Whewell that, as Leslie Stephen put it, the position of Master "gave little power of introducing reforms."[64] In 1844, Whewell was instrumental in revising the statutes of Trinity College; although his changes were minor, they signified, at least, some revision of regulations that had been on the books since Elizabethan times.[65] In addition, Whewell introduced new textbooks into the moral philosophy curriculum, in order to replace Paley's text, in use since 1787. In 1848 he proposed and helped to establish a natural and moral sciences tripos, which changed the options of study open to students who wished to graduate with honors. Yet, although he lobbied for the addition of the new exams, he argued that students must first take honors in mathematics and classics and only then go on to the new fields. (When the classical tripos was established in 1822, students were required to receive an honors degree in mathematics before sitting for the classics examinations. Whewell opposed all efforts to change this requirement, and was still in opposition when it was finally dropped in 1849.) In 1852, Whewell supported the revision of the university statutes recommended by the Senate syndicate and approved by the Royal Commissioners. The revised system was not a radical revision, but was more liberal than the existing one.[66] However, he was strongly opposed to the changes eventually enacted by the Act for an Executive Commission for Cambridge, which was passed in 1856. Whewell especially opposed the transferring of power from the traditional "caput"—composed of the heads of colleges and eight senior Fellows—to an elected council. The caput system had essentially given the Masters vetoing powers over university legislation. (Whewell's friendship with Adam Sedgwick suffered over this episode, as Sedgwick had been one of the members of the Royal Commission who called for statutes more democratic than those originally approved.)[67]

64. See Leslie Stephen, "Whewell, William"; D. A. Winstanley, *Early Victorian Cambridge*, p. 180; Harvey W. Becher, "Whewell's Odyssey," pp. 25–28; and Isaac Todhunter, *William Whewell*, *D.D.*, 1:215.

65. Robert Robson, "Trinity College in the Age of Peel," p. 332.

66. D. A. Winstanley, *Early Victorian Cambridge*, p. 43.

67. Yet Whewell seems to have been more upset by Sedgwick's deception of him than over his support of the new statutes, as he did not quarrel with Herschel, who was, with Sedgwick, one of the ex-members of the commission who sent a letter supporting more democratic reforms. See D. A. Winstanley, *Early Victorian Cambridge*, pp. 42–53. See also William Whewell, *Of a Liberal Education in General, and with Especial Reference to the University of Cambridge, Part III*, pp. 3–4.

To the dismay of Charles Lyell, Richard Owen, and other scientific acquaintances, Whewell rejected certain proposed changes to the curriculum at
Cambridge, including the addition of "modern scientific" topics such as comparative anatomy.[68] From the mid-1830s onward, he argued that mathematics
and classics should form the foundation of studies at the university. In his
Thoughts on the Study of Mathematics as Part of a Liberal Education (1835)
and "Remarks on Mathematical Reasoning" (1837), Whewell claimed that
the study of mathematics is the most effective way in which to learn to recognize truths. Indeed, he implied that such study is necessary for learning the
inductive method of science; as he explained in "Remarks on Mathematical
Reasoning," mathematics teaches one the "peculiar habits" necessary in
order "to follow a *chain* of reasoning."[69] Further, studying mathematical reasoning enables the student to see the importance of the distinct possession
of ideas—the Idea of Space, in the case of geometry, different ideas in the case
of the inductive sciences, such as mechanics. He urged that mechanics and
hydrostatics be added to the examinations for the B.A., so that students would
learn that mathematics has practical applications in the inductive sciences.[70]

In *On the Principles of English University Education* (1837), Whewell
introduced a distinction between "speculative" and "practical" teaching.
Practical teaching is active, in the sense that the student must not merely lis-
ten to lectures outlining the discoveries of others, but must do something
himself: solve problems, translate texts. Languages and mathematics can be
taught practically, but the physical sciences—especially those dealing with
new discoveries, where the foundations are constantly changing—can hardly
be taught except speculatively.[71] Practical teaching, Whewell noted, is applicable to sciences "in which principles having clear evidence and stable certainty, form the basis of our knowledge; and in which, consequently, a distinct possession of the Fundamental Ideas enables a student to proceed to
their applications, and to acquire the habit of applying them in every case
with ease and rapidity."[72] Gaining a facility for working with the clearest Fun-

68. In 1848 Richard Owen sent Whewell a strong letter, in which he bitterly complained that
his friend would not help him by advocating for a new chair in comparative anatomy. This letter
was discovered by Pietro Corsi in the Hawkins papers at Oriel College, Oxford, which suggests that
Whewell must have forwarded the letter to Edward Hawkins, Provost of Oriel. See Corsi, *Science
and Religion*, p. 268.

69. William Whewell, "Remarks on Mathematical Reasoning and on the Logic of Induction,"
in *Mechanical Euclid*, p. 148.

70. William Whewell, *Thoughts on the Study of Mathematics as Part of a Liberal Education*,
pp. 44–45.

71. William Whewell, *On the Principles of English University Education*, pp. 5–7.

72. Ibid., p. 9.

damental Ideas can enable the student to work well with the more indistinct ones. "It is only by the practical teaching of mathematics, that the Fundamental Ideas of science can become distinct among men in general."[73] Whewell especially incensed Lyell by noting that the role of geometry or mechanics in education could not be replaced by "sciences which exhibit a mass of observed facts, and consequent doubtful speculations, as geology." But Whewell—who at the time of writing the second edition of the *Principles* was serving as president of the Geological Society—did not mean to suggest that geology itself was merely a speculative science. Indeed, he went on to emphasize that although geology and the other sciences mentioned could not "do the work of mental cultivation," nevertheless they are "highly valuable acquisitions to the student, and may very beneficially engage his attention during the later years of his University career." Indeed, "some insight into the progressive sciences is an essential part of a liberal education."[74]

In *Of a Liberal Education in General, and with Particular Reference to the Leading Studies of the University of Cambridge*, first published in 1845, Whewell focused his attention less on the distinction between speculative and practical teaching, and more on a distinction between different types of knowledge: that which is "permanent," and that which is "progressive." On his view, students must learn the parts of knowledge that are "permanent," or known to be true, before they can move to those parts that are "progressive," or in the process of being debated. Otherwise, there is the risk that they will not come to recognize the nature of truth.[75] As Whewell had pointed out years earlier in an article in the *British Critic*, the primary function of the university is education, not research, and the main job of teachers is not discovering new truths, but transmitting knowledge of established truths and the proper method by which to learn to recognize and test new truths.[76] This is parallel to his claim, in *The Philosophy of the Inductive Sciences* and elsewhere, that the proper method of science must be taught by examining the accepted truths of physical science, not the more controversial propositions of morality and politics. Indeed, the educational policy based on the distinction between permanent and progressive knowledge grew out of Whewell's epistemological belief in a historical development of the Fundamental Ideas; he maintained that students should concentrate on those Ideas that have already been explicated, so that they will learn how to recognize correctly

73. Ibid., p. 20.
74. Ibid., pp. 41–2.
75. See, e.g., William Whewell, *Of a Liberal Education in General, and with Especial Reference to the University of Cambridge, Part III*, pp. 21 and passim.
76. William Whewell, "Science of the English Universities," pp. 71–72.

explicated ideas and the proper application of them in other fields (and, eventually, how to explicate ideas themselves). Thus, contrary to Lyell's complaint (in a lengthy letter to him), Whewell's educational writings were not inconsonant with his works on the history and philosophy of science.[77] Nor were they merely a sign of his conservatism, as some modern commentators have argued.[78] In light of his epistemology, Whewell was making a sound pedagogical point. Today, a student must master mathematics and classical physics before turning to quantum mechanics. Similarly, on Whewell's view, a student must learn to recognize and work with clear Fundamental Ideas—such as are present in mathematics—before studying sciences in which the Fundamental Ideas are only slowly coming into view.

To be sure, there were some politically conservative concerns underlying Whewell's views. He worried that the initial study of the progressive sciences could lead to radical forms of political thought among young men. One result of allowing students to begin with the progressive parts of knowledge would be to place them in the position of being critics instead of pupils; rather than learning that there are some truths which are unquestioned, they would immediately begin with topics where truth was contested, and thus where even the professor's point of view was contestable.[79] Criticism, Whewell believed, is an important part of the history of ideas, but it is for the already formed mind. He reported that he had been told that young men in Germany and France were extremely hostile toward the institutions of their country. If this is true, Whewell remarked, he believed that "such a consequence may naturally flow from an education which invokes the critical spirit, and invites it to employ itself on the comparison between the realities of society and the dreams of system-makers."[80]

For Whewell, education plays an important role in the development of the kind of "cultivated minds" that Mill also championed. Sounding a theme similar to Mill's, he noted that the business of the university is "the general cultivation of all the best faculties of those who are committed to their charge."[81] Indeed, his own translations of Plato's dialogues were intended for

77. Charles Lyell to William Whewell, April 17, 1847, WP Add.Ms.216 f. 57; quoted in Richard Yeo, *Defining Science*, p. 211.

78. See Richard Yeo, *Defining Science*, pp. 215–22, and Perry Williams, "Passing on the Torch."

79. William Whewell, *On the Principles of English University Education*, pp. 47–48.

80. Ibid., p. 51.

81. See ibid., p. 41. Whewell also stressed the role of the university in "the preservation and promotion of the general culture of mankind" (ibid). Mill, precisely because he feared that the present culture of mediocrity would be preserved, was nervous at the prospect of a uniform, government-provided secondary education (see John Stuart Mill, *On Liberty*, *CW* 18:302–4). But he agreed with Whewell, against the Benthamites, that the role of the university is not to teach professional

those who did not attend university. (He omitted those passages he felt were too difficult for the common reader.)[82] In addition to being important for the "mental cultivation" of young men,[83] Whewell maintained that education serves an important social purpose. He argued that a crucial result of university education is to inculcate in a student the feeling "that he *is* an Englishman" and the understanding of the principles on which his fellow citizens regulated their lives, so that he could acquire "a sympathy with their objects."[84] For this reason, men of all social and economic classes should be admitted to the university, allowing a "common participation in a liberal education," which was useful in placing all students in a position of equality and mutual respect. Indeed, this was one of the "most important and beneficial functions" of the university.[85] Surely his own experience of being raised from a probable career as a master carpenter or other similar trade to the elite levels of society must have influenced this view.

I would argue, then, that regarding university politics, Whewell was neither a liberal nor a staunch conservative. He loved the institution of Cambridge, and defended it vehemently from attacks—and perceived attacks—against it. Yet at the same time he recognized that improvement was needed. Change, though, would come from inside, and come slowly.[86] As Whewell ex-

---

or merely useful knowledge. Rather, like Whewell, Mill held that the university student should study both the ancients (classics) and the moderns (natural science). See "Inaugural Address to the University of St. Andrews," *CW* 21:218 and "Notes on the Newspapers": teachers should teach "those kinds of knowledge and culture, which have no obvious tendency to better the fortunes of the possessor, but solely to enlarge and exalt his moral and intellectual nature."

82. It has been suggested, with good reason, that although Whewell's university politics were somewhat progressive, "doubtless his personality more than any other single factor established his reputation as an uncompromising conservative" (Sheldon Rothblatt, *The Revolution of the Dons*, p. 104). At various junctures during his tenure as Master, his friends sent him letters chiding him for his arrogance and overbearing style of authority (see, for example, letters in Janet Stair Douglas, *The Life and Selections from the Correspondence of William Whewell, D.D.*, pp. 209, 235, 285–92). After his death, the young Henry Sidgwick—then a student at Trinity—expressed a view that was probably not uncommon, in a letter to his mother: "We are in a considerable state of agitation here, as all sorts of projects of reform are coming to the surface, partly in consequence of having a new Master—people begin to stretch themselves, and feel a certain freedom and independence" (*Life of H. Sidgwick* [London, 1906], p. 145; quoted in D. A. Winstanley, *Later Victorian Cambridge*, p. 241). It seems fair to point out, however, that at next year's General Meeting, the only new resolution was that Scholars should read Grace in Hall before rather than after dinner; thus it is not in fact so clear that many desired programs of reform were being held back by Whewell's "conservatism."

83. William Whewell, *On the Principles of English University Education*, p. 41.

84. Ibid.

85. Ibid., pp. 90–91. See also Richard Brent, *Liberal Anglican Politics*, pp. 195–96.

86. I agree with Martha Garland that Whewell's positions on university politics put him in the category of the "conservative reformers" she has characterized in her study of Cambridge in the first two-thirds of the nineteenth century. See Garland, *Cambridge before Darwin*, p. vii.

plained (perhaps intending the implicit reference to Mill's articles addressing the "spirit of the age," which had appeared several years earlier), "Universities and Colleges have for their office, not to run a race with the spirit of the
age, but to connect ages, as they roll on, by giving permanence to that which
is often lost sight of in the turmoil of more bustling scenes. . . . In order to introduce real improvements, we must bring to the task a spirit, not of hatred,
but of reverence for the past; not of contempt, but of gratitude towards our
predecessors. If we are able to go beyond them, it must be by advancing in
their track, not by starting in a different direction."[87]

Whewell's position on political issues outside the university was similar.
In 1843 he characterized himself as a "constitutional conservative," aligning
himself with the "old Whigs" rather than the "new"; but, tellingly, not with
the Tories.[88] By 1843 he was a member of the establishment, as Master of Trinity; at the same time, his marriage to Cordelia Marshall two years earlier had
joined him to the family of the "Marshalls of Leeds," who were very active in
Whig politics. As the author of one of his obituaries noted, "In politics he was
too independent, too fond of thinking for himself, to be a partisan. . . . As to
particular measures, such as Catholic Emancipation, he held with the Liberals, but his general sentiments and predilections were staunchly Conservative. He loved the historical traditions of England."[89] Whewell's stand on the
important issues of the day tended to be, like his views about the university,
on the side of "conservative reform": he supported change, but preferred it to
occur cumulatively and lawfully. In that way his view was conservative in the
sense defined by Peel—who had recommended Whewell for the Mastership
of Trinity—as the willingness to engage in reform by gradual means.[90]

For example, Whewell advocated the extension of the franchise that occurred with the Reform Bill of 1832, yet thought further change should come

87. William Whewell, *On the Principles of English University Education*, p. 137. In his 1845
*Of a Liberal Education in General, and with Particular Reference to the Leading Studies of the
University of Cambridge*, he similarly justified the need for changes in the university system to be
made slowly (but note that he justified as well the need to make changes, as opposed to maintaining the status quo) (see pp. vi–vii).

88. See William Whewell to Richard Jones, October 6, 1843, in Isaac Todhunter, *William Whewell, D.D.*, 2:318. See also 1:413.

89. W. G. Clark, "William Whewell," p. 551.

90. The Mastership of Trinity was a lifetime appointment made by the Crown, under the
advice of the prime minister. Whewell was recommended by Robert Peel, a fellow Lancastrian. On
the Lancaster connection, see Francis Espinasse, *Lancaster Worthies*. For this characterization
of Peel, see Nicholas Capaldi, *John Stuart Mill*, p. 118 and, more generally, Norman Gash, *Sir Robert Peel*. Jack Morell describes Whewell's politics as being Peelite in his "The Judge and Purifier
of All."

slowly; in 1845 he was still cautioning against further reform, at least for the present time.[91] Similarly, in his writings on morality, he strongly promoted the abolition of laws allowing slavery in the British colonies and in the United States, claiming, "Wherever Slavery exists, its Abolition must be one of the great objects of every good man." During the American Civil War, he refused to allow the *Times* in the Master's Lodge because it supported the slave-holding Southern states.[92] And we will see later in this chapter that Whewell had, for the time, fairly progressive views regarding the intellectual and moral capabilities of slaves. Yet at the same time, he cautioned that until laws allowing slavery were changed, they must be "submitted to."[93] In sum, Whewell's view of political change was consistent with his view of scientific change: there are no scientific revolutions—nor should there be political ones—completely discontinuous and unrelated to what came before. Indeed, as in the history of science, there must be "preludes" and "sequels" to any great "epochs" of change.

Whewell's political views, as we have seen, are more complex than they are generally considered. He was certainly no radical, fearing the kind of extreme changes called for by those who wanted to, in his words, "destroy the church and democratize the nation."[94] He disagreed with liberals in his equivocation on the admission of Dissenters to degrees at Cambridge. Yet unlike the most "orthodox" Tories, Whewell did not view current social arrangements and hierarchies as a predominant and unchanging feature of society; certainly, he believed that a clever boy from the lower ranks should be able to rise to social prominence and power. Nor did he believe that the current institutional status quo must be preserved for all time. He agreed that even his beloved Cambridge merited some improvement. Thus Mill's characterization of his adversary as an ardent supporter of the status quo was not completely accurate.

Indeed, in some ways Mill's political temperament was not so far from Whewell's. In later life, he seemed to concur with Whewell's preference for slow progression in political change, claiming that "reforms, worthy of the

91. Whewell explained, "In proportion to the largeness of the step made in the Reform Bill, should be the length of time which is allowed to elapse before any new Movement of an extensive nature is attempted. The New Part of the Constitution must have time to incorporate itself with the Old, before the body politic can bear with safety any new experiments" (*Elements of Morality, Including Polity,* p. 474; reference is to the fourth edition unless otherwise specified).

92. See Michael St. John Packe, *The Life of John Stuart Mill,* p. 425.

93. William Whewell, *Elements of Morality, Including Polity,* p. 235.

94. Whewell was describing the *Westminster Review* in a letter to Julius Charles Hare, October 15, 1838, Add.Ms.a.215 f. 43.

name, are always slow."[95] Whewell's impulses toward an intellectual aristocracy—he used to require undergraduates to stand in his presence at the Master's Lodge—were shared by Mill as well, though Mill rejected the institutional
"aristocracy" of the universities, so dear to Whewell. Both men supported an
aristocracy of the intellect, where the educated and cultivated minds would
lead the nation in the right way; thus they both disdained pure democracy, at
least for the present time. As we saw earlier, Mill came to support the idea of
plural voting. Whewell also, not surprisingly, approved of this idea. In 1859 he
wrote to his brother-in-law James Garth Marshall (whose pamphlet *Minorities and Majorities* had influenced Mill's view), "I have seen John Stuart Mill's
pamphlet on Reform [*Thoughts on Parliamentary Reform*], and do not much
disagree with him."[96] In Whewell's last publication he lauded the fact that the
people had elected a philosopher to Parliament: in his essay on Comte, he
wrote, referring to Mill, "It is no small glory of our times, that one of our
most popular constituencies has fully and practically adopted the great Platonic maxim, that it will never go well with the world till our rulers are
philosophers, or our philosophers rulers."[97] Both Mill and Whewell regarded
their task as reformers as including the extension of the right of education
and opportunities for self-improvement. Whewell exhorted governments to
take the lead in this reform, expressing the hope that nations would employ
their powers in "fully unfolding the intellectual and moral capabilities of
their members, by earlier education."[98]

Though Mill was more visionary than Whewell on the topic of extending
this right to women, Whewell (as we will see in chapter 5) was perhaps more
explicit and eloquent in expressing the right to education and upward class
mobility of the indigent poor. That each man achieved success because of his
education no doubt contributed to their emphasis on this method of self-
improvement. This position is motivated as well by their respective views of
moral philosophy. As we will now see, both Mill and Whewell emphasized
the moral importance of forming one's own virtuous character, in part by par-
taking of the proper kind of education.

95. John Stuart Mill, "Inaugural Address to the University of St. Andrews," *CW* 21:222.

96. William Whewell to James Garth Marshall, Good Friday 1859, in Janet Stair Douglas, *The
Life and Selections from the Correspondence of William Whewell, D.D.*, p. 510. Whewell specifi-
cally pointed to Mill's scheme for minority representation, in which each constituency elects three
representatives, and electors have three votes that they may give to one, two, or three candidates.
Mill noted in the pamphlet that he had taken this idea from Marshall's *Minorities and Majorities;
Their Relative Rights* (1853); see *Thoughts on Parliamentary Reform, CW* 19:330–31.

97. William Whewell, "Comte and Positivism," p. 353.

98. William Whewell, *Of the Plurality of Worlds*, p. 275.

Utilitarianism: Paley, Bentham, and Mill

The moral philosophy known as "utilitarianism" is the view that acts are right insofar as they increase human happiness, and wrong insofar as they decrease it. Though not yet called by this name, utilitarianism became a dominant position in Britain in the eighteenth century. In 1672, Richard Cumberland made the first statement of the Utilitarian principle by an English writer.[99] In 1725 Francis Hutcheson coined the phrase "the greatest happiness for the greatest number."[100] Later, Joseph Priestley, a leader of the Unitarians as well as a chemist, and the Anglican Paley, developed a theological form of utilitarianism. Paley's book, *Principles of Moral and Political Philosophy*, published in 1785, was quickly adopted as the required textbook in moral philosophy at Cambridge, and remained so until Whewell's efforts to dislodge it were finally successful in the 1840s.[101] Both Paley and Priestley emphasized that happiness is not only what humans desire for themselves, but also what God, a benevolent deity, desires for us.[102] God, therefore, places upon us the moral injunction to promote the general happiness.[103] We should promote the common happiness because it is the will of God, and by doing the will of God we will be rewarded in our next life.[104] It was by considerations of the future life that Paley and Priestley could account for the coinciding of our selfish interest with the interest of society; that is, they harmonize because God made us such that our own interest coincides with the good of society.[105] In its theological aspect, utilitarianism does not necessarily lend itself to any one political position, as evidenced by the fact that Priestley was a political radical, while Paley was a conservative.

Utilitarianism found its most important systematizer and publicist in Bentham, who developed a nontheological variant. In his 1776 *Fragment on Government*, Bentham explained that the only motivation for our actions is the desire for our own happiness. On the other hand, he claimed that the standard or criterion of morality is "the greatest happiness of the greatest

99. See Ernest Albee, *A History of English Utilitarianism*, p. 52.

100. See Geoffrey Scarre, *Utilitarianism*, p. 23. Yet Hutcheson himself was not purely a consequentialist, as he combined consequentialist reasoning with a moral sense in moral reflection (for more on this see ibid., p. 53, and Ernest Albee, *A History of English Utilitarianism*, p. 62).

101. See Henry Sidgwick, "Philosophy at Cambridge," p. 240.

102. Geoffrey Scarre, *Utilitarianism*, p. 60, and Ernest Albee, *A History of English Utilitarianism*, p. 172.

103. See Geoffrey Scarre, *Utilitarianism*, p. 61.

104. Ibid., p. 64; Ernest Albee, *A History of English Utilitarianism*, p. xiii; William Whewell, *Lectures on the History of Moral Philosophy in England*, p. 153.

105. Ernest Albee, *A History of English Utilitarianism*, p. xiv.

number." Thus he needed to account for how it is that our selfish interest in our own happiness or pleasure can coincide with the good of the greatest number.[106] He introduced various rewards and punishments (which he called sanctions) to effect this reconciliation. One's own happiness or pleasure is increased by augmenting the pleasure of others, due to various sanctions.[107] Bentham claimed the existence of four external sanctions: the physical (natural consequences), the political (legal rewards and punishments), the moral (the pleasure of favorable public opinion), and the religious (belief in God's judgments).[108] There is also an internal sanction, namely that of "sympathy"; Bentham acknowledged that even rulers could sometimes find their own happiness in providing happiness to their subjects.[109] Thus he allowed that people may act in benevolent or philanthropic ways, but only because of the pleasure that such actions give to them.

Bentham spent the remainder of his life attempting to draw out practical consequences from the "axiom" of utility. Having defined happiness in terms of the presence of pleasure and the absence of pain, he then claimed that there were seven circumstances which determine the value of a pleasure or a pain: its intensity, its duration, its certainty or uncertainty, its propinquity or remoteness, its fecundity (the chance of its being followed by other pleasures or pains of the same kind), its purity (the chance of its not being followed by a sensation of the opposite kind), and its extent (the number of people affected by it).[110] For all actions, the only criterion of worth was the degree to which they resulted in pleasure rather than pain, in accordance with these seven circumstances; thus Bentham, unlike Mill later, rejected the notion of pleasures that were *qualitatively* higher. (This resulted in his famous comment that pushpin—a child's game—has as much value as poetry, if it gives as much pleasure.)[111] He believed that numerical values could, in principle, be attached to the quantitative degrees of pleasure and pain in order to construct a "felicific calculus." Although he admitted that he could not at present assign such numerical values, he stressed that this calculus should be

---

106. Albee suggests that Bentham only introduced this idea of the societal good into his basically egoistic system merely because of his political views: "Bentham used the 'greatest happiness' formula because he was a reformer" (ibid., p. 180).

107. See W. D. Hudson, *A Century of Moral Philosophy*, p. 14, and Ernest Albee, *A History of English Utilitarianism*, p. 182.

108. As Ernest Albee notes, after once mentioning the religious sanction Bentham completely neglected it in his discussion of ethics (Albee, *A History of English Utilitarianism*, p. 167n.).

109. See W. D. Hudson, *A Century of Moral Philosophy*, p. 14; Geoffrey Scarre, *Utilitarianism*, pp. 76–77; John Stuart Mill, "Bentham," *CW* 10:97; and Jeremy Bentham, *Collected Works*, 1:14–15.

110. Geoffrey Scarre, *Utilitarianism*, pp. 74–75.

111. Jeremy Bentham, *The Works of Jeremy Bentham*, 2:253.

kept in mind as an ideal model of rational deliberation, even if it could not
yet be used for calculation.

Bentham argued for his utilitarian view by pointing out the critical flaw
he saw in the alternative theories. Most of these rest, he maintained, only
upon the opinion of the authors, not on any evidence or external standard,
and are thus only "arbitrary." He particularly criticized the intuitionist view
as leading to despotism, if any person's intuitions are considered binding
upon others, or anarchy, if they are binding only upon himself.[112] Whether an
author refers to a *moral sense*, or *common sense*, or the *understanding*, or to
natural law, Bentham claimed, it was all just a matter of the author's own
prejudices: this is what I say is right. Either that, or the author is making an
implicit appeal to utility, which would be better made explicit. In any case,
Bentham held that all these views merely evade the obligation to provide an
external, objective criterion of moral worth.[113]

Mill was trained by his father and Bentham to be the leader of the next
generation of utilitarians, and it is clear that at first they succeeded. Mill
recounted in his *Autobiography* that when he first read Bentham, "the feeling rushed upon me, that all previous moralists were superseded, and that
here indeed was the commencement of a new era in thought."[114] He was convinced that the greatest-happiness principle could bring about a new and
more enlightened age in morality and politics. However, by the time he was
twenty, in 1826, he had succumbed to his nervous breakdown. As we saw in
the introduction, the crisis was precipitated by his realization that he fit the
common—and derisive—characterization of the Benthamite "reasoning
machine."[115] "I now saw, or thought I saw, what I had before received with
incredulity—that the habit of analysis has a tendency to wear away the feelings"; indeed, he recognized that these analytic habits are "a perpetual worm
at the root both of the passions and of the virtues; and above all, fearfully
undermine all desires, and all pleasures."[116] By finding himself moved by
reading Romantic writers such as William Wordsworth, and listening to the
music of Mozart and Weber, Mill realized that his capacities for emotional
responses were not dead but merely slumbering. He thus was able to pull
himself out of his depression. But the experience left him with the strong
awareness of a principal error of Bentham's and his father's philosophy: "the

112. See W. D. Hudson, *A Century of Moral Philosophy*, p. 8.
113. Ibid., p. 9.
114. John Stuart Mill, *Autobiography*, CW 1:67.
115. See ibid., 1:111.
116. Ibid., 1:141–42.

neglect both in theory and practice of the cultivation of feeling."[117] Mill was determined to correct this fault in his own work, and thus his version of the utilitarian philosophy was meant to be based on a richer and fuller understanding of human capacities.

At first Mill wondered whether utilitarianism would have to be abandoned. In 1833, he published an essay highly critical of Bentham and his moral philosophy. It is noteworthy that, at the age of twenty-seven, he had it printed completely anonymously (as an appendix to Edward Bulwer's *England and the English*), to avoid offending his father. As we saw earlier, Mill came to believe that Bentham's view of human nature was excessively narrow. In his "Remarks on Bentham's Philosophy," he pilloried the recently deceased philosopher for his portrayal of man as a purely selfish creature, driven only by his pursuit of his own pleasure: "By the promulgation of such views of human nature, and by a general tone of thought and expression in keeping with them, I conceive Mr. Bentham's writings to have done and to be doing very serious evil."[118] He pronounced "fallacious" Bentham's assumption that men are alike in all times and places.[119] Already Mill had come to realize that men's happiness could not be increased merely by reforming institutions; rather, men themselves, particularly "the state of [their] desires," needed to be transformed.[120]

In 1835, however, Mill was provoked into a defense of utilitarianism by the publication of the third edition of Sedgwick's *Discourse on the Studies of the University of Cambridge*. In this work, Sedgwick attacked Paley's utilitarian moral philosophy, and argued that it should be displaced as the official teaching of the university. Mill was particularly annoyed that Paley's views were being presented to the public as *the* utilitarian position. (Of course, Sedgwick's concern was with the course of study prescribed at Cambridge, which meant that he would discuss Paley and not Bentham.) Mill and the other Benthamite radicals attacked Paley for using the principle of utility in defense of conservative doctrines; thus Mill did not want the public to associate radical utilitarians with the version of the doctrine promulgated by Paley.[121] More-

117. Ibid., 1:115.
118. John Stuart Mill, "Remarks on Bentham's Philosophy," *CW* 10:15.
119. Ibid., 10:16.
120. Ibid., 10:15.
121. A contemporary (nonutilitarian) commentator made this point in 1874, writing of Paley that "he had stolen into the camp of their reforming philosophy, and striven to carry off their best artillery, and then to use it in defense of doctrines to which they were wholly opposed; that is, the general excellence and merit of the British laws and constitution, and the Divine origin and authority of the Christian faith." See T. R. Birks, *Modern Utilitarianism*, p. 58.

over, as Mill explained in the *Autobiography*, he was unable to "speak out my whole mind" in this essay, because his father was aware he was writing it; indeed, he was obliged "to omit two or three pages of comment on what I thought the mistakes of utilitarian moralists, which my father considered as an attack on Bentham and on him."[122]

As a result, a comparison of this essay with the earlier one on Bentham leads to some disquieting results. In "Sedgwick's Discourse," Mill attacked Sedgwick for claiming that the principle of utility has a "debasing" and "degrading" effect, while he himself had argued two years earlier that "the effect of such writings as Mr. Bentham's, if they be read and believed and their spirit imbibed, must either be a hopeless despondency and gloom, or a reckless giving themselves up to a life of that miserable self-seeking, which they are there taught to regard as inherent in their original and unalterable nature."[123] Mill contemptuously rejected Sedgwick's argument that "waiting for the calculations of utility" is immoral, because "to hesitate is to rebel," when he himself had criticized Bentham on similar grounds, writing that "the fear of pain *consequent* upon the act, cannot arise, unless there be *deliberation*; and the man as well as 'the woman who deliberates' is in imminent danger of being lost."[124] Mill spent much of this essay arguing that utilitarianism as a philosophy should not be judged by Paley's work; "Of Paley's work," he sniffed, "we think on the whole meanly."[125] Yet many of Mill's criticisms of Paley's view here mirror his own criticisms of Bentham in the earlier piece. For instance, Mill here attributed the "lax morality taught by Paley" to Paley's conflation of utilitarianism with "considerations of expediency . . . of the most obvious and vulgar kind"; that is, he claimed that Paley was guilty of giving an overly narrow definition of "consequences," just as Mill had accused Bentham of doing in the 1833 essay.[126]

Mill published another essay on Bentham in 1838, a review of John Bowring's edition of Bentham's *Works*. Because his father was no longer alive, he felt more free to criticize Bentham openly. (Like all reviews in the *London and Westminster Review*, it was published anonymously, but it was widely known to have been written by Mill.) In this essay, he modified his earlier

122. See John Stuart Mill, *Autobiography*, CW 1:201.
123. See John Stuart Mill, "Sedgwick's Discourse," CW 10:68, and "Remarks on Bentham's Philosophy," CW 10:16.
124. See John Stuart Mill, "Sedgwick's Discourse," CW 10:66, and "Remarks on Bentham's Philosophy," CW 10:12.
125. John Stuart Mill, "Sedgwick's Discourse," CW 10:46.
126. See ibid., 10:55.

criticisms of Bentham to some extent. Nevertheless, as we saw in the introduction, by linking Bentham with the politically conservative S. T. Coleridge
as the "two great seminal minds of England in their age," Mill alienated many
of his old friends. In the essay he noted that Bentham did much good in advocating an objective, external standard for moral worth; while this standard,
by defining moral worth of actions purely in terms of their propensity to produce either pleasure or pain, was still inadequate, at least it improved upon
the appeal to moral intuitions. Yet Bentham was faulted once again for what
he ignored; Mill complained that "man is never recognized by him as a being
capable of pursuing spiritual perfection as an end; of desiring, for its own
sake, the conformity of his own character to his standard of excellence, without hope of good or fear of evil from other source than his own inward consciousness."[127] Now that he felt free to express his own view, Mill presented a
position rather different from the orthodox Benthamite doctrine. He rejected
the ideal of the felicific calculus as a means to choose the action with the
highest moral value. Rather, he claimed that the surest way to act morally
throughout one's life was by developing an excellent or "noble" character,
one which contains a sense of honor and personal dignity, the love of beauty,
of order and consistency, of power (in the sense of carrying out one's volitions), and of action, as well as an awareness of aesthetic and intellectual
value.[128] "Morality consists of two parts," Mill explained. "One of these is
self-education; the training, by the human being himself, of his affections
and will. That department is a blank in Bentham's system."[129] Instead, Bentham concentrated only on the regulation of man's actions.

A few years later, in the closing pages of the *System of Logic*, Mill
returned to this issue of the importance of developing an excellent character.
He noted that "the cultivation of an ideal nobleness of will and conduct,
should be to individual human beings an end" to which other goals should
"give way." He claimed that what characterized this "elevation of character"
was decided "by reference to happiness," both that of the individual himself
and that of the wider society.[130] Thus, in keeping with a broadly utilitarian
view, Mill argued that the acquisition of an excellent character is worthwhile
because it leads to greater happiness of the individual as well as those
affected by his actions. Yet he also suggested that at least some of the happi-

127. John Stuart Mill, "Bentham," *CW* 10:95.
128. See ibid., 10:95–96, 112–13. Interestingly, given the autobiographical details of his break-
down, Mill criticized Bentham specifically for his depreciation of poetry (see pp. 113–14).
129. Ibid., 10:98.
130. John Stuart Mill, *System of Logic, CW* 8:952.

ness one gains by having an excellent character is due to the recognition of this excellence.[131] In this way Mill's view can be seen as diverging from Bentham's utilitarianism, by recognizing an end for human action other than happiness itself.[132] In *System of Logic* he brought up an additional theme that he later developed further: the need for the proper type of moral education.

Mill's final and most complete statement of his moral philosophy is his long essay "Utilitarianism," which he drafted in the early 1850s but which was published in *Fraser's Magazine* in 1861. In his *Autobiography*, he admitted that he had gone through a period in which he rejected Benthamite utilitarianism, but claimed that by the time he wrote "Utilitarianism" he had returned to a more orthodox Benthamite position. He attributed this restoration of his early appreciation of Bentham to the influence of Harriet Taylor, whom he had married in 1851. Under her influence, he commented, "I had now completely turned back from what there had been of excess in my reaction against Benthamism."[133] He explained that Taylor had caused him to return to the radicalism of his younger days, after a period in which he was more conservative, and more "indulgent to the common opinions of society and the world." This renewed radicalism, Mill claimed, brought him back to the test of utility for all social and political institutions.[134] However, as we will now see, Mill's "test of utility" remained rather different from that which Bentham had endorsed.

Mill did begin "Utilitarianism" with a definition that could have been written by Bentham himself: "The Greatest Happiness Principle, holds that actions are right in proportion as they tend to promote happiness, wrong as they tend to produce the reverse of happiness. By happiness is intended pleasure, and the absence of pain; by unhappiness, pain, and the privation of pleasure."[135] However, Mill also introduced a new element that would have been anathema to Bentham, that of distinguishing between types or qualities of pleasures.[136] He believed that this distinction answered the criticism that utilitarianism was a "doctrine worthy only of swine"—or, as Thomas Carlyle

131. Geoffrey Scarre also makes this point; see *Utilitarianism*, p. 90.

132. Mill's position can in this way be seen as similar to that of neo-Aristotelians today who claim that the attainment of an excellent character is an intrinsically worthwhile end. For two rather different examples of this view, see Martha Nussbaum, *Cultivating Humanity*, and Alasdair MacIntyre, *After Virtue*.

133. John Stuart Mill, *Autobiography*, CW 1:237.

134. Yet, as we have seen, many of Mill's political positions in the 1850s and 1860s were quite at odds with those of the radicals—including his support for plural voting and his rejection of the secret ballot.

135. John Stuart Mill, "Utilitarianism," CW 10:210.

136. Ibid., 10:211.

had contemptuously put it, "pig philosophy."[137] Mill claimed that the Epi-
cureans—who had suffered the same charge—were right to answer that "it is
not they, but their accusers, who represent human nature in a degrading light;
since the accusation supposes human beings to be capable of no pleasures
except those of which swine are capable."[138] But since the human faculties
are more elevated than those of animals, it is not surprising that the gratification of these faculties requires pleasures which are also more elevated.[139]

The "higher faculties" are satisfied by intellectual and artistic pleasures,
as opposed to the lower, or "brutish," faculties, which are satisfied by the
pleasures of the flesh. Mill referred the question of quality of pleasures to the
"verdict of the only competent judges." Those who have knowledge of both
the higher and lower pleasures are those who are competent to judge which
type is most worth having, and to Mill it was "unquestionable" that those
who are acquainted with both the higher and lower pleasures give preference
to the former.[140] Since these experts would agree that "it is better to be a human being dissatisfied than a pig satisfied; better to be Socrates dissatisfied
than a fool satisfied," we can draw the conclusion that "if the fool, or the pig,
is of a different opinion, it is because they only know their own side of the
question."[141] Mill thus took a leaf from book 9 of Plato's *Republic*, in which
Socrates distinguished between lovers of wisdom, lovers of honor, and lovers
of gain, and claimed that lovers of wisdom have the greatest experience of all
these types of pleasure, and that therefore they are most qualified to decide
which sort of life is the most pleasant (the one spent seeking wisdom, of

137. Mill also paused to ridicule the misinformed view of those of "the common herd, including the herd of writers, not only in newspapers and periodicals, but in books of weight and preten-
sion," who make the mistake of conflating the word *utilitarian* with the rejection of pleasure in the
form of beauty, ornament, or amusement. (While on the contrary, the proponents of the theory of
utility "instead of opposing the useful to the agreeable or the ornamental, have always declared that
the useful means these, among other things" ["Utilitarianism," *CW* 10:209].) Here Mill may have
been referring in part to the recent novel *Hard Times*, in which Charles Dickens satirized utilitarianism as "Gradgrindism," which is characterized by the type of starvation of the feelings and imagination that Mill himself had experienced as a consequence of his complete embrace of Benthamism. Instead of explicitly noting this unfortunate aspect of Bentham's version of utilitarianism, and
describing his own way of avoiding it, Mill simply denied the problem. In a later passage he did,
however, admit, "Utilitarians who have cultivated their moral feelings, but not their sympathies
nor their artistic perceptions, do fall into" error (p. 221).

138. John Stuart Mill, "Utilitarianism," *CW* 10:210.

139. Mill thus rejected the reductionist psychology of Helvetius, which analyzed all pleasure
down to sensual pleasure, in favor of Hartley's, which recognized the qualitative hierarchy of pleasures which can arise from the process of association. For more on how Mill explained desire and
aversion as motives for action in terms of associationism, see Elie Halévy, *The Growth of Philosophical Radicalism*, pp. 46off.

140. John Stuart Mill, "Utilitarianism," *CW* 10:211–13.

141. Ibid., 10:212.

course). Pushpin was not, in Mill's analysis, as good as poetry. Moral development, therefore, requires the development of our higher faculties, especially
our intellect, in order that we can come to take more pleasure in poetry than
in pushpin. As Mill had noted in "The Spirit of the Age," in the normal condition of society it was "natural" for the people to put their trust in "the most
cultivated minds" for "finding the right and pointing it out."[142] Similarly, in
"Utilitarianism," he claimed that the people could trust those with experience of the higher as well as of the lower pleasures to decide which kind brings
the most happiness, and thus is the most "right," or moral, to pursue.

In "Utilitarianism," Mill connected this point about the quality of pleasures to his earlier claim about the importance of cultivating an excellent
character. He noted that a "nobler" character is one who prefers the higher
pleasures and who finds pleasure in the general good. He argued that such a
person is able to bring about more happiness to others, as well as to himself or
herself. Thus utilitarianism as an engine of social reform could only attain its
end by the "general cultivation of nobleness of character."[143] Mill emphasized
the need for proper education in order to bring about this cultivation. Suitable
training can create the "cultivated mind," or the "mental culture," which
appreciates the intellectual and artistic pleasures, and which finds one's own
personal happiness in the good of the whole.[144] Again, then, Mill emphasized
that moral development requires the improvement of our higher faculties,
such as the intellect. It also requires the development of a conscience. Such
progress builds on an innate foundation of natural sympathy.[145] According to
Mill, we innately have "social feelings," which he described as a "powerful
natural sentiment."[146] As civilization advances, this feeling becomes stronger,

142. See John Stuart Mill, "Spirit of the Age, V, pt. 2," *Examiner*, May 29, 1831, *CW* 22:312–
16. He has been criticized for introducing an inconsistency in utilitarian moral philosophy with
this distinction of the qualities of pleasures. Leslie Stephen claimed that this distinction "seems
to be making room for something very like an intuition" (*The English Utilitarians*, vol. 3, *John Stuart Mill*, p. 306), and Ernest Albee similarly noted that this aspect of his thought was one of his
"concessions to Intuitionism" (*A History of English Utilitarianism*, p. xvi).

143. John Stuart Mill, "Utilitarianism," *CW* 10:213–14. Geoffrey Scarre, on the other hand,
claims that Mill's earlier notion of developing one's character is not related to his distinction in
"Utilitarianism" between the higher and lower pleasures (see *Utilitarianism*, pp. 94–95).

144. John Stuart Mill, "Utilitarianism," *CW* 10:216. See also p. 218. In his review of Grote's
History of Greece [II], he praised "the potency of Grecian democracy in making every individual in
the multitude identify his feelings and interests with those of the state" (*CW* 11:325).

145. See John Stuart Mill, "Sedgwick's Discourse," *CW* 10:136, and "Inaugural Address to the
University of St. Andrews," *CW* 21:247.

146. John Stuart Mill, "Utilitarianism," *CW* 10:231. In his later essay "Nature," Mill claimed
that sympathy is a natural sentiment (indeed, he emphasized that "on that important fact rests the
possibility of any cultivation of goodness and nobleness"), but that what is natural is sympathy
toward a small group, such as one's own family. Education of the proper sort can lead to the extension of the natural sympathy to a wider social unit, even to the whole world (see *CW* 10:394).

and consists in "the desire to be in unity with our fellow creatures." By a proper education, this feeling can be strengthened even more, until we feel that we *are* in unity with others. This feeling becomes "deeply rooted in our character" and a "part of our nature."[147] Once this happens, we are capable of being motivated to act by the desire for the happiness of the whole.[148]

Mill explained how this occurs in terms of the associationist psychology he had inherited from Bentham and his father. The development of the conscience proceeds by creating an "indissoluble association" between the happiness of an individual and the good of the whole, such that the individual cannot even conceive of his own happiness without also conceiving of the general good.[149] This is brought about by making an individual think of virtue in a pleasurable light, and thus desire it.[150] Once these associations have been formed in our minds, our motive for acting—that is, to attain our own personal happiness—will cause us to act in ways that promote the happiness of the greatest number. This kind of education implants in us an inner sanction of duty that is "the essence of Conscience." Mill described this essence of conscience as "a feeling in our own mind; a pain, more or less intense, attendant on violation of duty, which in properly-cultivated moral natures rises, in the more serious cases, into shrinking from it as an impossibility."[151] Our conscience, then, is not an innate faculty, but is acquired by association—a fact that Mill claimed does not make our conscience less efficacious or less "natural." As he noted, "It is natural to man to speak, to reason, to build cities, to cultivate the ground, though these are acquired faculties."[152] In his posthumously published essay "Nature," he somewhat differently explained that the artificial was *preferable* to the natural, in the sense that what was created by man was often an improvement upon what was created by nature. "If the artificial is not better than the natural," he asked, "to what end are all the arts of life?"[153] The truly moral man must be artificially created: "This artificially

147. John Stuart Mill, "Utilitarianism," *CW* 10:227.
148. Mill pointed to the experience of ancient Sparta in showing the plasticity of the human mind under the proper education (see "Grote's *History of Greece* [I]," *CW* 11:302).
149. John Stuart Mill, "Utilitarianism," *CW* 10:218.
150. Ibid., 10:239.
151. Ibid., 10:228. In his "Inaugural Address to the University of St. Andrews," Mill linked this teaching of the love of virtue to aesthetic education: "If we wish men to practice virtue, it is worth while trying to make them love virtue, and feel it as an object in itself. . . . It is worth training them to feel, not only actual wrong or actual meanness, but the absence of noble aims and endeavors, as not merely blamable but also degrading: to have a feeling of the miserable smallness of mere self in the face of this great universe" (*CW* 21:253–54). The study of poetry can help us attain this feeling, and thus is valuable for moral education.
152. John Stuart Mill, "Utilitarianism," *CW* 10:230.
153. John Stuart Mill, "Nature," *CW* 10:381.

created or at least artificially perfected nature of the best and noblest human beings, is the only nature which is commendable to follow."[154]

Mill believed that the feeling of unity with our fellow beings, upon which the utilitarian morality depends, should be "taught as a religion, and the whole force of education, of institutions, and of opinion, directed, as it once was in the case of religion, to make every person grow up from infancy surrounded on all sides both by the profession and by the practice of it."[155] He referred to Auguste Comte's *Système de Politique Positive*, noting that he disagreed with Comte's particular moral and political system, but agreed with him on the possibility and efficacy of a kind of "religion of Humanity" that could take the place of supernaturalistic religions in inculcating an ethical and social philosophy. Mill developed this theme in his essay "The Utility of Religion," written between 1850 and 1858 but published posthumously in 1874. There, he again stressed that "the power of education is almost boundless."[156] Were people to be taught unity with mankind from their earliest days, as they are currently taught religious tenets, society could be reformed; for individuals in society would be motivated to act for the general good. The utilitarian morality would have "command over the feelings" as well as over our motives for acting.[157] Mill noted that people can be trained to feel an obligation toward the common good.[158] Indeed, he claimed that the "religion of humanity" he proposed would do better than supernatural religions in bringing about the proper association between one's own happiness and the happiness of the whole of society.[159]

This moral education—that is, "training the feelings and the daily habits"—is to be provided by one's family.[160] In "The Utility of Religion," Mill pointed to the power of the earliest education on religious topics that children receive from their parents. Similarly, the moral education required for the cultivated mind must begin early. Besides assigning this role to parents, Mill

---

154. Ibid., 10:396–97. Mill explicitly claimed that the capacity for developing the virtues is natural, though the virtues themselves are the result of education (p. 394).

155. John Stuart Mill, "Utilitarianism," *CW* 10:232.

156. John Stuart Mill, "The Utility of Religion," *CW* 10:409.

157. Ibid., 10:408–9.

158. Ibid., 10:421.

159. Ibid., 10:420. On this point, see also Michele Green, "The Religion of Sympathy," p. 1705.

160. See John Stuart Mill, "Inaugural Address to the University of St. Andrews," *CW* 21:247–48. Mill did note, however, that universities can help moral education by setting the proper tone: the university "should present all knowledge as chiefly a means to worthiness of life, given for the double purpose of making each of us practically useful to his fellow-creatures, and of elevating the character of the species itself; exalting and dignifying our nature" (ibid., 21:248).

also suggested that society at large can be influential in this regard.[161] He noted that public opinion should be "bent to the purpose" of enforcing the utilitarian standard of morality by the application of external sanctions of praise and blame, approbation and punishment.[162] In "Utilitarianism," he linked public opinion with education as having a "vast . . . power over human character."[163] In "The Utility of Religion," Mill even more strongly claimed that public opinion can be the "most overpowering" of motives for men's actions, and suggested that without the force of public opinion, even Christianity would not have the power it does in motivating people to consider the general good.[164] Thus, by applying approbation to actions that conduce to the general good, and disapprobation to those that are selfish, society can push the individual toward the proper actions, even if he is not motivated by the proper feeling of unity.

It seems, then, that Mill envisioned that the lessons conducive to morality would be taught by parents and exemplified in society at large, so that these lessons would be reinforced; moreover, those parents who did not provide the proper lessons would be motivated by feelings of shame to begin to do so. This proper education results in a particular type of character, considered "excellent" by Mill, one whose personal happiness is based on the higher rather than lower pleasures, as well as on bringing about the common good. Within these boundaries, there is some latitude for diversity, but not as much as commentators on Mill suggest. Indeed, Mill himself recognized that some might worry that this type of education and application of public opinion in order to mold character in a particular way might seem opposed to liberal political views. In response, he claimed that to those who have gained the social feelings, "it does not present itself to their minds as a superstition of education, or a law despotically imposed by the power of society, but as an attribute which it would not be well for them to be without."[165] Just as the person who has experienced both the higher and lower forms of pleasure is the one who can rightly judge which pleasures are more valuable, the person who has attained the "noble character" is the one who is qualified to decide whether the means necessary to gain this character were justified.

Because of such remarks, it has been argued that Mill's claims in "Utili-

161. John Stuart Mill, "The Utility of Religion," CW 10:408.
162. John Stuart Mill, "Utilitarianism," CW 10:228.
163. Ibid., 10:218.
164. John Stuart Mill, "The Utility of Religion," CW 10:410–12.
165. John Stuart Mill, "Utilitarianism," CW 10:233.

tarianism" conflict with those of his work *On Liberty*, even though the two were published within a few years of each other.[166] And it certainly does seem that the type of control to be exercised by education and public opinion on the formation of cultivated minds in the former work is at odds with the very strong plea for individuality and the liberty to form one's own character in the latter. One solution has been to point out that Mill drew a distinction between conduct that concerns others ("other-regarding" actions), and that which concerns only the individual ("self-regarding" actions); some commentators claim that the subject of "Utilitarianism" is the former type of actions, while the subject of *On Liberty* is the latter.[167] On this view, conduct that concerns others is the realm of morality, and is properly subject to control, while conduct concerning only the individual is the realm of personal liberty, and is not subject to control. Mill suggested this distinction in *On Liberty* when he explained that "in the part which merely concerns himself, his independence is, of right, absolute. Over himself, over his own body and mind, the individual is sovereign."[168] However, the use of education and public opinion in "Utilitarianism" suggests that the individual is not completely sovereign over his own mind, as his feelings, motivations, and conscience will be developed for him in a certain way. Moreover, even in *On Liberty* the scope of actions that are truly "self-regarding" on Mill's view appears quite small. I believe that if we are to follow Mill in viewing these two works as consistent, we must keep in mind the kind of society he wished to see created.

We have already seen that the goal of the liberty Mill endorsed is, in part, to allow individuals to rise above the custom of mediocrity, and to found custom on a better basis. But we have also seen that Mill did not trust the will of the majority to bring about this improved custom. Thus the liberty of the individual must be guided in certain ways. When Mill presented his "liberty principle" in *On Liberty*, he strongly qualified it by exempting certain societies from it. "Liberty, as a principle," he noted, "has no application to any state of things anterior to the time when mankind have become capable of being improved by free and equal discussion."[169] Minds must already be improved or cultivated to some degree before they are ready for this kind of lib-

---

166. See Gertrude Himmelfarb, *On Liberty and Liberalism*, and Isaiah Berlin, "Two Concepts of Liberty."

167. Alan Ryan, for example, describes Mill as drawing a distinction between the Art of Morality and the Art of Life (which encompasses prudence and excellence), and claims that Mill allows control in the sphere of morality but not the sphere of prudence and excellence. I do not agree that it is possible to draw such a clear distinction between the sphere of self- and other-regarding actions on which Ryan's analysis depends. See Alan Ryan, *John Stuart Mill*, chaps. 11 and 13.

168. John Stuart Mill, *On Liberty*, CW 18:224.

169. Ibid.

erty.[170] Such minds will feel sympathy with their fellow citizens, and will be at least inclined to act in ways conducive to the general good (or at least not opposed to it). These minds will still require further cultivation, and since this is the province of what concerns society, a certain kind of control exercised here is not ruled out by Mill's claims about actions and beliefs that concern only the individual. But minds that have been cultivated to some degree will also have begun to recognize that their own happiness is found in the higher pleasures rather than the lower ones. Such individuals can be left free to seek their own personal happiness each in their own way; but the cultivated mind will naturally prefer the higher to the lower pleasures, and so in societies in which minds have become cultivated, personal liberty will lead to a convergence toward lifestyles emphasizing intellectual and creative pleasures. Here is the kind of custom that Mill wished to see created by personal liberty.

Indeed, those who do not seek the higher pleasures may be pushed in that direction by societal pressure. Even in *On Liberty*, Mill made it clear that those who prefer the "lower," brutish pleasures of drink and fornication, even if their actions do not harm others, may and should be deterred in their conduct by shame and expressions of contempt, though not absolutely coerced by law. Thus he noted that "a person may suffer very severe penalties at the hands of others, for faults which directly concern only himself."[171] He specifically singled out as an example one "who pursues animal pleasures at the expense of those of feeling and intellect." If a man's behavior seems morally odious to us, even if it is not harmful to anyone else, we may show our disapproval to him by ignoring his society, and warning others to avoid him: a form of social pressure that, though Mill denied it amounted to "punishment," surely does not seem conducive to the free play of individual liberty. Rather, this social pressure imposes rules of conduct based on the kind of improved society Mill wished to see brought about. As he had explained at the start of *On Liberty*, "some rules of conduct . . . must be imposed, by law in the first place, and by opinion on many things which are not fit subjects for the operation of law."[172] Even the dispositions that lead to injurious acts, though not themselves injurious, may be socially sanctioned by our disapprobation and abhorrence.[173] In "Utilitarianism," Mill quite explicitly pointed

---

170. I disagree, however, with Joseph Hamburger's too-strong claim that Mill advocates—in chapter 3 of *On Liberty*—the liberation of only those with individuality or genius (see *John Stuart Mill on Liberty and Control*, p. 157).

171. John Stuart Mill, *On Liberty*, CW 18:278.

172. Ibid., 18:220.

173. Ibid., 18:279.

to the power of social pressure to enforce utilitarian morality as an "external standard," suggesting that legal penalties were not always required—and indeed, that they might be more costly in terms of resources than the ever-present and inexpensive use of social pressure.[174] He also indicated that "self-ishness"—what he sometimes referred to as "miserable individuality"—led to unhappiness, not only of the individual but of the greater society, and so was to be condemned and deterred.[175] Given Mill's strong statements about the "overpowering" nature of public opinion to influence action, allowing this kind of external sanction even on our seemingly self-regarding actions is not a trivial infringement of personal liberty. It is not surprising that, while writing *On Liberty*, he told Grote that he was "cogitating an essay to point out what things society forbade that it ought not, and what things it left alone that it ought to control."[176]

How can these types of interference in personal liberty be justified, given Mill's comment that "the sole end for which mankind are warranted, individually or collectively, in interfering with the liberty of action of any of their number is self-protection. . . . the only purpose for which power can be rightfully exercised over any member of a civilized community, against his will, is to prevent harm to others"?[177] Light is shed on this by a later passage in *On Liberty*, in which Mill restated this principle: "To individuality should belong the part of life in which it is chiefly the individual that is interested; to society, the part which chiefly *interests society*."[178] Mill here broadened the cases in which individual liberty may be curbed, from straightforward "harm to others" to what "concerns" or "interests" society. Society's interest lies in having citizens who are cultivated, who prefer the higher pleasures, and who are motivated to action by the desire to bring about the general good. To allow citizens to become cultivated in this way requires allowing them the liberty to rise above current custom, and to create more excellent characters. In both *On Liberty* and his later *Auguste Comte and Positivism*, Mill suggested that the best way to bring about excellent characters would be by allowing the freedom to develop in diverse ways, so that an individual could nurture what was best about him or herself.[179] But ultimately, he was not neutral on the question of what counted as being well developed. The liberty to

174. See John Stuart Mill, "Utilitarianism," *CW* 10:228. See also "The Utility of Religion," *CW* 10:410–11.

175. See, e.g., John Stuart Mill, "Utilitarianism," *CW* 10:215.

176. See Alexander Bain, *John Stuart Mill*, p. 103.

177. John Stuart Mill, *On Liberty, CW* 18:223.

178. Ibid., 18:276.

179. As Mill wrote to Thomas Carlyle in 1834, "Though I hold the good of the species . . . to be the *ultimate* end, (which is the alpha and omega of my utilitarianism) I believe with the fullest

develop oneself is circumscribed by the interest society has in a certain kind
of development, namely the kind that leads to an appreciation of the higher
pleasures, and the motivation to act for the common good.

In *On Liberty*, Mill emphasized the importance of allowing liberty in
the realm of the personal: that is, on actions that neither add to nor subtract
from the happiness of the whole. It is here that the individual is granted freedom, at least for the current time, to develop his or her own character in contradistinction to the prevailing custom of mediocrity. Yet even this freedom
has boundaries delineated by morality, in particular Mill's notion of the qualities of pleasures; thus education and social pressure can be applied to nudge
individuals toward forms of life that emphasize seeking the higher pleasures,
in order to ensure that each individual properly seeks the greatest quality of
his own happiness. In "Utilitarianism," Mill's emphasis is rather on the realm
of the moral, that is, the inherently social; thus the development of the individual that we find discussed there is one that must be guided by society, so
that each person develops the proper kind of conscience and therefore
becomes capable of choosing those actions that are most beneficial to the
whole. In "The Utility of Religion," Mill brought together the two realms under the rubric of Morality, showing that he regarded liberty and authority as
both being necessary to bring about the reformed society he sought:

> A morality grounded on large and wise views of the good of the whole, nei-
> ther sacrificing the individual to the aggregate nor the aggregate to the
> individual, but giving to duty on the one hand and to freedom and spon-
> taneity on the other their proper province, would derive its power in the
> superior natures from sympathy and benevolence and the passion for
> ideal excellence: in the inferior, from the same feelings cultivated up to
> the measure of their capacity, with the superadded force of shame.[180]

Note that Mill here explicitly distinguished between the "superior natures,"
those that have been fully cultivated, and the "inferior natures," those that
have not. The superior natures, or the individuals with "genius," are the ones
who are allowed to express their contempt for the inferior natures when they
indulge in the "brutish pleasures," and to thus attempt to control, by their disapprobation, even the self-regarding actions of the others.[181] In this way Mill
attempted to bring together the political insights of Bentham and Coleridge

---

Belief that this end can in no other way be forwarded but by the means you speak of, namely by
each taking for his exclusive aim the development of what is best in *himself*" (*CW* 12:207–8).

180. John Stuart Mill, "The Utility of Religion," *CW* 10:421.

181. On this point, see Joseph Hamburger, *John Stuart Mill on Liberty and Control*, pp. 171–180.

by allowing liberty of a sort while also advocating societal control—in the form of education and public opinion, especially the opinion of the intellectual class—in order to create a cultivated society. Indeed, Mill noted that is sometimes necessary to *force* a society to become cultivated. As he himself allowed in *On Liberty*, "The spirit of improvement is not always a spirit of liberty, for it may aim at forcing improvement on an unwilling people."[182] This force is needed because, unfortunately, a condition of stasis is "far more congenial to human nature" than a progressive one, in which there is "freedom and intellectual cultivation."[183] For this reason he praised Athenian imperialism: although it was a "blemish" according to universal standards of morality, it was "beneficial to the world" as "an organ of progress."[184]

It is obvious, by this point, that Mill's view of morality and politics ultimately diverged quite radically from the standard Benthamite utilitarian position. Mill, then—contrary to his assertion in the *Autobiography*—never really repaired the breach between his view and that of Bentham. This was recognized by many in Mill's own time: one commentator, soon after Mill's death, noted that "his divergence from the teaching of Bentham is here very manifest, and almost amounts to a surrender of the main position he professes to defend. . . . The changes . . . bear witness to the secret power of the antagonists he affected almost to despise."[185] John Grote similarly claimed that Mill should have given his system a new name, as it differed so radically from the "utilitarianism" of any other thinker.[186] As we will see now, in some important respects Mill's moral philosophy was similar to the intuitionism he derided Whewell for endorsing—and may even have been influenced by it.

Whewell's Attacks on Utilitarianism

In 1838, while preparing his first lectures as professor of moral philosophy, Whewell took note of Mill's essay criticizing Bentham in a letter to Julius Charles Hare:

---

182. John Stuart Mill, *On Liberty*, CW 10:272.

183. John Stuart Mill, "Grote's *History of Greece* [II]," CW 11:313.

184. Ibid., 11:321. In *Considerations on Representative Government*, Mill even argued that in earlier societies slavery had the important function of teaching the necessity of obedience, so that people would be ready for progress to a higher state. See G. R. Searle, *Morality and the Market in Victorian Britain*, p. 52.

185. T. R. Birks, *Modern Utilitarianism*, pp. 12–14. The author had been a student of Whewell's, and clearly saw Whewell's type of moral system as superior to Mill's.

186. John Grote, quoted in ibid., p. 38. Henry Sidgwick also noted that Mill's introduction of qualitative measures of pleasure was inconsistent with hedonistic utilitarianism (see *Methods of Ethics*, pp. 94–95).

It is certainly very encouraging to see on all sides strong tendencies to a reform of the prevalent system of morals. The article in the London review is an indication of this, and appears to me to be in many important points right, and at any rate right in the rigorous rejection of Bentham's doctrines and keen criticism of his character. But I confess I do not look with much respect upon a body of writers who, after habitually showering the most bitter abuse on those who oppose Bentham's principles, come around to the side of their opponents, without a single word of apology, and with an air of . . . complacency, as if they had been right both before and after the change.[187]

Hare wrote back defending Mill and his change of heart, claiming that "it does not seem to me that we ought to look with suspicion on a man who comes over in this way from the enemy's camp, openly, and by a change of which one may trace the progress. Converts have ever been the most powerful champions."[188] Whewell pointed out in response that if Mill were really a convert from the enemy camp, he would hardly have chosen to publish in the organ of the enemy's position, the *London and Westminster Review*.[189] And of course, he was right that Mill was not about to become a "champion" of the moral philosophy of Whewell. But it turns out that Mill's reformed utilitarian system resembled Whewell's moral philosophy in some of its central features.

Whewell and Hare, along with their friends Sedgwick, Thirlwall, and F. D.

---

187. William Whewell to Julius Charles Hare, October 15, 1838, WP Add.Ms.a.215 f. 43. Two years earlier, some years after the publication of Mill's 1833 criticism of Bentham, Whewell had noted, "At present I think we may discern in utilitarian writers a wish to conciliate rather than to shock the moral sympathies of their countrymen; and perhaps the opposite schools of moralists may thus be brought nearer to each other (preface to Mackintosh's *Dissertation on the Progress of Ethical Philosophy*, p. 32).

188. Julius Charles Hare to William Whewell, October 25, 1838, WP Add.Ms.a.206 f. 173.

189. See William Whewell to Julius Charles Hare, October 30, 1838, WP Add.Ms.a.215 f. 44. In 1834, annoyed at John Bowring's success in maintaining control over the *Westminster Review*, Mill and Sir William Molesworth (a wealthy young liberal) began the *London Review*, intending it to be an organ of radicalism. Two years later, Molesworth bought out the *Westminster Review* from T. Perronet Thompson, who had been keeping it afloat financially since 1828. Mill then became editor of the combined *London and Westminster Review* in 1836. (See G. L. Nesbitt, *Benthamite Reviewing*, pp. 130–63.) By this point, as he wrote to Edward Bulwer, he hoped to move the review in a new direction: "I hope you will believe that if the Review has hitherto been too much in the old style of Radical-Utilitarianism with which you cannot possibly sympathize very strongly (nor I either), it is because the only persons who could be depended upon as writers, were those whose writings would not tend to give it any other tone. My object will now be to draw together a body of writers resembling the old school of Radicals only in being on the movement side" (John Stuart Mill to Edward Lytton Bulwer, November 23, 1856, *CW* 12:311–13). Clearly by 1838 Mill had not succeeded in convincing some that the journal had moved away from its "Radical-Utilitarian" roots.

Maurice, considered utilitarianism the "enemy."[190] And because utilitarianism had become the dominant position at Cambridge, it was necessary to wage a coordinated battle to vanquish it. Even before Paley's book appeared, his main doctrine—"that actions are good in as far as they tend to pleasure"—had been taught at the university, and students there, as Whewell noted, "had accustomed themselves to look upon it as the only rational and tenable doctrine."[191] As we have seen, Sedgwick attacked Paley's moral philosophy and its role at the university in his *Discourse on the Studies of the University*. When the opportunity arose, Whewell decided to take on the professorship in moral philosophy in part to counter this ascendancy of the hated doctrine.[192] Around the time Whewell became professor, Hare wrote to him that "what you say . . . about the pernicious influence of our teaching Paley on the morals of the country, I have long thought. May you prosper in substituting something better for it."[193] In due course, Whewell wrote his own textbook, *Elements of Morality, Including Polity* (1845), in order to serve as this "substitute" for Paley's. Prior to this, he suggested that the university replace Paley's book with works of Bishop Joseph Butler and Sir James Mackintosh, which Whewell had edited for use as textbooks.[194] Eventually, Whewell managed to have Paley's *Moral and Political Philosophy* removed from the required reading of the curriculum.

Whewell attacked the utilitarian system directly in his preface to Mackintosh's *Dissertation* (1836), his four sermons in *On the Foundations of Morals* (1837), his *Elements of Morality, Including Polity* (1845), and his *Lectures on the History of Moral Philosophy in England* (1852). One of his worries about utilitarianism was that it appealed to many people because of its

---

190. On this point, see Boyd Hilton, *The Age of Atonement*, p. 171.

191. William Whewell, *Lectures on the History of Moral Philosophy in England*, p. 165.

192. In 1874 Whewell's former student T. R. Birks, at that time the successor to Whewell, John Grote, and F. D. Maurice as the Knightsbridge Professor of Moral Philosophy, remarked that the first signs of a new, anti-Paleyite era at Cambridge were the 1818 republication of S. T. Coleridge's *The Friend*, which contained an attack on Paley, and which was much read and discussed at the university; and the attacks of Sedgwick and Whewell (T. R. Birks, *Modern Utilitarianism*, p. 8).

193. Julius Charles Hare to William Whewell, January 4, 1838; WP Add.Ms.a.206 f. 171; quoted in Martha Garland, *Cambridge before Darwin*, p. 65.

194. Joseph Butler (1692–1752), author of *Sermons on Human Nature* (1726) and *The Analogy of Religion, Natural and Revealed* (1736), was a strong opponent of the hedonistic philosophy, which he refuted in an appendix to the latter work. In the 1820s, Whately had introduced Butler to the Oxford curriculum as an antidote to Paley (see Boyd Hilton, *The Age of Atonement*, p. 172). Sir James Mackintosh (1765–1832) was a Scottish, Whig member of Parliament who opposed the reactionary measures of the Tories, and played a role in the Catholic Emancipation and Reform Bills. At the end of his life he wrote *Dissertation on the Progress of Ethical Philosophy* and a *History of England* (for the *Cabinet Cyclopedia*).

supposed "scientific" nature. He lamented the fact that Paley's book had such an influence on so many generations of Cambridge graduates, in great part because it was taken to be "rational" and thus "superior ... to those vague and empty doctrines, of loftier sound, which had preceded the time of Locke, as the philosophy of Newton was to that of Aristotle."[195] Paley, for instance, professed to deduce all the commonly received rules of morality from his axiomatic principle, that actions are good insofar as they tend to pleasure. He contrasted this with the "moral sense" philosophy, which leads to "arbitrariness" in its doctrines as opposed to the "objective" standard offered by his philosophy.[196] Indeed, although Whewell considered Butler's the most adequate moral philosophy thus far, he criticized it for being too vague—for not specifying the exact nature and limits of the moral faculty, which left his system open to challenge by the supposedly "scientific" utilitarianism.[197] Mill, like Bentham, referred frequently to utilitarianism as "the scientific doctrine of ethics."[198] The Benthamite notion of a felicific calculus, whereby one could, in principle, scientifically and objectively determine the correct action to take, particularly appealed to the scientific-minded. Henry Sidgwick later pointed to this perception of utilitarian moral philosophy when he described how, in the early 1860s, under the spell of Mill, he and his contemporaries aimed at "a complete revision of human relations, political, moral, and economic, in the light of science."[199]

Whewell's attacks on utilitarianism in his *Elements of Morality* and his later *Lectures on the History of Moral Philosophy in England* moved Mill to write an angry response: his dual review of the *Lectures* and the *Elements* appeared in October 1852. As in the case with his earlier essay on Sedgwick, he attacked Whewell for making criticisms of Bentham similar to those that Mill had made himself. Whewell represented the supposed political enemy that Mill and Bentham's strictest followers were united against. Yet even in this review Mill continued, in a subtle way, to distance himself and utilitarianism from Benthamism, as when he noted, "It would be quite open to a defender of the principle of utility, to refuse to encumber himself with a defense of either of these authors [Paley and Bentham]."[200] By 1852, twenty years

195. William Whewell, *Lectures on the History of Moral Philosophy in England*, p. 166.

196. See Ernest Albee, *A History of English Utilitarianism*, pp. 169–70.

197. See William Whewell, *Lectures on the History of Moral Philosophy, 2nd edition, with Additional Lectures*, pp. 128–29.

198. See, e.g., John Stuart Mill, "Whewell on Moral Philosophy," *CW* 10:173.

199. Quoted in Stefan Collini, "The Ordinary Experience of Civilized Life," p. 344.

200. John Stuart Mill, "Whewell on Moral Philosophy," *CW* 10:173.

after Bentham's death, Mill had begun to believe that it was time to present his own view, the one that represented utilitarianism in its "best" form.[201] Indeed, around the time he wrote his review of Whewell's books, he began drafting "Utilitarianism."

Comparing the criticisms Whewell made against utilitarianism with the defense of utilitarianism offered by Mill, it is clear that to some extent, both men were arguing at cross-purposes: Whewell was critiquing the positions presented by Paley and Bentham, while Mill was responding with claims based on his own evolving doctrine, which differed from both of the others. Their arguing at cross-purposes is seen in the following case. Whewell disputed the claim that it is possible to perform the felicific calculus so central to Bentham's view. Although it would be good if there *could* be this kind of calculation, he contended that "the amount of happiness resulting from any action is not calculable."[202] One reason is that we can never know that we have taken into account *all* the consequences of an action, so it is impossible to calculate which action leads to the most happiness. Any action may have an infinite number of consequences of which we are ignorant.[203] Thus, since we can never calculate which action causes the greatest amount of pleasure, we cannot, in fact, apply the principle of utility as a scientific, objective method for deciding how to act. Not only this, but utilitarianism fails even to provide *any* method of moral choice. "If we cannot call our actions *good* or *evil* till we have performed this summation, till we have balanced against each other the positive and the negative quantities of such a calculation, we are surely thrown upon a task for which our faculties are quite unfit: we have the tangled course of life to run, and are blindfolded by the hand which is to assign the prize."[204]

In responding to this criticism of the notion of a felicific calculus, Mill made two points. First, he claimed that no one would deny that we cannot

201. In the essay on Sedgwick, Mill had noted, "A doctrine is not judged until it is judged in its best form," on which grounds he criticized Sedgwick for focusing on Paley's version of utilitarianism ("Sedgwick's Discourse," *CW* 10:52).

202. William Whewell, *Lectures on the History of Moral Philosophy in England*, p. 210. See also *Lectures on Systematic Morality, Delivered in Lent Term, 1846*, p. xx.

203. William Whewell, *Lectures on the History of Moral Philosophy in England*, p. 156.

204. Ibid., p. 171. Moreover, Whewell claimed that any system of morality must be able to answer the question, "What ought I do?" in a given situation. He contended that his system of morality was able to do so, while the utilitarian system was not. In his *Lectures on Systematic Morality Delivered in Lent Term, 1846*, he showed how his system can lead to answers to this question (see especially pp. 95–97). Indeed, as he wrote to his brother-in-law, Frederic Myers, one benefit of his system of morality was that with it he was able to find solutions to difficult cases of conscience, which had eluded him until he adopted his system (see September 6, 1845, in Janet Stair Douglas, *The Life and Selections from the Correspondence of William Whewell, D.D.*, pp. 325–30).

ever calculate *all* the consequences of an action, but this does not mean there is no point to considering those that we can know about.[205] We can have foresight, even though we cannot foresee everything. Mill was undoubtedly correct here, but to some degree he was missing the point. The question raised by Whewell was not so much whether we can have some idea of the consequences of our actions, but whether we can have a "scientific," precise, calculation of the happiness that would be caused by different possible actions. In making his second point, Mill claimed that in establishing moral rules it is necessary only to calculate the consequences of *classes* of actions, which he suggested is "a much easier matter" than calculating the consequences of individual actions. While Bentham had explicitly rejected this kind of "rule-utilitarianism" in his *Deontology*, Mill claimed that this work, published posthumously, did not in fact represent Bentham's views; rather, he believed that Bentham's friend and literary executor, Bowring, took "unwarranted liberties" with the papers from which he formed this book. (Mill and his father disliked Bowring, and resented the influence he had over Bentham, especially toward the end of Bentham's life.)[206] In any case, Whewell had anticipated the rule-utilitarian argument in his own discussion. He noted that the utilitarian may say that the consideration of consequences may be applied to general rules, and that to violate a general rule is itself an evil.[207] But as he pointed out, the principle of utility applies to actions, not to general rules; therefore on the utilitarian system general rules can only be good insofar as the individual actions that fall under that rule produce pleasure. Otherwise, something else is being appealed to besides pleasure, some higher virtue of rule-following.[208] Whewell claimed further that by appealing to general rules, the utilitarian is covertly adhering to an independent morality, that is, one which does not base the moral rightness of actions on their consequences but rather on some internal standard of virtue.[209]

Another criticism Whewell targeted against utilitarianism, both in the form expounded by Paley and by Bentham, is that it reduces men to the level of beasts. "If we go along with Paley," he cautioned in one of his sermons, "we plunge willingly into the slough of selfishness."[210] In his preface to Mackintosh's *Dissertation*, Whewell complained that the utilitarian philosophy may

205. John Stuart Mill, "Whewell on Moral Philosophy," *CW* 10:180.
206. Ernest Albee claims that there is "no conclusive evidence to this effect" (*A History of English Utilitarianism*, p. 177), and modern commentators seem to agree.
207. William Whewell, *Lectures on the History of Moral Philosophy in England*, p. 156.
208. Ibid., p. 157.
209. See ibid.
210. William Whewell, *On the Foundations of Morals*, p. 75.

be called brutish because "it recognizes no difference between the pleasures of man and those of the lower animals."[211] (In essence, Whewell was claiming that the Benthamites erroneously refuse to distinguish between the higher and lower pleasures. He was, therefore, anticipating Mill's own refinement—if not abandonment—of utilitarianism some twenty-five years later.) He drew the obvious consequence from Bentham's position: "If we are to promote human happiness in this sense, I do not see how the same obligation does not lie upon us to promote the happiness of brutes with the same care."[212] And indeed, as Whewell noted in the *Lectures on the History of Moral Philosophy*, Bentham did explicitly draw this conclusion, claiming that we are under equal obligation to take into account the pleasures of animals, as well as those of humans, when performing our felicific calculus.[213] Whewell disdainfully rejected this claim:

> Animals may indeed be the *objects* of morality. We may treat them with kindness or unkindness; and cruelty to animals is a vice, as well as cruelty to men. But cruelty to animals and cruelty to men stand upon a very different footing in morality. . . . We are bound to men by the universal tie of humanity . . . we have no such tie to animals. . . . The Morality which depends upon the increase of pleasure alone would make it our duty to increase the pleasures of pigs or of geese rather than those of men, if we were sure that the pleasure we could give *them* were greater than the pleasure of men.[214]

With this statement, Whewell obviously struck a nerve: Mill responded quite angrily, even though soon after writing his review he himself began to formulate a view distinguishing between higher and lower pleasures (only

211. William Whewell, preface to Mackintosh's *Dissertation on the Progress of Ethical Philosophy*, p. 31.
212. Ibid.
213. See Jeremy Bentham, *Collected Works*, 1:142n–43n, and William Whewell, *Lectures on the History of Moral Philosophy in England*, pp. 224–25. As Lea Campos Boralevi notes in her *Bentham and the Oppressed*, utilitarian views before Bentham did not count animals in the calculation of the greatest number, although Hume had claimed that men are "bound by the laws of humanity to give gentle usage to these creatures" (*Enquiry concerning the Principles of Morals*, chap. 3). Because Bentham's morality was not grounded in a rational faculty, but rather on the hedonistic psychology based on avoiding pain and seeking pleasure, it was not inconsistent to include animals into the calculus. As Bentham wrote in the *Introduction to the Principles of Morals and Legislation*, "The question is not, Can they *reason*? Nor, Can they *talk*?, but Can they *suffer*?" (see Campos Boralevi, *Bentham and the Oppressed*, pp. 165–67).
214. William Whewell, *Lectures on the History of Moral Philosophy in England*, p. 223.

the latter of which could be felt by animals). Mill called Bentham's claim that
the pleasures and pains of animals are of equal importance to those of men
"noble," and he accused Whewell of the same "superstitions of selfishness"
that lead some men to justify slavery or serfdom.[215] Indeed, he denounced
Whewell for justifying these superstitions and thus slavery itself: "According
to the standard of Dr. Whewell, the slave-masters and the nobles were right.
They too felt themselves 'bound' by a 'tie of brotherhood' to the white men
and to the nobility, and felt no such tie to the negroes and the serfs."[216] Mill
was correct in suggesting that the argument of Whewell's that we feel a "closer
tie" to other men than to animals is not a very compelling argument. (After
all, some people feel a closer tie to their pet cats than to other people.) Yet
Whewell's argument that animals cannot be the subjects but only the objects
of morality deserves more attention, and is also relevant to the issue of slavery—as Mill seems not to have realized.

Whewell's position is based on the claim that animals cannot be the subjects of morality, because they are not themselves moral agents. Animals do
not have a moral faculty; as we will see in the following section, Whewell
believed the moral faculty to be reason,[217] and animals—though they have
instincts—are not capable of reason. He explained, "Man alone is the Moral
Creature, for he alone can conceive Rules of Action;—can judge of actions as
right and wrong."[218] Because of this, *all* men, white and black, are moral subjects, and thus have moral rights; Whewell emphasized his claim that "the
same faculties of mind" are present in both races, that blacks share the "Universal Reason of man."[219] He asserted that "black men and white men are on
the same footing" morally, and thus that it is *immoral* to enslave a black man
and take away his moral rights.[220] Indeed, as Whewell noted (rather progressively for his time), those who forbade the education of slaves were implicitly
agreeing that black men share the same rational faculties as whites; otherwise
there would be no need to outlaw their education. In his *Of the Plurality of
Worlds*, Whewell argued, "We have good reason to believe that there is no
race of human beings who may not, by a due course of culture . . . be brought
into a community of intelligence and power with the most intelligent and
the most powerful races. This seems to be well-established, for instance,

215. John Stuart Mill, "Whewell on Moral Philosophy," *CW* 10:186.
216. Ibid.
217. William Whewell, *Elements of Morality, Including Polity*, p. 27.
218. Ibid., p. 593.
219. Ibid., p. 233.
220. Ibid., p. 594.

with regard to the African negroes; so long regarded by most, by some probably regarded still, as a race inferior to Europeans."[221]

Mill contemptuously (and somewhat bizarrely) claimed that he was willing to stake the whole defense of utilitarianism on this one question: "Granted that any practice causes more pain to animals than it gives pleasure to man, is that practice moral or immoral?" Parroting Whewell's own phrase, he asserted that anyone willing to raise his head from the "slough of selfishness" must answer "immoral."[222] Whewell was quite willing to accept these terms, as we saw in his letter to Richard Jones cited at the start of the chapter.[223]

Though Whewell did not mention it, we should attend to an important change in wording made by Mill here. Whewell had not claimed that we should *ignore* pain caused to animals in our consideration of the morality of actions. He had merely argued against Bentham's claim that the *pleasures* of animals and men were to be considered on equal terms. On such a view it would follow that if there were enough food to feed one elderly and seriously ill person or four healthy kittens for one year, and if it could be known that the four kittens would have the greater amount of pleasure, then the morally correct action would be to feed the kittens and to let the elderly invalid starve.[224] In the *Elements of Morality*, Whewell wrote that taking animal pleasures into account as equal to human pleasure "appears to me to be a reductio ad absurdum of the 'greatest happiness principle.' That the pleasures of pigs and geese are to be weighed against the pleasures of men, in order to obtain Moral Rules, is so wide a deviation from the general sentiments of mankind, that such a doctrine appears to be a sufficient refutation of the principles from which it flows."[225] On the other hand, Whewell did not deny that it is morally wrong to cause unnecessary pain to animals; he even argued that it is wrong to do so *even if* it causes pleasure in humans. Unnecessary cru-

221. William Whewell, *Of the Plurality of Worlds*, p. 81.

222. John Stuart Mill, "Whewell on Moral Philosophy," *CW* 10:187.

223. William Whewell to Richard Jones, October 10, 1852, WP Add.Ms.c.51 f. 273.

224. Lea Campos Boralevi notes that in his *Theory of Legislation*, Bentham claimed that animals did not have equal rights with men. Thus, for instance, as he wrote, "Men must be permitted to kill animals; but they should be forbidden to torment them" (see chap. 16; Jeremy Bentham, *Theory of Legislation*, ed. C. K. Ogden [London: K. Paul, Trench, Truber and Co., 1931], pp. 428–29; quoted in Campos Boralevi, *Bentham and the Oppressed*, p. 169). The life of an animal was not therefore put on a par with human life, and so Bentham avoided situations such as in the example of the elderly ill person and the young healthy kittens. But this point seems at odds with Bentham's claims in his *Principles of Morals*, for in *Theory of Legislation*, he suggested that we need *not* consider animal pleasures on a par with human ones. (For surely the amount of pleasure one human receives from a lifetime of consuming animals rather than vegetables cannot, on a purely felicific calculus, be greater than the pleasure all of those animals put together would have experienced if they had all lived full lives instead of being killed prematurely, however painlessly.)

225. William Whewell, *Elements of Morality, Including Polity*, p. 592.

elty to animals is odious, and a crime, and a sign of a bad disposition. Indeed, it fosters habits of disposition that can lead to immoral acts against persons, such as "regarding pain with indifference, or even with gratification." But, as he explained, "this is a very different matter from regarding the pleasures of animals as deserving the same kind of regard with those of men."[226] Thus ill-treating animals is wrong because of its consequences upon human character.

In his preface to Mackintosh's *Dissertation* published in 1836, Whewell had maintained[227] that in speaking of the importance of calculating the consequences of our actions, Bentham and Paley ignored their consequences upon our character. Yet, he asserted, "if among the most useful effects of actions, we conceive the most useful to be the improvement of man's moral character;—if we frame our rules so that they shall conduce as much as possible to virtuous feeling, as well as to beneficial action,—to purity of heart, as well as to rectitude of conduct;—if we aim at man's general well-being, and not merely at his gratification;—I know not what moralist would object to a criterion of morality so drawn from consequences."[228] Whewell believed that such a conception implies that moral good or virtue is a primary object, "valuable for its own sake," not as subservient to the purpose of utility.[229] He observed that "happiness consists rather in habits of the mind than in outward gratifications; and is to be sought rather by formal moral dispositions than by prescribing acts."[230] That is, happiness is sought in forming a good character. Thus, in the *Elements of Morality*, Whewell noted that laws protecting horses from harsh treatment by hackney drivers are enacted not to protect the rights of horses, but rather "to prevent ferocity, and hardheartedness" in people.[231] The problem with the utilitarians, he claimed, was

226. Ibid., p. 593. Interestingly, in the *Theory of Legislation* Bentham gave similar justification for why we may not torment animals. One reason to criminalize "gratuitous" cruelty to animals, Bentham claimed, was that "it is a means of cultivating a general sentiment of benevolence, and of rendering men more mild; or at least of preventing that brutal depravity, which after fleshing itself upon animals, presently demands human suffering to satiate its appetite" (Jeremy Bentham, *Theory of Legislation*, ed. C. K. Ogden [London: K. Paul, Trench, Truber and Co., 1931], pp. 428–29; quoted in Lea Campos Boralevi, *Bentham and the Oppressed*, p. 171). But, as Whewell pointed out, discussing the effects of our actions toward animals in terms of our own moral character is a different matter from claiming, as Bentham had in the other work, that we must take into account animal pleasure in our felicific calculus.

227. Incorrectly, in the case of Bentham's claim in the *Theory of Legislation*, as we saw in the previous note.

228. William Whewell, preface to Mackintosh's *Dissertation on the Progress of Ethical Philosophy*, pp. 26–27.

229. Ibid., pp. 27–28.

230. Ibid., p. 42. See also William Whewell, *Lectures on the History of Moral Philosophy in England*, p. 263.

231. See William Whewell, *Elements of Morality, Including Polity*, p. 594.

that they concentrate only on the calculation of outward gratifications, and not on the creation of an inner, moral, character.

It is interesting to compare this claim to Mill's criticism of Bentham in his 1838 essay: that Bentham ignored the creation of an excellent character, which contains the dispositions to act morally even without having to perform the felicific calculus. Mill's criticisms of Bentham here, and the later development of his own moral philosophy, seem rather similar to Whewell's criticisms of Bentham in 1836, in his preface to Mackintosh's *Dissertation*. It is possible that Mill's reading Whewell's preface influenced this aspect of his refinement of utilitarianism. Mill surely read the *Dissertation* when it originally appeared in 1830, as it contained a lengthy criticism of James Mill and the Benthamite "sect" (to which James Mill responded in his 1835 *Fragment on Mackintosh*). But while Mackintosh stressed the need to cultivate the "sentiments" (that is, feelings) that lead to virtuous actions, it was Whewell who called for a general improvement of man's moral character, which is close to the view that Mill later adopted. Interestingly, Whewell noted in his preface that if all sides in the debate over morality would adopt such a position, "perhaps the opposite schools of moralists may thus be brought nearer to each other."[232]

Whewell's Inductive Morality and Mill's Critique

Whewell consistently expressed his belief that morality shares the same inductive methodology as science. As early as 1827 he wrote to John Herschel that one benefit of his view of induction was that it "brings moral considerations exactly in their right place."[233] In a notebook entry of July 1831, Whewell was somewhat more guarded: "It remains yet to be determined whether the Inductive method can be applied to studies which deal with facts of consciousness (subjective sciences)," such as morality.[234] Yet by the time he began his draft of *The Philosophy of the Inductive Sciences* in a notebook dated December 1833, Whewell had recovered his confidence. He indicated that his subject, the philosophy of science, is interesting enough when considered only as related to the physical sciences. "But," he added,

> it is a subject of a far higher and deeper interest when we include in our survey all branches of human knowledge, those which concern his moral

232. William Whewell, preface to Mackintosh's *Dissertation on the Progress of Ethical Philosophy*, p. 32.

233. William Whewell to John Herschel, November 23, 1827, in Isaac Todhunter, *William Whewell, D.D.*, 2:86.

234. William Whewell, notebook, WP R.18.17 f.15, p. 46.

and religious condition as well as those which refer us to the material world. And it is not only allowable but necessary, to consider all the branches of human knowledge as having before them the same prospect of improvement, progression, and perfection, till we have discovered how and why the rules and processes under which the physical sciences flourish and advance are incapable of being applied, with some modifications, to other parts of our knowledge.[235]

Ultimately, however, he did not discuss the moral sciences in the *Philosophy*, and he even criticized Mill for including book 6, on the moral sciences, in the *Logic*. But the reason was not because Whewell no longer believed that moral science is inductive; rather, it was because he held that to make his strongest case for induction he must begin from those sciences which are most certain, and which everyone would accept as true knowledge.[236] Indeed, in the preface to the published version of the *Philosophy*, he noted that the doctrines in the book are "suited to throw light upon Moral and Political Philosophy, no less than upon Physical."[237]

Throughout his career, Whewell continued to express his belief in the inductive nature of moral philosophy. In numerous letters to Jones in 1834 and 1835, he discussed his plan to describe the analogy between moral and physical science.[238] In his "Introductory Lectures on Moral Philosophy" given in 1839 and 1841, he argued that "inquiries into the nature of truth, the

235. William Whewell, notebook, WP R.18.17.f.8, pp. 12–13. Whewell's renewed certainty may have been inspired by a sermon he heard that same month: on Commemoration Day at Trinity, his student (and second Wrangler in 1834) Thomas Birks preached on the thesis "that there is a moral truth which in its own way is as certain as mathematical truth." Whewell called it "very profound and consistent" in a letter to Julius Charles Hare on Christmas Day (WP Add.Ms.a.215 f.29).

236. Thus Whewell explained, "I think it will be allowed that by taking, as I have done, the Physical Sciences alone, in which the truths established are universally assented to, and regarded with comparative calmness, we are better able to discuss the formal conditions and general processes of scientific discovery, than we could if we entangled ourselves among subjects where the interest is keener and the truth more controverted" (William Whewell, *The Philosophy of the Inductive Sciences*, 1:vii).

237. Ibid., 1:x.

238. For example, see William Whewell to Richard Jones, August 2, 1834 (WP Add.Ms.c.51 f.175). Giving an outline of his proposed *Philosophy of the Inductive Sciences*, Whewell noted that he wanted to describe not only the physical but also the "hyperphysical" sciences: "morals, taste, politics, language," etc. He claimed that he intended to do the same for morals as for natural science, i.e., "showing that the ideal relation that there are things we *ought* to do is inevitably involved in all our thinking on this subject, and tracing the fundamental principles which have been assumed in the principal moral philosophies." See also letters of July 27, August 5, and August 21, 1834, WP Add.Ms.c.51 f.173, 174, 175; and May 9, May 26, August 21, and September 1, 1835, in Isaac Todhunter, *William Whewell, D.D.*, 2:210–11, 213–15, 222–25.

means and methods of its discovery, and the philosophy of science, even though they set out from the study of physical science, . . . cannot fail to exercise a strong and favourable influence upon our studies with regard to moral truth, moral science, and the true philosophy of human life."[239] Moreover, he claimed that just as physical sciences start out as practical and then become speculative, ethics was currently at a practical stage but would eventually progress to a more speculative one.[240] Nevertheless, in the first edition of the *Elements of Morality*, Whewell compared moral philosophy to geometry, leading some to believe that he saw morality as being a deductive science.[241] For example, his brother-in-law, Frederic Myers, in a lengthy letter full of praise for the *Elements*, pointed out his only major disagreement with Whewell: "To claim for morality the rank of a Science of any kind is in my judgment scarcely permissible . . . but to compare it in rigour of demonstration with the very proofs from mathematical science is to my mind even scarcely tolerable."[242] In a later letter, Myers suggested that Whewell alter this analogy: "I don't think that you will lose much . . . if you will permit us to consider 'Systematic Morality' as a kind of 'Moral Mechanics' instead of 'Moral Geometry.'"[243] Whewell appears to have followed this advice; in the second edition of the book he altered the analogy to one comparing morality to mechanics.[244] Yet this does not mark any change of opinion regarding the inductive nature of morality. As we have seen, Whewell had consistently spoken of the analogy between morality and physical science, even before the first edition of the *Elements of Morality* was published in 1845. Indeed, he explained to Myers that he had intended to use the scientific analogy and not the geometrical one all along; only he had worried that since not many people

239. William Whewell, *Two Introductory Lectures to Two Courses of Lectures on Moral Philosophy*, p. 28.

240. Ibid., pp. 39–42. Indeed, though he does not quite put it in this way, Whewell gave a similar reason for removing "Moral Casuistry" from the title of his professorship (see *Lectures on the History of Moral Philosophy in England*, preface).

241. Whewell had also drawn this analogy in his 1837 collection of sermons, *On the Foundations of Morals*, pp. 35–42.

242. Frederic Myers to William Whewell, July 31, 1845, WP Add.Ms.a.205 f. 137.

243. Frederic Myers to William Whewell, undated letter [but clearly written after the letter of July 31, 1845], WP Add.Ms.a.205 f. 139.

244. Thus, in the preface to the second edition of the *Elements of Morality, Including Polity*, Whewell wrote, "In my first edition, I said a few words implying an analogy between the relation of Truths to each other in Morality, and in that subject in which the nature and foundation of truth is supposed to be best studied, Geometry. The suggestion of any such analogy appears to be received by many readers with great impatience; and as none of my conclusions depend upon the analogy, I willingly withdraw all discussion of it" (p. 9). However, it should be noted that Whewell did continue to use the analogy in the text (see, e.g., pp. 11 and 94).

were familiar with mechanics, they would find the analogy too obscure. Mechanics, he agreed, works better in the analogy (for reasons that we will return to later).[245]

Mill, as we have seen, had criticized Whewell's moral philosophy as *not* being inductive; instead, he considered it an exemplar of the "a priori" or intuitionist view of morality. But just as his analysis of Whewell's philosophy of science as a priori ignored one crucial member of the antithetical pair—the empirical part—so too his characterization of Whewell's morality missed a crucial element. Whewell's morality *is* intuitionist in the sense of claiming that humans possess a conscience—a faculty that enables them to discern directly what is morally right or wrong. His view differs from that of earlier philosophers such as the Earl of Shaftesbury and Francis Hutcheson, who claimed that this faculty is akin to our sense organs and thus spoke of conscience as a "moral sense." Whewell's position is more similar to that of intuitionists such as Ralph Cudworth and Samuel Clarke, who claimed that our moral faculty is reason. He maintained that there is no *separate* moral faculty, but rather that conscience is reason exercised on moral subjects. Accordingly, he referred to moral rules as "principles of reason."[246]

Although he sometimes used the term *intuition* to describe the way in which we come to discover the moral rules,[247] it is clear that for Whewell, intuition is a process of reasoning. Thus he more often described the discovery of the moral rules as an activity of reason.[248] His morality does not have one problem associated with the moral-sense intuitionists. For this sort of intuitionist, the process of decision-making is nonrational; just as we feel the rain on our skin by a nonrational process, we feel what the right action is. This is often considered the major difficulty with the intuitionist view: if the decision is merely a matter of intuition, it seems that there can be no way to settle disputes over how we ought to act.[249] However, as we have seen, Whewell is not a moral-sense intuitionist; he never suggested that decision-making in morality is a nonrational process. Indeed, he pointed out in *Elements of Morality* that reason leads to common decisions about the right way to act (although our desires/affections may get in the way): "So far as men

245. William Whewell to Frederic Myers, September 6, 1845, WP Add.Ms.a.214 f. 57.
246. William Whewell, *Elements of Morality, Including Polity*, p. 3.
247. See, for example, William Whewell, *Lectures on Systematic Morality, Delivered in Lent Term, 1846*, p. 11.
248. See William Whewell, *Elements of Morality*, pp. 23–24; see also p. 45. On pp. 12–13, it is clear that Whewell saw no contradiction between his claims that the moral rules are discovered by intuition and that they are known by reason.
249. See John Rawls, *A Theory of Justice*, p. 34.

decide comformably to Reason, they decide alike."[250] Thus the decision on
how we ought to act should be made by reason; thus disputes can be settled
rationally on Whewell's view.

Reason leads us to certain moral rules. First, there is a "supreme" rule of
human action, which is "to do what is right and to abstain from doing what
is wrong."[251] Whewell noted that "with regard to the Supreme Rule, the question *Why?* admits of no further answer. Why must I do what is right? Because
it *is* right."[252] This is why his system of morality is an "independent morality," which he defined as a system that "look[s] upon moral goodness and
rightness as in themselves sufficient and supreme ends of human action." He
contrasted independent morality with "dependent systems," which "make
moral goodness and rightness derive their value and force from subservience
to some other ulterior end;—as pleasure or gain of some kind."[253] Utilitarianism, of course, was his prime example of the latter. Notably, in 1820
Whewell read in Madame de Stael's *De l'Allemagne* (1810) that Kant opposed
the utilitarian reductionism in ethics, and wanted instead to establish morality on necessary and universal grounds.[254] Whewell's "independent morality"
follows the same approach.

In Whewell's system, particular moral rules, which follow from the supreme rule, determine what is right. Further, "in order that Moral Rules may
exist, Men must have Rights."[255] For example, in order that there be a moral
rule "Thou shalt not steal," there must be a right to property, which I would
be subverting by taking for my own something that "belonged" to another. In
the *Elements,* Whewell claimed that five "primary and universal" rights correspond to the five main kinds of desires found universally among humankind: (1) the right of personal safety, which corresponds to the desire for personal safety; (2) the right of property, which corresponds to the desire for
having; (3) the right of contract, which corresponds to the desire for mutual
understanding; (4) family rights, which correspond to the desire for family
society; and (5) rights of government, which correspond to the desire for civil
society.[256] Because rights are based on these "natural desires," which encompass more than mere desire for pleasure and the avoidance of pain, Whewell

250. William Whewell, *Elements of Morality, Including Polity,* p. 43.
251. Ibid., p. 49.
252. Ibid.
253. William Whewell, *Lectures on Systematic Morality, Delivered in Lent Term, 1846,* p. 1.
254. This has been pointed out by Pietro Corsi, *Science and Religion,* pp. 155–56.
255. William Whewell, *Elements of Morality, Including Polity,* p. 51.
256. Ibid., pp. 51–52.

believed his system of morality was founded in a richer conception of human nature than the utilitarian system.[257] What he meant by this "correspondence" between the natural desires and these rights is that the rights are required so that man can live as a moral agent. For instance, the universal desire for personal safety requires that there be a right to personal safety. Without such a right, individuals would live in constant fear of actual or expected violence from others. The existence of the right causes men to be more calm, which in turn diminishes the number of assaults on others. Whewell explained that "in this calm, man, free from extreme agitations of Fear and Anger, can act with a reference to Rules founded on other men's Rights; and can thus, and no otherwise, exercise his rational and moral nature."[258]

Whewell noted further that these rights must be realized in positive law. For example, to realize the right of property (and thus the moral rule "Thou shalt not steal") in civil society, there must be laws forbidding the unwarranted removal of another's property.[259] Moreover, laws function in another important way. Whewell explained, "Law supplies the Definitions of some of the terms which morality employs; and without these definitions, Moral Rules would be indefinite, unmeaning, and inapplicable. Morality says, You shall not seek or covet another man's property: Law defines what *is* another man's property."[260] He claimed that one must follow the moral rules expressed in positive law.[261] But these laws are not to be taken as themselves being equal to the moral rules. Indeed, he clearly noted that "laws . . . may be themselves immoral." For instance, a law may define property in such a way as to include ownership of other human beings, as in the case of slavery.[262] Over time, reason will reveal more of the moral rules. Once that happens, we can alter our laws to be in closer accord with the moral rules. But in the meantime, we must obey the positive laws of our society, because they define for us the "pri-

257. Whewell suggested this by his praise of the moral system of Hugo Grotius. In the preface to his translation of Grotius's *De Jure Belli et Pacis*, Whewell wrote: "It is, [Grotius] says . . . too narrow a view to say that Utility is the Mother of Rights; the Mother of Rights is Human Nature, taken as a whole, with its impulses of kindness, pity, sociality, as well as its desire of individual pleasure and fear of pain. Human Nature is the Mother of Natural Law, and Natural Law is the Mother of Civil or Instituted Law. . . . By thus founding Morality and Law upon the whole compass of man's human and social, as well as animal and individual nature, Grotius, as I conceive, makes his system more true and philosophical than many of the more recent schemes of the philosophy of morals" (William Whewell, *Hugonis Grotii, De Jure Belli et Pacis, libri tres, Accompanied by an Abridged Translation, with the Notes of the Author, Barbeyrac, and Others,* 1:v).
258. William Whewell, *Elements of Morality, Including Polity,* p. 51.
259. Ibid., p. 54.
260. Ibid., p. 584.
261. Ibid., p. 67.
262. Ibid., p. 68.

mary and universal rights of man"; without such definitions the moral laws are too vague to enable us to live as fully moral agents.

Mill strongly objected to two aspects of Whewell's morality, both of which led, in his opinion, to the support of the status quo and therefore the currently obtaining unjust political and social arrangements. First, he criti-́ cized Whewell's claim that there are moral rules that are necessary truths. Second, he complained about Whewell's conception of the strong connection between moral rules and the dictates of positive law. (I will return to this point later.)

Mill believed that the claim of necessity in morality leads to the conclusion that there can be no ethical progress. As we saw earlier, Mill conceived self-evidency as a nonhistorical property; a self-evident principle must always be recognized as such. For this reason he interpreted Whewell as suggesting that the rules of society then in place were self-evident necessary truths. If Whewell had held such a simple-minded view, he would be guilty of supporting the status quo, as Mill accused him of doing. However, although Whewell did claim that moral rules are necessary truths, and attributed to them the epistemological status of "axioms" which are self-evident, his notions of necessity and self-evidency do not obviate the possibility of progress in our view of the moral principles. This is not surprising, given his view of necessary truth in physical science.

Necessary Truth in Morality

As in the case of physical science, morality contains necessary truths, according to Whewell. In *Elements of Morality*, he explained that "moral Rules must be necessary truths."[263] Indeed, he characterized the main controversy in moral philosophy in England as concerning "the arbitrary or necessary nature of moral truth."[264] (Like Bentham and Mill, Whewell characterized the opposition as being in favor of arbitrary moral truth.) Moreover, moral knowledge, like scientific knowledge, is ordered by certain Fundamental Ideas. The realm of physical science requires and is structured by the Ideas of Space, Time, Causation, and others. The realm of morality is structured by the Ideas of Benevolence, Justice, Truth, Purity, and Order.[265] As we

263. Ibid., p. 58.
264. William Whewell, *Lectures on the History of Moral Philosophy in England*, p. 110.
265. Ibid., p. xxiii. In a notebook dated (by archivist) 1831–32, Whewell noted, "Contrary to the error of applying moral conceptions to physical phenomena is that of applying physical conceptions to moral phenomena." As an example of the former error, Whewell explained that Aristotle "applies moral or notional conceptions to physics—*violent* motion etc." and that Alchemy

saw earlier, without an Idea of Space, we could not make sense of our perceptions of objects in the external world. This Idea is required to perceive objects as existing with spatial relations to other objects, and with spatial characteristics of size and shape. Similarly, moral Ideas enable us to perceive actions as being in accordance with the demands of morality. Whewell drew this comparison between the moral Ideas and the physical Ideas as early as 1834. He told Jones, "When I have shown, as I hope to do, that the relations of duty and of the affections are as fundamental a part of man's thought as the relations of time and space . . . I think I shall have assigned man a moral and social constitution on firm grounds."[266]

We saw in chapter 1 that while the physical Ideas are thus regulative of our thoughts, they also accurately describe the relations of physical objects as they really exist, external to our experience; we do not merely project them onto the world. Objects in the world really do exist in spatial relation to each other, and with spatial characteristics of size and shape; moreover, the Idea of Space enables us to have knowledge of these relations and characteristics as they exist in the world. The moral Ideas function in a similar way. There is, for instance, an Idea of Justice, such that acts or laws that are in accordance with the Idea are truly just. Whewell's explanation for why it is that our physical Ideas correspond to properties and relations that exist in the external world, and our moral Ideas correspond to true moral properties, was a theological one. Specifically, the Ideas correspond to the world because both Ideas and world have a common origin in a divine creator. According to Whewell, God's Mind contains Divine Ideas, some of which were used by God as archetypes in creating the universe. For example, God exemplified an Idea of Space in his universe by creating all physical objects as having spatial characteristics, and as existing in spatial relation to each other. Similarly, Whewell explained, "our moral Ideas, the Ideas of Benevolence and Justice in particular, must also be realized in the universe as a scheme of Divine Government."[267] He recognized the difference in this regard between physical and moral Ideas: the moral Ideas are as yet imperfectly realized in the present world.[268] Indeed, the progress of man as a moral agent consists in his

---

arose from the latter error, e.g., by claiming that "gold is a *perfect* metal" (WP R.18.17 f. 13, p. 50). But by making this point Whewell did not intend a separation between the types of Ideas similar to that proposed by Coleridge between "Physical" and "Metaphysical" Ideas; rather, he was simply expressing the claim he made later in his published works, that the conceptions and Ideas applied to certain sets of facts must be "appropriate" to them.

266. William Whewell to Richard Jones, August 21, 1834, WP Add.Ms.c.51 f. 175.

267. William Whewell, *On the Philosophy of Discovery*, p. 393.

268. Whewell readily admitted that these moral Ideas are not uniformly realized in the world of man existing in his present life; seemingly unjust incidents happen daily. But from this fact he

constant approach toward the extension of the moral Ideas into the realm of fact.[269]

Again, as is the case with the physical Ideas, God has created us so that our minds contain the moral Ideas. We can use our moral faculty to know how to act morally, and thus exist as moral agents, because God has given us the "germs" of our moral Ideas. Before we can accurately use these Ideas, however, we must unfold and render more distinct the "germs," through the "explication of Ideas." Only when the Ideas become distinct can we derive axioms from them, or recognize these axioms as necessary truths. There is, Whewell believed, a "progressive intuition" of necessary truth in morality as well as in science.[270] This is why he explained of moral principles that "there is nothing inconsistent with their being Axioms, in their requiring calm reflection, steady thought, and a development of the moral ideas, in order to a full apprehension of their evidence and generality."[271] Indeed, in discussing our faculty of conscience (which, remember, is the name for reason applied to moral subjects), Whewell explained that "such a faculty, although it originally exists in us, requires to be enlightened and instructed, in order that it may be a safe and salutary guide."[272]

Although Mill did not realize it, this aspect of Whewell's moral philosophy is significantly similar to Mill's own. Recall that for Mill, our conscience must be developed by education, out of our innately existing "social feelings." Thus, for both men, humans are born with the germs or foundation of conscience, but this innate foundation must be developed over time before an individual has the capacity to judge correctly the morality of actions. We have seen that Mill prescribed an associationist education—the kind that we

---

did not conclude that the moral Ideas, unlike the scientific Ideas, are not exemplified in the universe. Rather, he saw this as a justification for the view that that there must exist an afterlife, in which the Ideas of Justice and Benevolence are already realized.

269. See William Whewell, "Second Memoir on the Fundamental Antithesis of Philosophy," p. 10.

270. William Whewell, *On the Philosophy of Discovery*, p. 388. See also his *Two Introductory Lectures to Two Courses of Lectures on Moral Philosophy*, pp. 43–44. Elsewhere, Whewell explained: "Tardily and gradually, no doubt, do the principles of moral truth emerge into view. . . . But has not this been so with abstract truths of the plainest kinds? Even those portions of human knowledge to which we here turn men's eyes, as the very type and exemplar of evident and indisputable speculative truth;—the properties, I mean, of space and number;—were not these too, brought into view, late and slowly and partially, among the most acute and luminous intellects of the ancient world?—. . . Yet who among us holds that therefore these doctrines are precarious? And who does not see that the faculties by which we apprehend the properties of space and number are not the less real, or the less trustworthy, because they require to be unfolded and expanded by exercise and by teaching" (*On the Foundations of Morals*, p. 35).

271. William Whewell, *Elements of Morality, Including Polity*, p. 11.

272. William Whewell, *On the Foundations of Morals*, p. 29.

use to come to our knowledge of the physical world—in order to bring about the proper kind of conscience. Whewell's prescription was different, of course; but, as in the case of Mill, he believed that the education necessary in order to perceive moral truth is the same as the education necessary to recognize physical truth.

To claim that moral truths are axiomatic and "self-evident," on Whewell's view, does not entail that we currently know these truths. He claimed instead that "it requires a culture of the human mind" to know which moral rules are self-evident.[273] Hence, "truths may be self-evident when we have made a certain progress in thinking, which are not self-evident when we begin to think."[274] As he explained in a letter to his brother-in-law, in mechanics,

> although we have now axioms, defined conceptions and vigorous reasonings, we can point to persons, and to whole ages and nations, who did not assent to those axioms because the mechanical ideas of their minds were not sufficiently unfolded. In this we have, it seems to me, an answer to the objection that what I assert as moral axioms are not evident to all men. They are as much evident to all men as the axiom that a body will not alter its motions without a cause, or the like. They become evident to men in proportion as all men have their conceptions of the terms of the science rendered clear and distinct.[275]

This is why mechanics is actually a better choice than geometry for the analogy with morality; the progressive intuition of the axioms of the former is more *obvious* than that of the latter. However, the geometrical analogy was never *wrong*; in geometry, no less than in mechanics, it is necessary to explicate the Fundamental Ideas and their conceptions before the axioms can be seen to follow necessarily from the Ideas. Moreover, once we have clearly explicated the moral Ideas, the laws that follow from them can be arranged in a deductive system, just as is the case for any completed science such as geometry or—in principle, in the future—any of the inductive sciences.

We see, then, that on Whewell's view there is no contradiction between his viewing the moral principles as axioms, or necessary truths, and his belief that we do not currently know them. This is why he explained that "to test self-evidence by the casual opinion of individual men, is a self-contradiction." We cannot decide which moral principles are necessary truths

273. William Whewell, *Lectures on Systematic Morality, Delivered in Lent Term, 1846*, p.39.
274. William Whewell, *On the Foundations of Morals*, p. 38.
275. William Whewell to Frederic Myers, September 6, 1845, in Janet Stair Douglas, *The Life and Selections from the Correspondence of William Whewell, D.D.*, pp. 325–32.

by taking a "Promiscuous Jury from the mass of mankind," just as we cannot determine the axioms of geometry by polling schoolchildren.[276] Contrary to Mill's accusation, his morality does not serve as a justification for any principles that are currently established in society. Our knowledge of moral truth is progressive, requiring the development and clarification of our moral Ideas. Whewell explained, accordingly, that his system of morality is consistent with the notion of morality as "a living and growing body of truth."[277] Moreover, this need for clarification of the Ideas sheds light on his claim that we can learn how to construct a moral system of necessary truths by observing common moral principles as expressed by the dictates of positive law, a claim Mill also found quite objectionable.

Morality and Law: Explication of Moral Ideas

Whewell asserted that we can look to the dictates of positive law as a "starting point" in framing a morality. Mill believed that this position amounted to making law the basis or foundation of morality, and this is another reason for his claiming that Whewell's morality served to justify the status quo. By making his criterion of right what is right according to law, Mill argued, Whewell appeared to be basing his morality on the very kind of "external standard" he criticized in moralists such as Hobbes, Bentham, and Paley. Moreover, he seemed to be using as an external standard the laws passed by those in power, people such as himself. Mill charged that "with Dr. Whewell, a strong feeling, shared by most of those whom he thinks worth counting, is always an ultima ratio from which there is no appeal."[278] For this reason Mill called the *Elements of Morality* "a catalogue of received opinions."[279] (A similar criticism could be made against Mill's utilitarianism, given his insistence that the "higher" pleasures are those that he, and his friends, would count as having the most value.)[280]

However, Whewell did not make law the criterion of morality. He allowed that law may be judged to be not in conformity with morality.[281] In fact, law is *subject* to morality; hence he held that "morality is the Standard of

---

276. William Whewell, *Lectures on Systematic Morality, Delivered in Lent Term, 1846*, pp. 34–35.

277. Ibid., p. 7.

278. John Stuart Mill, "Whewell on Moral Philosophy," *CW* 10:194.

279. Ibid., *CW* 10:169.

280. This is why Leslie Stephen claimed that Mill's own distinction between qualities of pleasures "seems to be making room for something very like an intuition" (*The English Utilitarians*, vol. 3, *John Stuart Mill*, p. 306).

281. See William Whewell, *Elements of Morality, Including Polity*, p. 68.

Law."[282] He also claimed that "laws may be unjust, and when unjust, ought to be changed."[283] Whewell specifically pointed to laws allowing slavery as cases of such unjust laws (though, as we have seen, he believed that until they are abolished, such laws ought to be "submitted to").

At this point, we might be tempted to sympathize with Mill's exasperation with Whewell. If we can know whether a law is in conformity with the moral Ideas, does this not suggest that we have an independent method for knowing what these Ideas are? If we did have this kind of method, then we would not need to look at positive law in order to determine moral truth; we would derive these truths directly from the moral Ideas. However, this is not Whewell's claim. To understand what he is getting at here, we must return to his notion of necessary truth. In order to perceive moral truths, the moral Ideas must be clarified, just as to see truths in mechanics or geometry requires the explication of the Ideas of Force and of Space. As we saw in chapter 1, the explication of Ideas is a partially empirical process, requiring the observation of relevant facts. In mechanics, for instance, Whewell claimed that the first law of motion was recognized to be a necessary truth only in the course of empirical experience. Observation of the facts of motion led to a clarification of the Idea of Cause, which in turn led to further more precise observations, and so on, back and forth in a kind of bootstrapping process, until the Idea of Cause was distinct enough for the law of motion to be recognized as following necessarily from the Idea and as therefore being a necessary truth.[284] In the case of morality, the facts to be observed are the facts of positive law and the history of moral philosophy. Observation of the facts of law over the course of history can, Whewell believed, help us to clarify our moral Ideas. As our Ideas become more distinct, we can observe the laws in light of the Ideas and then reconsider whether the laws are still in conformity with the moral Ideas.[285]

282. William Whewell, *Lectures on Systematic Morality, Delivered in Lent Term, 1846,* p. 105.

283. William Whewell, *Lectures on the History of Moral Philosophy in England,* p. xvii.

284. For more details about this example, see my "It's *All* Necessarily So."

285. Mill believed that this argument constituted a "vicious circle" (see John Stuart Mill, "Whewell on Moral Philosophy," *CW* 10:189). Later, he seems to have modified his opinion about the usefulness of studying the history of law and morality, tying it in with his own belief in "many-sidedness." In his 1867 "Inaugural Address to the University of St. Andrews," Mill held that these subjects should be taught in the universities in order to show what the "best and the wisest" have thought on the subject. He suggested that this could be useful in helping to discover moral truth, because "there is not one of these systems which has not its good side" (*CW* 21:248). Interestingly, Whewell's idea here is not unlike John Rawls's later use of the notion of "reflective equilibrium." Our initial, commonsense or intuitive notion of a moral principle such as justice may not be the best notion; Rawls notes that the most acceptable theory is the one which best survives a process

Thus understood, Whewell's moral system, which proposes deriving moral truths from this bootstrapping process involving the observation of positive law, does not justify the status quo, nor evil practices such as slavery. Indeed, in the supplement to his second edition of the *Elements*, Whewell rather indignantly rebutted Mill's accusation that his system does justify such evils.[286] In the first edition, Whewell had written that "the Laws of the State are to be observed, even when they enact Slavery"; in the second edition, he changed "observed" to "submitted to."[287] Moreover, he explicitly noted,

> And let it be observed, that when we say that such Laws are to be submitted to, while they exist; we say this in conjunction with the strongest expressions which we can use, asserting the duty of all States to abolish Slavery; and of all persons to do all in their power to promote such a result. . . . Moreover we hold a belief that such a submission to the Laws, while they exist, combined with such a constant effort to produce a reform in the laws, is the most hopeful mode of promoting the abolition of slavery, and the course most beneficial to the slaves themselves. This may be a mistake; but at any rate it is an opinion not arising from any love of slavery or toleration for it.[288]

Indeed, Whewell stressed the need for granting political power to the lowest classes of society, in order to prevent slavery from occurring; he explained that "social freedom can hardly exist without political freedom: the lowest class can hardly have and retain Rights, without possessing some political power of maintaining them."[289] Again, though, this power had to be given lawfully, not seized by illegal revolution or other means.

Mill rejected Whewell's requirement for lawful means of political change. In the section "Justice and Utility" in "Utilitarianism," he claimed that the justice of laws should be determined by their correspondence to the principle of utility. Laws are unjust if they fail to conform to this criterion. Mill noted, however, that some people argue that such unjust laws should be repealed, but not broken, because it is in the common interest to keep intact the value of following the law. But Mill contended that this opinion "condemns many

---

of "reciprocal modification" in our general principles, commonsense rules, and judgments about particular cases (see Rawls, *A Theory of Justice*, pp. 19–21). Like Whewell, Rawls stressed the importance of studying the history of moral philosophy (see ibid., p. 51n).

286. William Whewell, *Elements of Morality, Including Polity*, p. 591.

287. Ibid., p. 590.

288. Ibid., p. 591.

289. Ibid., p. 237.

of the most illustrious benefactors of mankind, and would often protect pernicious institutions against the only weapons which, in the state of things
existing at the time, have any chance of succeeding against them."[290] Such a
view rejects illegal tactics as a means to repealing an unjust law. Mill believed
his project of reforming the world required such tactics; he explicitly allowed
the use of illegal force when the cause is just and force has a reasonable
chance of success.[291] Whewell was, indeed, more of a political conservative
than Mill on the question of the morality of disobeying unjust laws. He too
wanted to reform the world, but by lawful means only.

As we saw in the prologue, Mill eventually did realize that Whewell was,
in fact, personally opposed to slavery; he expressed his pleasure at learning
this in a letter he sent him in 1865.[292] Although he was thus finally convinced
that Whewell personally supported the abolition of slavery, Mill never
recorded any change in his opinion of Whewell's moral philosophy. His misunderstanding of Whewell's epistemology and how it underlies his moral philosophy seems to have precluded this.

A Reformist Morality

As we have seen, Mill was wrong to criticize Whewell's moral philosophy for
being irredeemably conservative in justifying the status quo. Indeed, had he
understood the epistemology underlying Whewell's view of physical and
moral science—especially his notion of the progressive intuition of necessary truth and the explication of ideas—Mill might have seen Whewell's
moral philosophy as potentially a *reformist* one. Advocates of reform could
use Whewell's doctrine to argue that what is currently believed to be moral
is not, in fact, and that a slow progression of our understanding is necessary
before we come to realize the true moral axioms.[293] This doctrine is not a *radical* one; remember that on Whewell's view the steps toward truth must be
slow and steady, not abrupt and discontinuous. Yet there is nothing inherently conservative about his moral philosophy. Whewell himself, though

290. John Stuart Mill, "Utilitarianism," *CW* 10:252.

291. See Geraint Williams, "John Stuart Mill and Political Violence."

292. John Stuart Mill to William Whewell, May 24, 1865, WP Add.Ms.a.209 f. 48 (1). Whewell's
original note was brief: "I thank you for your kindness in sending me your Examination of Sir
William Hamilton's Philosophy. Perhaps I may venture to say that your criticisms on this cele-
brated writer seem to me to be just; and that there are other points besides those which you have
noticed, on which his services to philosophy seem to me to have been overrated" (May 15, 1865,
WP Add.Ms.a.209 f. 48 (2), copy in hand of Helen Taylor).

293. Indeed, John Rawls has similarly argued against the claim that reflective equilibrium is
conservative, noting that, on the contrary, the method allows that our present positions are subject
to revision. See his *Collected Papers*, pp. 288–89.

offering a defense against Mill's incorrect depiction of him as a proponent of
slavery, may not have fully recognized—nor, perhaps, fully desired—the progressive implications of his own view. When his *Elements of Morality* first
appeared, he wrote to Jones, "Both the *Athenaeum* and the *Spectator* have
taken up my *Morality,* somewhat hostiley [*sic*]. I suppose that was to be expected; for they are both of them strong movement [that is, reform] journals,
and my speculations are, upon the whole, conservative."[294]

The political consequences of Whewell's moral philosophy, then, were
not as far from the political consequences aimed at by Mill's morality as Mill
imagined. Indeed, their conceptions of morality were quite similar in some
important respects. Both men eschewed the utilitarianism of Bentham, which
asserted that pleasure was the sole determinant of virtuous action. Instead,
both erected moral philosophies that stressed the importance of creating
morally excellent characters that would find happiness in acting virtuously.
Both believed that a proper education—one aimed at "cultivating minds"—
would help in creating this kind of moral character. Moreover, both had hopes
that a widening of the scope of this type of education could lead to an improved society.

We have seen in this chapter that in the case of both morality and politics, Whewell's view was not as far from Mill's as the younger man supposed.
Given that Mill's motivation for developing a philosophy of science was to
defeat an epistemology he was certain led to moral and political consequences completely opposed to those he supported, it is important to note
that his certainty was not fully justified. Had Mill realized this, perhaps he
would not have written *System of Logic,* or would not have argued for such
an extreme phenomenalism in its pages.

294. William Whewell to Richard Jones, August 19, 1845, in Isaac Todhunter, *William Whewell, D.D.,* 2:327–28. Whewell's former student and biographer Isaac Todhunter, however, did
notice this implication. He shrewdly noted that Whewell's own doctrines might have been urged
against his resistance to certain changes in the university (ibid., 1:414).

CHAPTER FIVE

Reforming Political Economy

We find, that when we look at the world as it is, with its elements of good
and evil, of progress and permanence, we have no longer the interests of
the proprietor in constant and necessary opposition to those of the rest of
the community, but that the common welfare of each class resides in the
common advance of all, and that the moral and social elevation of the lowest orders are inseparable conditions for any considerable and permanent
prosperity of the highest.
—*William Whewell, "Jones—On the Distribution of Wealth and the
Sources of Taxation," 1831*

At present I expect very little from any plans which aim at improving even
the economical state of the people by purely economic or political means.
We have come, I think, to a period, when progress, even of a political kind,
is coming to a halt, by reason of the low intellectual and moral state of all
classes: of the rich as much as of the poorer classes.
—*John Stuart Mill to Edward Herford, 1850*

We have seen that both Mill and Whewell believed that reforming society
required the renovation of science. This connection between reforming
society and science is most explicit in their writings on political economy.
Whewell, along with his friend Richard Jones, endorsed an inductive methodology in political economy, which fell in with their general "induction project." But this methodology also conformed to their moral and political views:
they saw it as a means to combat the utilitarian moral philosophy and the
political consequences the utilitarian radicals drew from it. Moreover, unlike
the utilitarians, they believed, following Adam Smith and T. R. Malthus, that
political economy was a branch of morality and politics, and so must concern

man not only as an economic abstraction but also as a moral, social, and intellectual being.

Mill's case is somewhat more complicated. His writings on political economy spanned the early 1820s through 1848 (with later editions of relevant works appearing until 1872), culminating in his magisterial two-volume work, *The Principles of Political Economy.* Over the course of these decades, Mill's political interests had shifted from the concern of the Philosophical Radicals—namely, the reform of institutions—to that of his middle and later years, the reform of individuals. At the same time, there was a shift from his youthful Benthamite view of man to a richer conception derived from S. T. Coleridge. Mill's earliest works in political economy, not surprisingly, conform to a view of reform and of man's nature similar to that of his father and Jeremy Bentham; hence they present a view of political economy as having a narrow scope, one concerned with man only in his economic aspect. On the other hand, his final works on the topic reflect his shift away from Bentham, and mirror his interest in creating a cultivated society: in *The Principles of Political Economy,* for instance, we find him concerned with man as a moral, social, and intellectual, as well as an economic, being.

We have also seen that by the late 1830s, Mill came to believe that the inductive method of Coleridge and the deductive method of Bentham needed to be combined. In the physical sciences, Mill constructed a method that was both inductive and deductive, at least in the case of the more complex laws, which require the "deductive method" of discovery and proof. He also wanted to create a combined inductive and deductive methodology for the social sciences. However, precisely because of the richer notion of man's nature that he derived from Coleridge, Mill found himself unable to do so. Ironically, then, this important insight of Coleridge's led him back to Bentham's deductive method. A way out of this methodological impasse was determined by Mill while writing his *Principles of Political Economy:* he reached back to an older distinction he had drawn between the Science and Art of political economy, in order to delineate proper domains for deductive and inductive methods in the social sciences.

Methodology and Morals in Political Economy: Ricardo and Malthus

In the early Victorian period, political economy gained an increasing amount of attention in Britain. By the end of the Napoleonic Wars, various political and economic threats to stability were perceived: postwar economic distress, popular unrest throughout Europe (especially worrisome to writers was the

1830 Revolution in France) and at home, the advancement of a manufacturing system with a consequent decrease in living and working conditions of the urban poor, the rise of a wealthy capitalist class and the related increase in the gap between rich and poor.[1] Certain changes in agriculture had also adversely affected farm laborers: the enclosure of open fields, for example, caused laborers to lose the use of that land as garden plots or as wasteland for fuel and game (thus they became dependent on daily wages for subsistence).[2] Because there was increasingly a lack of employment in winter months—in years of bad harvests or low market prices, farmers could not afford hired help to make repairs or fertilize fields—many laborers were without means of subsistence at all for part of the year.[3] For these reasons it became important practically as well as intellectually to define the science of political economy, and to determine whether and to what extent the science could be applied to solve the problems faced by society.

It is not surprising, therefore, that many of the day's leading writers on logic and scientific method (including not only Mill and Whewell, but also John Herschel, Richard Whately, and Charles Babbage) were interested in the "dismal science," as Thomas Carlyle famously dubbed political economy.[4] During this period, political economy became a topic of popular interest as well. For example, as we have seen, Harriet Martineau's series *Illustrations of Political Economy*, in which principles of Ricardian economics were taught through tales of common experience, reached monthly sales of over ten thousand by 1834.[5] Before turning to the debate between Mill and Whewell on political economy, it would be useful to look back to the very start of the century, when the issues and methods of that science were being defined.

In 1798, T. R. Malthus anonymously published his classic work, *An Essay on the Principle of Population, as It Affects the Future Improvement of Society, with Remarks on the Speculations of Mr. Godwin, M. Condorcet, and Other Writers*. This book was intended as an anti-Utopian tract. Malthus argued against the view of William Godwin and Marie-Jean-Antoine Condorcet that human society could reach a perfect state if only social and political institutions were reformed. Malthus claimed, on the contrary, that misery and vice were caused by a fundamental law of nature, which therefore could not be contravened by institutional innovation. His argument began with two

1. See Donald Winch, *Riches and Poverty*, p. 226.
2. See Karl Polanyi, *The Great Transformation*, pp. 96–97.
3. Raymond Cowherd, *Political Economists and the English Poor Laws*, p. xi.
4. On Carlyle's original use of the term, see Nicholas Capaldi, *John Stuart Mill*, pp. 225–26, and David Levy, *How the Dismal Science Got Its Name*.
5. See Mark Blaug, *Ricardian Economics*, p. 129n.

assumptions about mankind that seem indisputable: humans require food to survive, and have a natural sex drive.[6] Malthus then added two other less obvious assumptions, namely, that an unchecked population will grow at a geometric rate (in a sequence which can be represented by the numbers 1, 2, 4, 8, 16), while the means of subsistence can only increase at an arithmetic rate (1, 2, 3, 4, 5).[7] From these assumptions he drew the conclusion that there will always be pressure between the population and the available resources. Indeed, he argued that a natural and necessary oscillation keeps the population in check. When population outstrips food supply, many people at the lower end of society will suffer: the group death rate will be higher and individuals will tend to marry later and produce fewer children, causing a decrease in population. But as the population decreases, food supply will increase relative to the size of the population, which will in turn lead to an increase in the number of people (because of fewer deaths from starvation, larger numbers of earlier marriages, and more children).[8] Eventually, population will again outstrip the available resources, and the process will begin again.

On Malthus's view, the number of people in a nation is held down by two types of checks: positive (those that raise the death rate, such as hunger, disease, war) and negative (those that lower the birthrate, such as delayed marriages, birth control, abortion, prostitution). Malthus called these checks, respectively, "misery" and "vice." He believed it was impossible to put a complete end to this cycle of misery and vice, and thus a large segment of society—those individuals who live at the precarious edge of subsistence—would always suffer. In subsequent editions of *An Essay on the Principle of Population*, he presented a somewhat more optimistic view, introducing the option of "moral restraint"—purposely delaying marriage until one could support a family—for keeping population in check. This provided greater scope to human agency in avoiding misery and vice, and is reflected in the new subtitle given to these editions: *An Essay on the Principle of Population; or, A View of Its Past and Present Effects on Human Happiness; With an Inquiry into Our Prospects Respecting the Future Removal or Mitigation of the Evils Which It Occasions*. Yet Malthus continued to dismiss the views of Godwin

---

6. T. R. Malthus, *An Essay on the Principle of Population*, pp. 14–15.

7. Malthus rightly credited numerous eighteenth-century thinkers, such as David Hume, Richard Price, Adam Smith, and Robert Wallace, who had earlier claimed that population depends on the availability of subsistence, and varies accordingly (see ibid., pp. 7–8). Malthus's innovation was to put this commonplace in terms of geometric and arithmetic rates, which added a misleading precision to the point.

8. Ibid., pp. 21–29.

and Condorcet, who claimed that institutional change could banish poverty and conflict.

In 1811, Malthus, already famous as the author of the "population principle" (as his doctrine was known), became acquainted with David Ricardo, who had just begun to publish newspaper articles on monetary policy. The two men remained friends, correspondents, and adversaries until Ricardo's death in 1823. Their rivalry in issues of political economy set the terms of the debate between Mill and Whewell in three important ways. The first area of contention between Malthus and Ricardo involved methodology in political economy. Ricardo's *Principles of Political Economy*, published in 1817, endorsed a deductive method.[9] He claimed that political economy is "a strict science like mathematics."[10] In this science it is necessary, Ricardo argued, to start from certain axiomatic principles, such as the axiom that "Men desire to obtain as much wealth with as little effort as possible."[11] These axioms are not discovered by systematic or extensive experience; rather, they are axioms of "universal" experience about human and physical nature, known by introspection and casual observation. From these axioms and certain idealized situations based on explicitly unrealistic simplifying assumptions (which Ricardo called "strong cases"), we can deduce laws regarding the production of wealth. The laws of production are expressed as tendency statements, such as "The market price of labor has a tendency to fall to the natural price."[12]

9. I disagree with the view of Walter Bagehot, Joseph Schumpeter, and others who claim that Ricardo was a "methodological neophyte" who had no explicit methodological views at all. The discussion of methodology that takes place over years of correspondence with Malthus is, I believe, ample proof of the error of this view. See Bagehot, *Biographical Studies*, and Schumpeter, *Economic Doctrine and Method*.

10. David Ricardo, *The Works and Correspondence of David Ricardo*, 3:181.

11. See Neil DeMarchi, "The Empirical Content and Longevity of Ricardian Economics," and Sergio Cremaschi and Marcelo Dascal, "Malthus and Ricardo on Economic Methodology." DeMarchi characterizes the three central statements considered axiomatic by Ricardo in the following way: first, that the successive application of fixed amounts of capital and labor to land of a given quality and limited extent will eventually result in less-than-proportionate returns; second, that population tends to increase to the limit set by the level of wages; and third, that "men desire wealth, without limit to the range of their desires" (see "The Empirical Content and Longevity of Ricardian Economics," pp. 54–55; for the statement of these principles in Ricardo see *Principles of Political Economy*, in *The Works and Correspondence of David Ricardo*, 1:70–72, 120; I:78, 93–94, 96–97, 100–101; 6:133–35, 148 and 8:102–3).

12. Whately distinguished between two uses of the term *tendency*: "the existence of a cause which, if *operating unimpeded*, would produce [some specified] result" and "the existence of such a state of things that the result *may be expected to take place*" (see Richard Whately, *Introductory Lectures on Political Economy*, pp. 249–50). The first corresponds to a *ceteris paribus* proposition, the second to a prediction about probable outcomes. As Neil DeMarchi notes, Ricardo shifted between these two uses of the term ("The Empirical Content and Longevity of Ricardian Economics," p. 56).

The natural price of labor is the amount of wages necessary to enable the laborer to subsist and replace himself in the population (that is, to support a family).[13] Natural price therefore depends on the price of food and other necessary items. The market price of labor is the amount of wages actually paid, and this depends on the operations of supply and demand: "labour is dear when it is scarce, and cheap when it is plentiful."[14] Thus market price depends on the size of the laboring population. Ricardo, implicitly using Malthus's population principle, maintained that there is a natural tendency for market price to conform to natural price. This is so because when market price exceeds natural price, the laborer can rear a "healthy and numerous family." This will result in the increase in laboring population; when this happens market price will fall, even to a point below natural price, as the supply of labor will exceed demand. When market price falls below natural price, the condition of the laborer becomes "wretched," for he will not be able to support his family. In drawing the conclusion that the market price will always tend toward the natural price of labor, Ricardo adduced no empirical evidence, either from the case of labor in England or elsewhere. Moreover, it does not seem possible, by any empirical information, to disconfirm this conclusion. Because there are other factors besides decreasing population that can lead to an increase in market price, such as the increase of capital, Ricardo noted that "notwithstanding the tendency of wages to conform to their natural rate, their market rate may, in an improving society, for an indefinite period, be constantly above it."[15] Thus it appears to be impossible to disconfirm the "tendency" of market price to conform to natural price. When pressed by Malthus to explain why apparent deviations from his tendency statements did not disconfirm them, Ricardo claimed that he was interested in effects that would occur over the long run, not in the short run. But his definition of this period of "short run" was vague, suggesting that it was adjustable to take account of new facts. Thus he specified that the short-run period for the operation of technical progress to check diminishing returns in agriculture was twenty-five years, while the short-run period for the numerous influences operating upon supply and demand was only five years.[16]

Malthus's own *Principles of Political Economy* was published in 1820, in part to counter the orthodoxy coalescing around his friend's work. Malthus

13. David Ricardo, *Principles of Political Economy*, in *The Works and Correspondence of David Ricardo*, 1:93.

14. Ibid., 1:94.

15. Ibid.

16. See Neil DeMarchi, "The Empirical Content and Longevity of Ricardian Economics," p. 58.

presented a rival inductivist approach.[17] He criticized Ricardo's claim that the axioms of political economy are known by introspection and casual observation of a few instances; rather, he argued, first principles must be "carefully founded on an experience sufficiently extended."[18] He claimed that the inductive methods of Newton are applicable to the moral and social sciences.[19] Like Newton, Malthus rejected conjectures that are not inferred from what has been observed, such as Condorcet's claim that the human life span will increase indefinitely.[20] In his own work, he used statistical data, such as life expectancy tables, to draw inferences to principles; he then spent the rest of his life collecting and interpreting a range of historical, anthropological, and other empirical evidence that he hoped to incorporate into future editions.[21] Malthus also argued that, after discovering his theories inductively, the political economist must be willing to bring them to the test of experience; he criticized Ricardo's unwillingness to give up his theories even if they were found to be "inconsistent with general experience."[22]

Interestingly, Ricardo's deductivism was generated by his skepticism regarding the possibility of knowledge in any realm involving human behavior. He argued against Malthus that "there are so many combinations—so many operating causes in Political Economy, that there is a great danger in appealing to experience in favour of a doctrine, unless we are sure that all the causes of variation are seen and their effects duly estimated."[23] Ricardo did

17. See Stefan Collini et al., *That Noble Science of Politics*, p. 65. As James Henderson has noted, the methodological issues involved in the transition from classical to neoclassical economics date to the late 1600s; followers of two rival methods, induction and deduction, were characterized by Dugald Stewart as "political arithmeticians or statistical collectors" and "political economists or political philosophers." Deductive approaches proceeded from laws of human nature, while inductive approaches started from evidence of the historical record. Adam Smith's *Wealth of Nations* was seen as having combined both approaches (see Henderson, "Induction, Deduction and the Role of Mathematics," p. 2, and Collini et al., *That Noble Science of Politics*, p. 21).

18. T. R. Malthus, *Principles of Political Economy*, p. 518.

19. See Donald Winch, *Riches and Poverty*, p. xxii.

20. T. R. Malthus noted of Condorcet's claim that there has not been "the smallest perceptible advance in the natural duration of human life since first we had any authentic history of man" and remarked that if we can infer Condorcet's claim without any inductive evidence, "there is at once an end of all human science. The whole train of reasonings from effects to causes will be destroyed. We may shut our eyes to the book of nature, as it will no longer be of any use to us" (Malthus, *An Essay on the Principle of Population*, 1st ed., pp. 50–51).

21. See Donald Winch, *Riches and Poverty*, p. 236. Malthus used the life expectancy tables of Leonhard Euler, Johann Peter Sussmilch, and others (though, as many have noted, he was often careless with this data and distorted statistics from Sussmilch's table). On this point, see Deborah Redman, *The Rise of Political Economy as a Science*, pp. 296–97.

22. T. R. Malthus, *Principles of Political Economy*, p. 10. Note that what Malthus required for disconfirmation is not just one "isolated fact" but "experience sufficiently extended" (ibid., p. 518).

23. David Ricardo to T. R. Malthus, October 7, 1815, in *The Works and Correspondence of David Ricardo*, 6:295.

not think it possible to be certain that we have accounted for all the causes of human behavior. Because of this, he thought that in searching for "natural and constant" causal relations, it is necessary to abstain from the "accidental causes" and "temporary effects" that confuse the "empirics" and "practical men."[24] The way to do this is to consider only one of those causes, namely the desire to maximize material advantages. Ricardo admitted that this is an "abstraction"—that in reality men have other motivations as well—but claimed that in order to have a science of political economy, it is necessary to assume that in their economic dealings, men act exclusively with a view toward attaining a maximum value with a minimum sacrifice.[25] He distinguished between "questions of fact" and "questions of science," the former concerning the actual cause of a behavior, the latter concerning the interest rationally served by a behavior. Political economy, according to Ricardo, is only concerned with "questions of science."[26]

On the other hand, Malthus was less skeptical about our possibilities for at least probable knowledge of man's many motivations. Thus he believed that political economy must seek *all* the causes of man's economic behavior. He criticized Ricardo's "unwillingness to acknowledge the operation of more causes than one in the production of particular effects."[27] He appealed again to Newton: "It is certain that we cannot too highly respect and venerate that admirable rule of Newton, not to admit more causes than are necessary to the solution of the phenomena we are considering, but the rule itself implies, that those which really are necessary must be admitted."[28] In order correctly to discover causal laws regarding man's economic behavior, Malthus claimed, we must be concerned with how man actually does behave, and what are all the real causes of his actions; and this, he argued, requires an inductive method of generalization from systematic observations.[29]

24. See Neil DeMarchi, "The Empirical Content and Longevity of Ricardian Economics," p. 54.

25. On this point, see John Neville Keynes, *The Scope and Method of Political Economy*, p. 116. James Fitzjames Stephen responded to the criticism that Ricardian political economists were guilty of reducing men to abstract profit-maximizers: "The world at large rather unjustly view political economists as an iron-hearted race, believing in nothing but statistics, and a set of iron-hearted calculations founded upon them; a charge founded on the fact that they are addicted to what many people consider the bad and even wicked habit of thinking and speaking about one thing at a time, and so arriving at definite results" (James Fitzjames Stephen, "Money and Money's Worth," p. 106).

26. David Ricardo to T. R. Malthus, October 22, 1811, in *The Works and Correspondence of David Ricardo*, 6:63–65.

27. T. R. Malthus, *Principles of Political Economy*, p. 6.

28. Ibid., p. 7. Malthus was invoking Rule One of Newton's four methodological rules in book 3 of his *Mathematical Principles of Natural Philosophy* (namely, that "we are to admit no more causes of natural things than such as are both true and sufficient to explain their appearances").

29. Although Malthus and Ricardo were thus generally opposed on the question of whether methodology in political economy should be inductive or deductive, Malthus at times struck a

This brings up a second difference between the views of Ricardo and Malthus relevant to the later debate between Mill and Whewell. On Malthus's view, because we must consider all the causes of man's behavior, political economy must encompass moral and political considerations as well as economic ones. Like Smith, then, he regarded political economy as a branch of the science of politics and morals. In the *Essay on Population,* he pointed out that legislators were then facing a choice between promoting "wealth" (political economy strictly defined) and "happiness" (morality and politics more generally).[30] Malthus believed that the correct choice is obvious, as he explained in the *Principles:* "If a country can only be rich by running a successful race for low wages, I should be predisposed to say at once, perish such riches."[31] On the other hand, since Ricardo held that political economy must deal only with the motivation of economic self-interest, he denied that political and moral considerations have a place in economic reasoning. In his *Notes on Malthus,* Ricardo specifically cautioned that in political economy, one should avoid consideration of the moral usefulness of the factors being discussed.[32] He complained of Malthus that he conflated the role of the political economist with that of the moral philosopher or theologian. The role of

---

position of compromise. As we will see later on, when Whewell and Jones attempted to conscript Malthus in their fight against the deductivist approach, Malthus refused to fully support them, claiming that while Ricardo erred on the side of being overly deductive, Jones too erred, by being overly inductive. See T. R. Malthus, *Principles of Political Economy,* pp. 6–7. See also Stefan Collini et al., *That Noble Science of Politics,* p. 80, and Neil DeMarchi and R. P. Sturges, "Malthus and Ricardo's Inductivist Critics." Indeed, Malthus, a Cambridge-trained mathematician (and ninth Wrangler in his day), attempted to introduce mathematical precision into political economy by expressing his population principle in terms of arithmetic and geometric ratios; he also later told Whewell that he thought there would be some value to expressing principles of political economy using the method of fluxions. (See T. R. Malthus to William Whewell, May 26, 1829, WP Add.Ms.a.209 f.10.) At the same time, Ricardo often used empiricist language, seeking laws expressing "natural and constant" causes, rather than "accidental causes," and comparing his laws to Newton's law of universal gravitation. See David Ricardo, *Principles of Political Economy,* in *The Works and Correspondence of David Ricardo,* 1:108; see also Sergio Cremaschi and Marcelo Dascal, "Malthus and Ricardo on Economic Methodology," pp. 500–501. Nevertheless, it would be wrong to conclude, as Deborah Redman does, that the methods of Malthus and Ricardo are not contradictory (*The Rise of Political Economy as a Science,* p. 259). Over their long friendship, much of their correspondence involved argument over this very issue.

30. See Stefan Collini et al., *That Noble Science of Politics,* p. 65.

31. T. R. Malthus, *Principles of Political Economy,* p. 236. As the result of a series of bad harvests between 1794 and 1800, Malthus saw that it was possible for industrial production to rise while agricultural production remained stationary—thus leading to an increase in wealth of the nation but a decrease in the happiness of the laboring classes. This led Malthus to his view that "I really cannot conceive anything more detestable than the idea of knowingly condemning the labourers of this country to the rags and wretched cabins of Ireland, for the purpose of selling a few more broadcloths and calicoes. The wealth and power of nations are, after all, only desirable as they contribute to happiness" (ibid., pp. 299–300).

32. See David Ricardo, *The Works and Correspondence of David Ricardo,* 2:337–38.

the political economist, Ricardo argued, "is to tell you how you may become rich, but he is not to advise you to prefer riches to indolence, or indolence to riches."[33]

The third relevant point of contention between Malthus and Ricardo has to do with the question of the relation between the different classes of society. Both men developed a theory of rent that led to a similar conclusion, namely that, assuming no improvement in agricultural technology, the price of food is determined by the cost of growing it either on the least fertile soil or on land which was already cultivated (thus requiring more labor and capital in increasing dosages to achieve the same result). Rent, for Malthus and Ricardo, was regarded as the aggregate surplus paid to landowners as a class after wages and profits on capital (that is, the necessary costs of producing food) had been met. Thus rents were determined by the price of food. When rents were high, it was because the price of food was high; it was not the case that food prices were high because rents were high.[34] Both Malthus and Ricardo agreed that the same type of correlation applied between rents and price of food. Where they disagreed, however, was on the consequence of this for the interests of the various classes of society. In his pamphlet on the Corn Laws, Ricardo famously noted that "the interest of the landlord is always opposed to the interest of every other class in the community. His situation is never more prosperous, as when food is scarce and dear; whereas, all other persons are greatly benefited by procuring cheap food."[35] On the other hand, in his *Principles of Political Economy*, Malthus agreed with Smith that "the interest of the landholder is closely connected with that of the state; and that the prosperity or adversity of the one involves the prosperity or adversity of the other."[36] Indeed, he elsewhere more strongly claimed of landowners that "there is *no class in society* whose interests are more nearly and intimately connected with the prosperity of the state."[37]

This disagreement between Malthus and Ricardo resulted in their views of political economy being seen as falling under different political categories. Malthus was taken as being more supportive of the interests of the landowning class, and thus was viewed by his contemporaries as a political conserva-

33. Ibid.
34. See Donald Winch, *Riches and Poverty*, p. 350, and T. R. Malthus, *Principles of Political Economy*, p. 134.
35. David Ricardo, *The Works and Correspondence of David Ricardo*, 4:20–21.
36. T. R. Malthus, *Principles of Political Economy*, pp. 204–5.
37. T. R. Malthus, *The Grounds of an Opinion on the Policy of Restricting the Importation of Foreign Corn*, p. 162; emphasis added.

tive.[38] Indeed, Karl Marx would later call him a "bought advocate" for the landowning classes.[39] Although James Mill accepted the population principle, he made "crushing" attacks on Malthus at the meetings of the Political Economy Club, of which they were both original members.[40] On the other hand, Ricardo's economic views were seen by Mill as being supportive of the political program of the Philosophical Radicals. He and Ricardo had become acquainted in London, and met frequently between 1811 and Ricardo's death in 1823.[41] Mill encouraged Ricardo to expand the views expressed in his *Essay on Profits* into the *Principles of Political Economy and Taxation*.[42] Eventually, he persuaded Ricardo to become the spokesperson for the utilitarian radicals in Parliament.[43] Ricardo had probably already been exposed to radical politics and utilitarianism through his close friendship with Thomas Belsham, who, like Joseph Priestley, was a leading Unitarian minister. (Ricardo had converted to Unitarianism soon after his marriage.)[44] On his part, Mill had already stud-

38. See Stefan Collini et al., *That Noble Science of Politics*, p. 77, where it is argued that this is why Malthus's economic views were not given more positive support by the Whig *Edinburgh Review.*

39. Karl Marx, quoted in Donald Winch, *Riches and Poverty*, p. 12. In a letter to Ricardo, John Ramsay McCulloch referred to Malthus's *Principles of Political Economy* as a "textbook . . . of a few landlords" (quoted in Salim Rashid, "Malthus' *Principles* and British Economic Thought, 1820–1835"). In a sense the charge of conservatism was a case of "guilt by association." Because Malthus argued against the claims of those who welcomed the French Revolution as a means toward the perfectibility of society, he was associated with the counterrevolutionary views of Edmund Burke (as expressed in his *Reflections on the Revolution in France* [1790]), even though Malthus saw himself as an arbiter between these extreme views. Indeed, he included two chapters in *An Essay on the Principle of Population* expressing his dismay that the war with France had lent a pretext for diminishing civil liberties at home (see bk. 4, chaps. vi and vii).

40. See William D. Grampp, "Malthus and His Contemporaries," p. 281. The antipathy went in both directions: when Malthus heard that there was going to be an article on political economy in the Supplement to the *Encyclopedia Britannica*, he advised the editor, Macvey Napier, that it should *not* be written by either James Mill or John Ramsay McCulloch, because either of them would present the doctrines of Ricardo as if they were the received view, which they were not. Napier ignored the advice, and McCulloch wrote the article (see Grampp, "Malthus and His Contemporaries," p. 285). Malthus then complained to Napier that the view of Ricardo and his followers "will not stand the test of experience" (see Donald Winch, *Riches and Poverty*, p. 354).

41. Nevertheless, it is surely an exaggeration to claim, as Elie Halévy does, that Mill gave Ricardo "lessons in method" of political economy, which Mill had learned from J. B. Say (Halévy, *The Growth of Philosophical Radicalism*, p. 277; for the opposing view see Sergio Cremaschi and Marcelo Dascal, "Malthus and Ricardo on Economic Methodology," p. 494; Samuel Hollander, *The Economics of John Stuart Mill*, 1:24–28; and Stefan Collini et al., *That Noble Science of Politics*, pp. 112–26).

42. See Donald Winch, *Riches and Poverty*, p. 354.

43. See Stefan Collini et al., *That Noble Science of Politics*, p. 105. Murray Milgate and Shannon Stimson argue strongly against the "received view" that Ricardo was politically naïve, with no political views of his own (see *Ricardian Politics*).

44. Thomas Belsham's book *Elements of the Philosophy of the Mind, and of Moral Philosophy* (1801) incorporated many of Joseph Priestley's views. For example, in the book Belsham defined

ied political economy, having attended Dugald Stewart's lectures on the subject in Edinburgh in 1798. In 1821, Mill united Ricardian economics with Bentham's science of legislation in his own *Elements of Political Economy*.[45]

What exactly is it about Ricardo's work that appealed to James Mill and the Philosophical Radicals?[46] They especially appreciated his suggestion that the landowning classes were opposed to the interests of the rest of society, which tied in nicely with their demand for the end to aristocratic privilege.[47] A specific instance of the congruence of the utilitarian radical position with Ricardo's political economy occurred in the debate over the Corn Laws. In 1814 and 1815, Parliament had passed laws that raised duties on foreign corn to the extent that it gave domestic farmers a virtual monopoly on the home market.[48] Ricardo claimed, in his 1815 pamphlet *An Essay on the Influence of a Low Price of Corn on the Profits of Stock*, that this would cause the price of food to increase, and thus would raise money wage costs while depressing real wages. This result would hurt both those who employed labor (manufacturers and farmers) and the laborers themselves. Landowners alone would benefit, by the higher rents they could charge for the use of their land. Ricardo argued for abolishing these duties and allowing free trade, a position that seemed to the Philosophical Radicals to benefit the people while stripping the landowning aristocracy of an unfair privilege.

On the other hand, Malthus (in his 1814 *Observations on the Effects of the Corn Laws*) argued inductively for the opposite position. In doing so he objected to what he called Ricardo's "doctrine, that we are to pursue our general principles without ever looking to see if they are applicable to the case before us."[49] He claimed that although in the abstract free trade was prefer-

---

virtue as "the voluntary production of the greatest sum of happiness" (p. 379). For more on the connection between Ricardo and Belsham, see Sergio Cremaschi, "Ricardo's Philosophy." James Mill, however, had mixed feelings about Belsham's philosophy, especially its relation to Hartleyan associationism (on this point see Elie Halévy, *The Growth of Philosophical Radicalism*, p. 439).

45. See Pietro Corsi, "The Heritage of Dugald Stewart," and Salim Rashid, "Dugald Stewart, 'Baconian' Methodology, and Political Economy."

46. I am not claiming, of course, that all followers of Ricardo were political radicals. John Ramsay McCulloch is a notable exception. But James Mill certainly believed that Ricardianism led to his political views: thus, as Donald Winch notes, he reproved McCulloch for failing to follow Ricardian logic "to its proper [political] consequences" (Donald Winch, *Riches and Poverty*, p. 356).

47. As Jeff Lipkes notes, "the profound animus [of the Philosophical Radicals] toward the aristocracy was both rationalized and intensified by Ricardian economics" (*Politics, Religion, and Classical Political Economy in Britain*, p. 18).

48. In 1814 Parliament abolished the outdated bounty on exportation; in 1815 it ended the sliding scale of duties on imports, substituting absolute prohibition up to a certain fixed (high) price and free importation above that level (see Mark Blaug, *Ricardian Economics*, pp. 6–10).

49. T. R. Malthus, *The Grounds of an Opinion on the Policy of Restricting the Importation of Foreign Corn*, p. 149.

able, in the actual conditions then obtaining (including the fact that France had just passed laws limiting the free exportation of corn), some type of protectionism would actually benefit most classes of society, including the laboring classes.[50] Malthus's argument for the protectionist position was also related to his claim that political economy must consider more than purely economic factors. As we have seen, he believed that the political economist must take into account moral and social factors as well as purely economic ones. In the case of the Corn Laws, he noted that free trade could have results disadvantageous to "national quiet and happiness."[51] This is so because if England imported corn, more capital (that amount now spent on growing corn) would be channeled into manufacturing, and this seemed contrary to what was best for the happiness of the greater number of citizens. He concluded that England should not abandon protectionist restrictions on the free trade of corn, because more people would be injured than helped by free trade.[52] Interestingly, Malthus argued on utilitarian grounds for the maintenance of the Corn Laws, yet his view was rejected by the utilitarian radical reformers. In part this was because he was a follower of William Paley, having been taught at Cambridge by Paley's student William Frend. Malthus's form of utilitarianism, therefore, was the theological one that the radical reformers rejected.[53]

Inductivist Critics of Ricardian Political Economy: Whewell and Jones

Whewell and Jones targeted Ricardian political economy as part of their project to fight against the "downwards mad" deductivists and promote inductivism in all areas. They were not the only ones engaged in this campaign. The circle of inductive political economists with ties to Whewell have been dubbed the "Whewell Group,"[54] but they are more commonly known as the "Cambridge Inductivists," and include not only Jones but also John Caze-

50. As Donald Winch points out, there is some evidence that Malthus later changed his opinion about the Corn Laws (see *Riches and Poverty*, pp. 334–35).

51. See T. R. Malthus, *The Grounds of an Opinion on the Policy of Restricting the Importation of Foreign Corn*, p. 117.

52. Ibid., pp. 168–69.

53. In his 1815 Corn Law pamphlet, Malthus appealed implicitly to William Paley's principle of utility by examining in detail the effect of the abolition of duties on imported corn on the interests of various categories of citizens, and concluding that this abolition would be contrary to the happiness of the greatest number. Laborers, farmers, manufacturers not involved in foreign trade, landholders—all would be disadvantaged; the only winners would be manufacturers involved in foreign trade and stockholders and others on fixed salaries (such as himself). See ibid., pp. 158–68.

54. See James Henderson, "Induction, Deduction, and the Role of Mathematics."

nove,[55] Babbage,[56] Herschel,[57] and others.[58] This group was instrumental in founding, in 1833, Section F of the British Association for the Advancement of Science and the Statistical Society of London, both of which were formed with the express intent to counter the hegemony of the Political Economy Club of London.[59] The explicit goal of starting the Statistical Society was to establish generalizations based on empirical data, which could be used in inductively inferring theories of political economy.[60] The first meeting of Section F took place in rooms being used by Jones at Cambridge during the third British Association gathering.

Whewell himself had been interested in political economy since the early 1820s, and corresponded with Jones on the topic while Jones was beginning work on his economic treatise.[61] Whewell argued against Ricardo's deductivism in a series of papers presented to the Cambridge Philosophical Society beginning in 1829. In these papers, he translated the axioms and conclusions of Ricardian political economy into mathematical terms in order to launch three types of attack against Ricardo's methodology. First, he argued that deduction is not the proper form of reasoning in political economy. Next, he

55. John Cazenove (1788–1879) was a friend of Richard Jones's who agreed with his view of rent and also supported Malthus's position against Say's Law. A member of the Political Economy Club from 1821 to 1831, Cazenove wrote several works in political economy before assisting Whewell with the editing of Jones's posthumously published *Literary Remains* in 1859.

56. Charles Babbage wrote a study of factory organization and the division of labor, in which he reproached Ricardian political economy for "too small a use of facts and too large an employment of theory." See Babbage, *On the Economy of Machinery and Manufactures*, p. 119.

57. Although Herschel did not publish a treatise on political economy like the others, he marked an interest in the topic in his *Preliminary Discourse on the Study of Natural Philosophy*, when he noted the importance of proper scientific method to the practical sciences such as political economy. He claimed that in political economy, "great and noble ends are to be achieved . . . by bringing into exercise a sufficient quantity of sober thought, and by a proper adaptation of means" (*Preliminary Discourse*, p. 74). Mill had pointed to this very aspect of Herschel's book when he reviewed it (see John Stuart Mill, "Herschel's *Preliminary Discourse*," CW 22:284–87). Years later, Herschel sent Richard Jones a plan to modify property taxation procedures to decrease the burden on the poorer classes (see abstract of letter in *Calendar of the Correspondence of Sir John Herschel*, p. 219). Herschel was, for a time, the head of the Mint, so he had a professional interest in monetary policy.

58. Even T. Perronet Thompson (a Fellow of Queens College, Cambridge, and for a time the publisher of the *Westminster Review*) expressed similar views on the importance of considering political economy a branch of physical (i.e., inductive) science in his work *The True Theory of Rent in Opposition to Mr. Ricardo and Others*.

59. See James Henderson, "Induction, Deduction, and the Role of Mathematics," pp. 22–23. Indeed, Lawrence Goldman claims that Susan Cannon's "Cambridge Network" was actually bound most tightly by the members' contempt for the method of Ricardian economics (see "The Origins of British 'Social Science,'" p. 594).

60. See Lawrence Goldman, "The Origins of British 'Social Science,'" p. 599.

61. See letters of William Whewell to Richard Jones, August 16, 1822, WP Add.Ms.c.51, f. 14; May 24, 1825, WP Add.Ms.c.51 f. 21; and September 10, 1827, WP Add.Ms.c.51 f. 41.

claimed that Ricardo's reasoning was unsound not only because of this, but also because the premises upon which Ricardo based his deductions were ascertained by only the most casual observation, and were, in fact, false or not universal. At best, then, Ricardo's deductions were applicable only in special cases.[62] And finally, Whewell criticized the particular deductions made by Ricardo. Even *if* it were appropriate to use deductive reasoning in political economy, and even *if* the assumed starting premises were correct, Ricardo's conclusions would still be invalid.[63]

Whewell's papers were misunderstood from their first appearance: he was criticized in both *Fraser's Magazine* (April 1830) and *British Critic* (January 1831) for having allegedly claimed that it was possible to completely mathematize political economy.[64] Jones also worried that Whewell had not made his intentions clear enough: "You should, I think . . . repudiate more distinctly and earnestly than you have done any belief in the practical truth of the axioms and definitions of which you are tracing the necessary results."[65] Yet Whewell believed he had clearly expressed his disagreement with both the deductive method and the particular postulates of Ricardo.[66]

In the 1831 paper, Whewell explicitly explained the role of mathematiza-

62. On this point, see Neil DeMarchi and R. P. Sturges, "Malthus and Ricardo's Inductivist Critics," pp. 381–82.

63. Whewell's papers present a formal version of the type of argument that had been made by T. B. Macaulay against James Mill's deductive *Essay on Government*. Interestingly, Whewell read his first paper to the Cambridge Philosophical Society on March 2 and 14, 1829. The same month, Macaulay, who was elected a Fellow of Trinity in 1824, published his first paper attacking Mill in the *Edinburgh Review*. Although I have not found evidence to confirm this, it is certainly possible that Whewell and Macaulay may have discussed their respective papers while working on them.

64. See James Henderson, "Induction, Deduction and the Role of Mathematics," p. 17.

65. Richard Jones to William Whewell, April 18, 1831, WP Add.Ms.c.52 f. 16; quoted in Pietro Corsi, "The Heritage of Dugald Stewart," p. 119. Some modern commentators have similarly misinterpreted the point of Whewell's papers. See, for instance, Samuel Hollander, "William Whewell and John Stuart Mill on the Methodology of Political Economy," pp. 130 and 142ff. S. G. Checkland also mistakenly claims that Whewell grew impatient and abandoned the inductive method for political economy, turning instead to deductive speculations ("The Advent of Academic Economics in England," p. 66); but this is not the case: even in *On the Philosophy of Discovery* of 1860 and his *Six Lectures on Political Economy* of 1862, Whewell continued to describe political economy as an inductive science. On the other hand, J. A. Cochrane correctly notes that Whewell was criticizing Ricardian deductivism, and points out that his mathematizing of Ricardianism was useful and important in the development of twentieth-century mathematical economic models (see "The First Mathematical Ricardian Model").

66. William Whewell to Richard Jones, April 2, 1829, WP Add.Ms.c.51 f.63. Others followed Whewell in applying mathematics to critique the Ricardian model, such as Edward Rogers and J. E. Tozer. Some, such as T. Perronet Thompson, Whewell's friend and former student John Lubbock, and Dionysus Lardner used mathematics as part of an inductive methodology to develop new, anti-Ricardian theories (see James Henderson, "Induction, Deduction and the Role of Mathematics," pp. 12–15). Thompson's was the first mathematical model of the profit-maximizing rule (i.e., marginal revenue equals marginal cost) employed today in the theory of the firm. Lardner developed

tion in his work: "The mathematical investigation proceeds from these [Ricardian] principles as *postulates,* and has no concern with their truth or falsehood. . . . For my own part, I do not conceive that we are at all justified in asserting the principles which form the basis of Mr. Ricardo's system, either to be steady and universal in their operation. . . . Some of them appear to be absolutely false in general, and others to be inapplicable in almost all particular cases. Perhaps, however, to trace their consequences may be one of the most obvious modes of verifying or correcting them."[67] For example, he discussed Ricardo's postulate of rent, which relies on "suppositions" that, Whewell noted, had been shown by Jones to be inapplicable 99 percent of the time. He remarked that Ricardo's postulate of wages depends on an assumption about the "necessary and universal operation of wages upon population" that is "entirely gratuitous and unfounded." Indeed, Whewell pointed out that many of the assumptions made by Ricardo were so far from being adequate to represent the known facts, that deductions from them bore almost no similarity to the "actual state of things."[68]

Whewell clearly indicated his view that mathematical reasoning, and deductive reasoning in general, can help in deducing conclusions from axioms, but not in arriving to axioms in the first place. Only induction can lead to axioms. And the required induction involves a broader scope of information than merely economic aspects of man's behavior: it must be "concerned with the moral and social principles of man's actions and relations."[69] Indeed, for this reason Whewell doubted that political economy would ever be at the state in which it could usefully be mathematized.[70] Thus he explicitly agreed with Malthus's notion of a broader scope of political economy, and rejected

---

the profit-maximizing rule in such a way as to make an important contribution to the theory of the behavior of monopolies. Lubbock is seen as making an important contribution to monetary analysis (ibid., pp. 14–15).

67. William Whewell, "Mathematical Exposition of Some of the Leading Doctrines in Mr. Ricardo's *Principles of Political Economy and Taxation,*" pp. 2–3.

68. See ibid., pp. 4, 6, 14. In his 1862 *Six Lectures on Political Economy,* Whewell similarly described Ricardo's position on rent, noting, "It will be my business to show that this proposition is altogether erroneous" (p. 65). After making his argument, Whewell addressed the question, How did this erroneous position become so prominent in political economy? "To this I reply, that this happens because this proposition was an ingenious deduction from the doctrine of rent of a certain hypothesis; namely on the hypothesis of a constantly decreasing return to agricultural labour," and then went on to show that this hypothesis is itself false as a universal assumption: "There seems to be no reason whatever to suppose that this is the rule of the general case" (p. 72).

69. William Whewell, "Mathematical Exposition of Some Doctrines of Political Economy," p. 3.

70. "The most profitable and philosophical speculations of Political Economy are . . . those which are employed not in reasoning *from* principles, but *to* them: in extracting from a wide and patient survey of facts the laws according to which circumstances and conditions determine the progress of wealth, and the fortunes of men. Such laws will necessarily at first, and probably always, be too limited and too dependent on moral and social elements, to become the basis of mathemat-

the Ricardian use of the abstract, economic man. It was wrong, he believed, to represent man as "a mere aggregate of physical powers, a mere instrument for the production of wealth." Instead, as he put it in his review of Jones's book, "all parts of his nature, all the elements of his progress, all the conditions of his well-being, are indissolubly united—his energy and activity as a cultivator bound up with his civil position—his social comforts and prospects associated with his moral qualities and duties."[71]

In these papers Whewell accused the Ricardians of having neglected the effect upon principal forces of the many disturbing causes that account for man's actions.[72] He made this point explicitly when arguing against Ricardo's development model, the center of which was the law of diminishing returns. Whewell claimed that diminishing returns in agriculture were likely to be offset by the technological changes occurring both in agriculture and other related fields. Jones's historical analysis of agriculture in England and other countries had led both of them to this conclusion.[73] While agreeing that there had been a historical rise in (absolute) rents in England, Whewell argued that this was most plausibly explained as the effect of increased capital and improved methods of agriculture, rather than as an instance of the law of diminishing returns.[74] He claimed that this problem in Ricardo's analysis resulted from faulty deduction.

Whewell noted another case in which Ricardo's deduction from premises was invalid. He considered Ricardo's claim that "a tax on wages must fall on labourers; because if it did not so fall, wages would rise, and in consequence of this rise of wages, the price of manufactured goods would rise; and in consequence of this rise of goods, wages would again rise; and so on, without any assignable limits."[75] Whewell went on to show, using an arithmetic series, that wages and prices would actually increase by determinable amounts—

---

ical calculation" (William Whewell, "Jones—*On the Distribution of Wealth and the Sources of Taxation*," pp. 42–43).

71. See ibid., p. 60.

72. This is similar to another of T. B. Macaulay's criticisms of James Mill; Mill had argued that despots and aristocracies will always "plunder and oppress" the people, *if nothing checks their behavior*. Macaulay claimed in response, "It is quite clear that the doctrine thus stated is of no use at all, unless the force of the checks be estimated." See T. B. Macaulay, "Utilitarian Theory of Government," p. 209.

73. See William Whewell, "Mathematical Exposition of Some of the Leading Doctrines in Mr. Ricardo's *Principles of Political Economy and Taxation*," p. 15, and James Henderson, "Induction, Deduction, and the Role of Mathematics," p. 8.

74. See Neil DeMarchi, "The Success of Mill's *Principles*," p. 134; see also William Whewell, Prefatory Notice to *Literary Remains, Consisting of Lectures and Tracts on Political Economy, by the Late Rev. Richard Jones*; and Richard Jones, *An Essay on the Distribution of Wealth*, pp. 192–96, 205.

75. William Whewell, "Mathematical Exposition of Some Doctrines of Political Economy," p. 2.

they are not quantities with "no assignable limit."[76] So, on the reasonable
assumptions that wages are less than the whole cost of production, and that
only part of a laborer's wages is spent on manufactured goods, there is in fact
no wage-price spiral of an unassignable limit, but rather a finite series of wage
and price increases.[77]

At the same time that Whewell took on the negative task of showing
Ricardo's premises as well as his reasoning to be in error, he insisted that
Jones take on the positive task of producing more accurate doctrines, by the
correct inductive approach. Jones was cajoled, chided, encouraged, and supported by Whewell while he was slowly working on his book. Whewell also
provided material help in the form of obtaining from John Lubbock data on
mortality rates and on the relationship between age of marriage and number
of children.[78] In addition, he petitioned the University Press to offer a contract to Jones for publication of the book.[79] When the book finally appeared,
Whewell sent a copy to Herschel, punning that "in the long run, its failure is
not possible."[80] He then wrote three reviews, two of which appeared in print.
He had a short notice of Jones's book in the London *Literary Gazette* (February 1831) and a longer review in the *British Critic* (July 1831). Whewell also
submitted a piece to the *Quarterly Review*, but the editor, John G. Lockhart,
had already asked for a notice of the book from George Poulette Scrope.[81]
Although the review by Scrope was quite positive, Whewell was disappointed;
he had hoped that his review of Jones's book would appear in the same issue
of the *Quarterly* as his review of Herschel's *Preliminary Discourse*. He
hinted at this in a letter to Herschel: "I hold myself as one of the most fortunate men of the age in having before me at the same moment, just published
by two intimate friends, two such books as yours and his. I am also bold

76. See ibid., p. 3.

77. See Neil DeMarchi and R. P. Sturges, "Malthus and Ricardo's Inductivist Critics," p. 382n.

78. See Harvey Becher, "Whewell's Odyssey," p. 9. Herschel provided help to Jones as well, by
finding for him a book in Amsterdam on Dutch and Portuguese commerce (see letter from John
Herschel to Richard Jones, September 11, 1822, JHP 19.31) and by reading much of Jones's book in
proofs.

79. See William Whewell to John Herschel, November 23, 1827, in Isaac Todhunter, *William
Whewell, D.D,* 2:86. In a letter to the Reverend Wilkinson, he expressed his exasperation with
Jones: "Jones's book will I expect . . . *faire époque* in the science when it does appear . . . but he is
heinously dilatory" (in Janet Stair Douglas, *The Life and Selections from the Correspondence of
William Whewell, D.D.,* p. 137).

80. William Whewell to John Herschel, February 10, 1831, in Isaac Todhunter, *William Whewell, D.D.,* 2:115.

81. For details, see ibid., 1:52–53. Scrope's review appeared in November 1831 (see Mark Blaug,
Ricardian Economics, p. 152). In the review Scrope praised Jones's book as having "dealt the finishing stroke to the miserable 'Theory of Rent' of the Ricardo school." McCulloch defended Ricardo
against Jones's criticisms in his own review, which appeared in the *Edinburgh Review* September 1831.

enough to think that I can point out the meanings and merits of these same books better than most people."[82]

Jones began his book, *An Essay on the Distribution of Wealth*, by criticizing deductive methodology in political economy.[83] The problem with the method of Ricardo, he explained, is that it did not take account of the way the world is and how humans in the world actually behave. Thus he noted that "Mr. Ricardo was a man of talent, and he produced a system very ingeniously combined, of purely hypothetical truths; which, however, a single comprehensive glance at the world as it actually exists, is sufficient to shew to be utterly inconsistent with the past and present condition of mankind."[84] Jones explicitly appealed both to Francis Bacon and Herschel in advocating for an inductive, empirical approach to the science. Bacon was right, Jones claimed, in warning us against anticipations of nature.[85] Herschel, a more recent authority on both philosophy of science and science, had also pointed out the way that modern inductive scientists could proceed.[86]

By neglecting inductive method, Jones argued, Ricardo was led into numerous absurdities. For instance, he pointed out that Ricardo based his pos-

82. William Whewell to John Herschel, February 10, 1831, in Isaac Todhunter, *William Whewell, D.D.*, 2:115. Herschel looked forward to Whewell's review of both books; he wrote to Jones of Whewell's intention: "I am glad he is going to review us—let us put him to the torture if he does not butter us both well" (February 17, 1831, JHP 19.80).

83. The neglect of Jones by many historians of economics may be due to the view, expressed by William Miller, that his notion of induction was essentially the "naïve" view of induction as the mere enumeration of instances. As Miller puts it, "He seems to have thought that he could collect facts and from these a comprehensive theory would drop of its own weight" ("Richard Jones," p. 201). But Jones, like Whewell, did not have such a naïve view of induction. Writers in the nineteenth century were more favorable to Jones: Marx devoted seventy pages in the *Theorien über den Mehrwert* to praising Jones's book (whereas, as we saw earlier, he considered Malthus a "bought advocate" of the landowning classes) (see Donald Winch, *Riches and Poverty*, p. 12). Alfred Marshall thought Jones's methodology "had its defects," but nevertheless noted that he "largely dominated the minds of those Englishmen who came to a serious study of economics after his works had been published [posthumously] by Dr. Whewell in 1859" (A. G. Pigou, ed., *Memorials of Alfred Marshall* [reprint, New York, 1956], p. 296; quoted in Miller, "Richard Jones," p. 198). In *Palgrave's Dictionary of Political Economy*, originally published in 1894 and reprinted in 1925, F. Y. Edgeworth claimed that Jones "deserves to be regarded as the founder of the English historical school," and as the "Bacon" of political economy (pp. 490–91).

84. Richard Jones, *An Essay on the Distribution of Wealth*, p. vii.

85. On Bacon, see, for example, ibid., pp. xxii–xxiii, xl.

86. See ibid., p. xx, n.1, also the appendix, which was praised by Whewell as "giving a very inductive character to the book" (see William Whewell to Richard Jones, January 13, 1831, WP Add.Ms.c.51 f.95, and Jones, appendix, *An Essay on the Distribution of Wealth*, pp. 3–4). In a letter of December 16, 1830, Jones asked Herschel's permission to quote passages from his *Preliminary Discourse*, in case Herschel's book had not yet appeared by the time Jones's was published (JHP 10.348). When the book was published, Herschel praised its inductive character. He wrote to Jones: "It is really comfortable and consolatory to see a subject that seemed to have been raked into only to bring up noisome conclusions and bewildering paradoxes once more put on the footing of a rational enquiry and treated on broad inductive principles" (February 17, 1831, JHP 19.80).

tulate of rent on a type of landlord-tenant relationship (which Jones called "farmers' rents") that accounted for only about 1 percent of cases on the globe.[87] Jones adduced numerous other types of relationships—the serf, métayer, ryot, and cottier as well as the farmer relationship—and showed that Ricardo's definition of rent does not apply to these. Thus he noted that the conclusions that Ricardo drew from his definition or "postulate" of rent are false as universal principles.[88]

In an early letter to Whewell, Jones had described his work in economics as having, in part, a political aim: namely, "the establishment of sound political economic views in a subject which has hitherto only called itself a science to enforce a dogmatical philosophy of the most pernicious kind."[89] This is because Jones believed, like Whewell, that an inductive method in political economy leads to different political consequences than does Ricardian economics. For example, because Ricardo's so-called universal postulate of rent was shown to apply to only a very small portion of the world, the consequence Ricardo had drawn from this postulate—namely, that the interest of the landlord is *always* opposed to the interest of the rest of society—was demonstrably false. Jones indicated the importance of this point by claiming, in his preface, that one of the goals of his book was to show the interconnectedness of all classes of society, and thus "to demonstrate the error of those gloomy notions of a perpetual discord between rival interests in society."[90] In particular, he wished to refute the "repulsive doctrine" that the landowners' interests are permanently in conflict with those of the other classes. In examining the facts about rent (in this volume) and wages (in his planned, but never completed, second volume), Jones claimed he would at last be able to demonstrate the error made by Ricardo and his followers. Thus political consequences drawn from this conclusion, such as Ricardo's position on the Corn Laws, were not supported. Nor were the land reform proposals of the utilitarian radicals necessary.

Moreover, Jones argued that an inductive approach to political economy showed that the members of the lower classes are not doomed to a precarious and wretched existence. Both Whewell and Jones rejected the view that the laboring classes must inevitably suffer. Whewell had argued against this

87. On this point, see Neil DeMarchi and R. P. Sturges, "Malthus and Ricardo's Inductivist Critics," pp. 382–83. In his second edition of the *Principles*, Malthus redefined rent so as to include "all the different kinds of rent referred to by Mr. Jones" (*The Works of T. R. Malthus*, 5:124n18).

88. Yet Jones did grant that farmers' rent constitutes a necessary stage of progress in supposing the growth of a class of capitalists who direct labor and introduce more productive methods of cultivation.

89. Richard Jones to William Whewell, September 27, 1827, WP Add.Ms.c.52 f. 15.

90. Richard Jones, *An Essay on the Distribution of Wealth*, p. xxxix.

claim in a sermon he drafted in 1826.[91] In his later *Of the Plurality of Worlds*, he claimed that "the normal condition of man is one of an advance beyond the mere means of subsistence, to the arts of life and the exercise of thought in a general form."[92] Indeed, Whewell and Jones agreed with those nonutilitarian reformers who argued that since the economic interests of all classes are intertwined, it is incumbent upon the wealthy to help improve the economic situation of the poor, for the good of all. In his review of Jones's *Essay on the Distribution of Wealth*, Whewell highlighted this claim:

> We find, that when we look at the world as it is, with its elements of good and evil, of progress and permanence, we have no longer the interests of the proprietor in constant and necessary opposition to those of the rest of the community, but that the common welfare of each class resides in the common advance of all, and that the moral and social elevation of the lowest orders are inseparable conditions for any considerable and permanent prosperity of the highest.[93]

One way in which Jones argued for this position was by rejecting Ricardo's "iron law of wages," that is, the claim that wages cannot permanently rise above what is sufficient to provide bare necessities. As we have seen, Ricardo used Malthus's population principle and its conclusion that population grows much more quickly than do the means of subsistence to draw the conclusion that the market price of labor tends constantly toward the natural price. Since the natural price is that at which the laborer and his family could barely just survive, this "law" of wages implied that the laborer would tend always to a condition of struggle and bare survival. Ricardo did suggest that laborers could rise above mere subsistence and thus avoid distress, but only by restricting their population or by the increase in capital accumulation.[94] But he was not very optimistic about either of these possible solutions, and so it seemed to

91. See William Whewell, "Drafts of Sermons [1827]," WP R.6.17, f. 22–26; quoted in Harvey Becher, "Whewell's Odyssey," p. 11. In this draft, Whewell rejected the view of the Ricardians that "those portions of society that win their bread by the labor of their hands and eat in the sweat of their brows, are destined to eternal and irredeemable degradation." He claimed that this doctrine "shook and startled the mind of pious and benevolent men," because it cast doubt on "God's goodness." In his delivered sermon, Whewell omitted these passages, because Jones had objected to being "scooped" by him. See letter from William Whewell to Richard Jones, December 15, 1826, WP Add.Ms.c.51 f. 33.

92. William Whewell, *Of the Plurality of Worlds*, p. 83.

93. William Whewell, "Jones—On the Distribution of Wealth and the Sources of Taxation," p. 60.

94. See David Ricardo, *Principles of Political Economy*, in *The Works and Correspondence of David Ricardo*, vol. 1, chap. 5.

follow from his analysis that poverty would always be extreme for the lower classes. Jones, on the other hand, was rather optimistic about the possibility for laborers to rise above mere subsistence. In his book, he admitted that the fact that humans are capable of so much reproducing can cause suffering. However, he expressed confidence about the ability of "human effort" to overcome this suffering. Further, he believed that "no one community is necessarily doomed to endure any portion of such suffering at all."[95]

Jones criticized the Ricardians not only for their deductive methodology and their political conclusions, but also for their claim that the proper subject of political economy is wealth alone. Like Whewell, Jones believed that political economy must encompass moral, social, and political concerns.[96] He argued that subjects in political economy can never "be successfully pursued if such [moral, political, and social] subjects be wholly eschewed by its promoters. There is a close connection between the economical and social organisation of nations and their powers of production."[97] For this reason, years earlier, Whewell had told Jones that he belonged to the "ethical school of political economists."[98]

In 1831, after he had completed *An Essay on the Distribution of Wealth*, Jones happened to read Whately's *Elements of Logic*. He was particularly incensed by the appendix, which had been written by Whately's colleague Nassau Senior, and which argued for the importance of starting with correct definitions in the science of political economy. Jones was surprised that the Oriel dons supported the Ricardian view of deductive method in political economy. Whewell's initial response was more phlegmatic: "I quite rejoice to have such specifics of what is to be avoided."[99] Jones inquired with his publisher about the status of his book, which had been printed but not yet bound.[100] It is therefore likely that the unnumbered page placed between the

---

95. Richard Jones, *An Essay on the Distribution of Wealth*, p. xxxv. This echoes the omitted parts of Whewell's 1827 sermon, which Jones found too close to the views he hoped to publish first: "It has been maintained . . . that the fiat of His will by which the Creator ordained the increase and multiplication of men, impelled them . . . to want and degradation, to vice and misery. It has been passed on . . . that the tendency of mankind to replenish the earth ever pushes them on till the sharp discipline of pain, the iron hand of want . . . drive them back" (William Whewell, "Drafts of Sermons [1827]," WP R.6.17, f. 22–26; quoted in Harvey Becher, "Whewell's Odyssey," p. 11).

96. Richard Jones, *An Essay on the Distribution of Wealth*, pp. 4–5.

97. Richard Jones, *Literary Remains, Consisting of Lectures and Tracts on Political Economy, by the Late Rev. Richard Jones*, pp. 44–45.

98. William Whewell to Richard Jones, August 16, 1822, WP Add.Ms.c.51 f. 14.

99. William Whewell to Richard Jones, February 25, 1831, WP Add.Ms.c.51 f. 99.

100. See Richard Jones to William Whewell, February 25, 1831, WP Add.Ms.c.52 f.21; quoted in Pietro Corsi, "The Heritage of Dugald Stewart," p. 120.

table of contents and the first page of text was inserted at the last moment, and was aimed at the Oriel school.[101] This page is blank but for the following disclaimer:

> It has been mentioned to me, that I have given no regular definition of the word *Rent*. The omission was not undesigned. On a subject like this, to attempt to draw conclusions from definitions, is almost a sure step towards error. A dissertation, however, on the use and abuse of definitions would be out of place here. I have pointed out the origin of payments made to the owners of the soil. I have tracked their progress. If any reader, during this enquiry, is really puzzled to know what we are observing together, I shall be sorry: but I am quite sure that I should do him no real service by presenting him in the outset with a definition to reason from.

Meanwhile, Whewell took on the task of criticizing the Oxford school in several of his works from 1831 to 1832, though these criticisms were generally implicit rather than explicit. In his review of Jones's book, for example, Whewell made ironic reference to a comment made by Senior.[102] In his review of Herschel's *Preliminary Discourse*, he expressed the hope that the day would come soon "when 'definitions' shall no longer be considered as the fountains of what we can know with regard to things altogether independent of the operations of human thought."[103]

These attacks did not go unnoticed by Whately, who in 1831 had ascended to the bishopric of Dublin. The next year he put out the second edition of his *Introductory Lectures on Political Economy*. To this edition Whately added a ninth lecture aimed at Whewell, in response to his review of Jones's book. In it, Whately argued against the "mistake of beginning by a crude collection of facts."[104] He appealed to Bacon, claiming that Bacon himself would have agreed that starting from a mass of facts was dangerous.[105] Whewell counterattacked with an article, "On the Use of Definitions." He contended explicitly here that the exact definition of terms in science follows the advance of knowledge, rather than causes it. With Whately his clearly intended target—since the only references in this essay are to writ-

---

101. I owe this point to Pietro Corsi, "The Heritage of Dugald Stewart," p. 120.

102. See William Whewell, "Jones—*On the Distribution of Wealth and the Sources of Taxation*," p. 56.

103. William Whewell, "Modern Science—Inductive Philosophy," p. 377.

104. Richard Whately, *Introductory Lectures on Political Economy*, p.224.

105. Ibid., p. 237.

ings of Whately's—Whewell noted that people who argue over definitions at earlier stages of scientific growth are omnipresent, "but truth has generally passed rapidly forwards, and left them behind to enjoy their favorite amusement."[106]

While arguing against Ricardian economics and its use by Whately and the Oriel school, Jones and Whewell attempted to enlist Malthus in their fight against the "downwards mad" political economists. In 1829, Whewell sent Malthus his paper "Mathematical Exposition of Some Doctrines of Political Economy," and received an enthusiastic response. Malthus was present for the first meeting of the Section F of the BAAS, and Whewell visited him several times in his final years.[107] After being introduced to Jones by Whewell, Malthus apparently recommended the younger political economist to be his successor at the Haileybury College.[108] However, while Malthus agreed that Ricardo went too far in the direction of "abstract" reasoning, he seems to have thought that Whewell and Jones went too far in the opposite direction.[109] Thus he wrote to Whewell that he was gratified to have the support of Whewell and Jones against Ricardo, but said, "My apprehension at present is that the tide is getting too strong against him; and I even think that Mr. Jones is carried a little out of the right course by it."[110] Moreover, in response to Whewell's paper on definitions, Malthus agreed with Whewell that "the most exact definitions are not so much the causes as the consequences of our advances in knowledge." However, he went on to add in this letter that definitions can sometimes be useful as starting points, if taken provisionally and with the admission that "they might subsequently give way to others more complete." (Malthus had originally worried that Whewell's paper on definitions was a criticism of his own 1827 paper, "Definitions of Political Economy," but then realized that it was meant as an attack on Whately.)[111] Malthus had other, substantive disagreements with Whewell and Jones as well. For example, his population principle entailed that there would always be poverty and misery at the bottom levels of society. In addition, he was not

106. William Whewell, "On the Uses of Definitions," p. 2.
107. See Neil DeMarchi and R. P. Sturges, "Malthus and Ricardo's Inductivist Critics," p. 379.
108. See letter from Richard Jones to John Herschel, JHP 10.410. The letter is not dated, but it was probably sent in 1837 or 1838, for it contains a reference to Whewell's newest book, which appears by Jones's description of it to be the *History of the Inductive Sciences*.
109. See Neil DeMarchi and R. P. Sturges, "Malthus and Ricardo's Inductivist Critics," p. 386.
110. T. R. Malthus to William Whewell, May 31, 1831, WP Add.Ms.a.209 f.11.
111. T. R. Malthus to William Whewell, April 1, 1833, WP Add.Ms.a.209 f.12. Whewell had explicitly exempted Malthus from his criticism of those who seek definitions of political economy in his review of Jones's book (see "Jones—*On the Distribution of Wealth and the Sources of Taxation*," p. 54).

willing to abandon the basic theory of rent based on diminishing returns that he and Ricardo had developed, and that Whewell and Jones rejected.[112]

Neither did Whewell and Jones accept all of Malthus's principles. They rejected his population principle and the consequences of it drawn by many political economists. While Whewell was preparing a sermon on the topic of the poor in 1826, he remarked to Jones that he would not have time "for more than a very short attack on the principle of population," which he called a "false induction."[113] In a later letter he gave examples of inductive method in subjects other than natural philosophy, noting that "a good deal of Malthus' [essay on] population is a beginning of such a process *excluding of course his anticipatory thesis*, which is the only thing usually talked of."[114] On their view, Malthus arrived at his population principle by ignoring his own inductive methodology. Jones, who discussed this issue explicitly in his book, claimed there that Malthus was correct in noting that we possess a physical power of multiplying such that if it were exerted unchecked, our numbers must eventually outstrip any possible increase in food. Jones also agreed with Malthus that much of the happiness or misery of a large portion of the population depends on the extent to which this power is controlled. But he believed that Malthus gave a defective division of the checks to population.[115] Jones held that there are various motives for self-restraint of our reproductive capacity, not all of which are resolvable into vice or misery.[116] Notably, he stressed the increase in the artificial desires, which starts with the upper classes but then trickles down to the lower; he calls this the "gradual spread of refinement." This causes a check to population, as people realize they will not be able to spend money on satisfying such desires if they need instead to feed a large family.

Their rejection of Malthus's population principle led Whewell and Jones to argue against the New Poor Law of 1834. This law amended the previous poor laws by abolishing outdoor relief to able-bodied laborers and their families, requiring instead that all assistance be administered in workhouses, in which conditions would be harsh and families separated.[117] Malthus's popu-

112. See letter from T. R. Malthus to William Whewell, May 31, 1831, WP Add.Ms.a.209 f.11.

113. William Whewell to Richard Jones, December 15, 1826, WP Add.Ms.c.51 f.33. See note 95 above.

114. William Whewell to Richard Jones, February 25, 1831, WP Add.Ms.c.51 f. 99.

115. Richard Jones, *An Essay on the Distribution of Wealth*, pp. viii–ix.

116. Ibid., pp. xxxvi–xxxviii.

117. The Philosophical Radicals were enthusiastic supporters of the law, and especially its workhouse requirement. Indeed, to many people, the workhouse requirement of the new Poor Law was associated with utilitarian radicalism, even though it had numerous supporters of other political stripes as well, including the Whigs (the party which introduced the bill) and many clergymen.

lation principle had provided a justification for altering the current system of relief to the poor.[118] In the first edition of his *Essay on Population*, Malthus suggested (as follows from his population principle) that there is a limited quantity of resources, and everything given to unproductive paupers was given at the expense of deserving independent laborers.[119] Because of this, a "labourer who marries without being able to support a family, may in some respects be considered as an enemy to all his fellow-labourers."[120] Further, the expectation of the right to relief removed one of the incentives to "moral restraint," encouraging early marriages even by those without the means to support children by their own labor. Thus Malthus concluded that the Poor Laws "in some measure create the poor which they maintain."[121] Moreover, the Poor Laws have negative effects on the moral condition of the people they are supposed to be helping. The Poor Laws "are strongly calculated to eradicate" the spirit of independence of the lower classes. They further "diminish both the power and the will to save among the common people, and thus . . . weaken one of the strongest incentives to sobriety and industry, and thus to happiness."[122]

In the first edition of his *Essay on Population*, Malthus argued for a workhouse requirement and other remedies; by the 1803 second edition he condemned such remedies and claimed that relief to the poor should be abolished completely. He argued against the payment of children's allowances, which he believed promoted population increase by lessening the preventa-

---

Thus, for instance, Dickens satirized the "utilitarian economists" for their support of the workhouse in *Hard Times*. More generally, the whole field of political economy became associated with the utilitarian moral and political philosophy. For example, of the utilitarian radical reformers, Macaulay wrote: "There is not, and we firmly believe that there never was, in this country, a party so unpopular. They have already made the science of Political Economy—a science of vast importance to the welfare of nations,—an object of disgust to the majority of the community"("Westminster Reviewer's Defense of Mill," pp. 203–4). Even William Wordsworth wrote a short poem, "To the Utilitarians," which he sent to H. Crabb Robinson in 1833: "Avaunt this oeconomic rage! / What would it bring?—an iron age, / Where Fact with heartless search explored / Shall be Imagination's Lord, / And sway with absolute controul / The god-like functions of the Soul."

118. Thus the economist John Neville Keynes noted, "The reform of the English Poor Law in 1834 was largely due to the direct and indirect influence exerted by the writings of Malthus" (*The Scope and Method of Political Economy*, p. 291).

119. See J. R. Poynter, *Society and Pauperism*, p. 153.

120. T. R. Malthus, *An Essay on the Principle of Population*, 1st ed., p. 86.

121. Ibid., p. 100.

122. Ibid., p. 101. Harriet Martineau expressed these opinions in *Cousin Marshall*, part of her hugely popular series *Illustrations of Political Economy*. The tendency of the poor laws, she explained, was "to encourage improvidence with all its attendant evils.—to injure the good while relieving the bad,—to extinguish the spirit of independence on one side,—and of charity on the other,—to encourage speculation, tyranny and fraud,—and to increase perpetually the evil they are meant to remedy" (*Cousin Marshall*, p. 191).

tive check on marriage. (Indeed, he held that such payments strengthened the positive incentive to early marriages.)[123] He proposed a short phase-out period such that no legitimate child born from a marriage taking place one year from the start of the law, and no illegitimate child born two years from the start, should ever receive parish support.[124] A sermon to this effect should be read at all marriages, lest any young couple arrive at the altar without considering this information. Only the moral restraint of the poor could rescue them from pauperism.[125] Malthus advocated a national system of education for the laboring classes, to teach them the benefit to themselves as well as to the whole of society of this moral restraint. Otherwise, all that would be accomplished by giving paupers the means to support their growing families was the creation of a supply of laborers greater than the demand for labor, which Malthus claimed as the primary cause of poverty.[126]

Though Malthus's work provided a new justification for the abolitionist position on relief to the poor, it was a position typical of the tradition of Christian political economy out of which Malthus himself arose. Many pioneer writers on political economy in the eighteenth century were clergymen, and economics was often considered a branch of natural theology. Paley, for example, believed that the study of human nature, as well as the natural world, leads to a greater devotion to God. The study of human nature includes the laws of social behavior, such as economic behavior. (However, Paley came to the opposite conclusion than did Malthus later about the Poor Laws; in his *Principles of Moral and Political Philosophy*, he defended the notion that the poor have a right to relief from the state.)[127] These writers (including Paley) often believed that privation and differences in rank and wealth are part of

123. See J. P. Huzel, "Malthus, the Poor Law, and Population in Early 19th Century England," p. 139.

124. T. R. Malthus, *An Essay on the Principle of Population*, 1st ed., p. 118.

125. J. R. Poynter, *Society and Pauperism*, pp. 157–58.

126. In the end, the rationale for the legislation (from the 1834 *Commissioners' Report*) couched the need for the amendment in explicitly Malthusian terms: "It appears to the pauper that the Government has undertaken to repeal the ordinary laws of nature: to enact, in short, that the penalty which after all must be paid by someone for idleness and improvidence is to fall, not on the guilty person or his family, but on the proprietors of lands and houses incumbered by his settlement. Can we wonder if the uneducated are seduced by a system which offers marriage to the young, security to the anxious, ease to the lazy, and impunity to the profligate?" (Poor Law Commissioners, *Report of H. M. Commissioners for Inquiring into the Administration and Practical Operation of the Poor Laws* [London, 1834], p. 27; quoted in G. R. Searle, *Morality and the Market in Victorian Britain*, p. 169).

127. See William Paley, *Moral and Political Philosophy*, pp. 413–50. Elie Halévy notes that Paley's belief that collective happiness consists in the sum of individual happinesses led him to the view that in order to increase the collective happiness, the number of people capable of happiness must be increased; thus the happiness of a society will increase in direct proportion as the population (*The Growth of Philosophical Radicalism*, p. 500).

God's plan, as necessary for human improvement and the exercise of free will.[128] As Paley had written, "In a religious view, privation, disappointment, and satiety, are not without the most salutary tendencies."[129] Malthus later argued, against socialist radicals such as the Owenites, that inequality is necessary, because it is required to improve human virtue. Indeed, economic well-being was often taken by these Christian political economists as a sign of moral and religious worth. Writing in his Bridgewater Treatise of 1833, the Scottish Evangelical divine Thomas Chalmers explained that there is an "inseparable connection between the moral worth and the economic comfort of a people. . . . Political economy is but one grand exemplification of the alliance, which a god of righteousness hath established, between prudence and moral principle on the one hand, and physical comfort on the other."[130] Thus he told the Select Committee on the Irish Poor in 1830, "I think there is a great deal of sound political economy in the New Testament."[131] John Bird Sumner, later archbishop of Canterbury, similarly claimed that wealth is a visible token of virtue.[132]

Archbishop Whately was also in favor of abolishing the Poor Laws. In the appendix to his *Introductory Lectures on Political Economy*, he rejected the notion of applying some kind of poor law to help relieve the distress then building in Ireland, noting that such relief "would tend to make [the poor] work as little as possible. . . . I have seen that operate a great deal in England, and I think it would operate with much more rapid and destructive effect in Ireland."[133] Such interference in God's natural order upset God's plan; hence the increase in poverty and misery since the passage of these misguided laws in England. During the Potato Famine of 1846–49, Whately blamed the situation on the extension since 1838 of the Poor Law in Ireland, claiming that if the government had not interfered, the people would have seen that harder

---

128. See R. A. Soloway, *Prelates and People*, p. 112.

129. William Paley, *Natural Theology*, p. 274.

130. Thomas Chalmers, *On the Power, Wisdom, and Goodness of God as Manifested in the Adaptation of External Nature to the Moral and Intellectual Constitution of Man*, Bridgewater Treatise; quoted in G. R. Searle, *Morality and the Market in Victorian Britain*, p. 12. Malthus considered Thomas Chalmers his "best ally" (see Donald Winch, *Riches and Poverty*, p. 372). For more on the connection between Malthus and Chalmers, see Salim Rashid, "Malthus' *Principles* and British Economic Thought, 1820–1835."

131. Thomas Chalmers, *Evidence to the Select Committee on the State of the Poor in Ireland*, 1830, qu. 3584; quoted in Patricia Hollis, "Anti-Slavery and British Working Class Radicalism," p. 304.

132. See Peter Mandler, "Tories and Paupers," pp. 86–87.

133. Richard Whately, *Introductory Lectures on Political Economy*, p. 279.

work was necessary.[134] In 1859 he produced an annotated edition of Paley's *Moral and Political Philosophy*, in which he argued against Paley's claim that the poor have the right to relief. Even individual charity leads to distress: the particular effect of almsgiving is the "waste of a few shillings: the general effect would be to reduce thousands to beggary, and to paralyze the industry of a whole nation."[135] (Leslie Stephen was led to remark, "Archbishop Whately is said to have thanked God that he had never given a penny to a beggar. This view suggests some confusion between the Political Economy Club and the Christian Church.")[136] Indeed, the established church itself supported the Poor Law Amendment Act, leading to negative implications for church relations with the lower classes, and to the frequent satirizing of church authorities by novelists such as Anthony Trollope.[137] Church officials—such as Bishop Blomfield, the chair of the commission that had proposed the new law—had self-interested reasons for opposing the current poor law system, in which landowners paid rates based on the number of paupers needing relief in the parish. Since the Church of England was the greatest landowner of early nineteenth-century England, it was in its interest to abolish the payment of these rates.

However, some clergymen—including Whewell and Jones, as we shall now see—were disturbed by this antagonism toward the poor, and felt that the workhouse provision was contrary to Christian charity.[138] Jones's *Essay on the Distribution of Wealth* was published during the debates that led up to the formation of the commission to study the Poor Laws. In his work, he noted that the present system of relief constituted an "economic evil." By paying paupers wages even for unproductive labor, or for no labor at all, the Poor Laws add to the price of agricultural produce in Britain, which was bad for every-

134. See R. A. Soloway, *Prelates and People*, p. 163.

135. Richard Whately, *Paley's "Moral and Political Philosophy,"* p. 102.

136. Leslie Stephen, *The English Utilitarians*, vol. 3, *John Stuart Mill*, p. 170n.

137. See R. A. Soloway, *Prelates and People*, p. 163. Donald Winch claims that Malthusianism became part of "official Anglican doctrine" (see *Riches and Poverty*, p. xxix).

138. Moreover, a number of Christian writers felt that Malthus's population principle raised the problem of evil: why would God allow the creation of more mouths than could be fed? In the first edition of *An Essay on the Principle of Population*, Malthus included two chapters in which he developed the view that the principle of population gives rise to misery and vice, which is not a sign of evil but actually further evidence of the goodness of God, because this gives us the necessary opportunity for forming our spirit as God wishes us to do. However, he deleted these chapters in the second edition, and introduced the alternative of moral restraint. M. B. Harvey-Phillips sees this move as an abandonment of the theodicy in favor of a Paleyite theological utilitarianism ("Malthus' Theodicy," p. 605), whereas E. N. Santurri believes the addition of the alternative of moral restraint is evidence, on the contrary, that Malthus retained his theodicy ("Theodicy and Social Policy in Malthus's Thought," pp. 403, 416).

one, including the paupers themselves. He also adduced an additional reason for reforming the Poor Laws: the "moral evil" they have produced. Because of "bad laws" and "bad management," Jones believed, "the honesty of the labourers, their self-respect, their value for their character as workmen, all hope of bettering their condition in life by good conduct, industry and prudence; their sense of their mutual duties and claims as parents and children, all feelings and habits in short, that contribute to make men good citizens, and good men, have been undermined and impaired, and utterly destroyed."[139] Yet Jones strongly rejected the workhouse requirement. Any new plan for relief, he insisted, must result in the improvement of the pauper's position, not only ultimately but *at every step of the change.*[140] This suggestion was, Jones admitted, partly a matter of fairness. The poor did not invent nor desire this "miserable system" that has robbed them of their self-respect and independence; therefore they are not to be blamed and should not be punished for it.[141] Jones did not offer a solution, but suggested that there might be benefit to adopting a land-allotment scheme.[142]

Whewell publicly addressed the treatment of the poor in his *Elements of Morality, Including Polity,* published eleven years after the adoption of the New Poor Law. In outlining the "Duties of the State," he explained that "it is a Principle of Humanity, and, in an extended sense of Justice, a Principle of Justice also, that all man should possess the Natural Rights of Man; namely the Rights of Personal Security from violence; of Sustenance and Property so far as is requisite for Moral agency; and of Marriage."[143] Concerning the Right of Subsistence, he noted,

> there are large bodies of the people, even in States conspicuous for their general freedom, who hold these necessary means of moral being very precariously, and occasionally lose them altogether. Men perish of hunger in

139. Richard Jones, *An Essay on the Distribution of Wealth*, p. 317.

140. Ibid., p. 318.

141. In a later letter to Herschel, Jones signalled that he still held this view, as well as suggesting another, more economic reason for treating the laboring poor in a better way: "Liberty and comforts of the *mass* are the conditions of obtaining great productive power—and their virtuous habits and elevations of character the conditions of retaining it—and these and many like propositions arise out of the facts dealt with as plainly and with as little effort as any general law in physics" (undated, but clearly after 1837 and before 1841, as Jones referred to Whewell's new book on history of science and his own plans [which did not materialize] to review it for the *Quarterly Review;* JHP 10.410).

142. Richard Jones, *An Essay on the Distribution of Wealth*, pp. 317–18. After seeing the proofs of Whewell's *Elements of Morality, Including Polity,* Jones told Whewell, "I find no fault with any of your practical views about poor laws" (March 15, 1845, WP Add.Ms.c.52 f. 98).

143. William Whewell, *Elements of Morality, Including Polity*, p. 482.

opulent cities. Many are mendicants, who . . . depend for sustenance upon the casual bounty of their fellow-citizens. Many, belonging to the industrious classes, are frequently destitute; though willing to work, they can find no one to hire them. . . . Does the Duty of Humanity in the State admit of its tolerating the existence of such things?[144]

Whewell answered this question by noting that one "Duty of the State" is to teach the rich to act with benevolence. The state, therefore, should encourage, rather than outlaw, private charity to the poor—contrary to the views of Whately and Harriet Martineau, who had both called for an end to all private philanthropy.[145] Moreover, Whewell believed, the destitute who are not helped sufficiently by private charity must be provided for by the state. Unlike Whately, he was in favor of the imposition of the Poor Laws—and hence of public relief—in Ireland.[146] Yet, because another duty of the state is to teach "Foresight and Thrift," the state should not "maintain in comfort" all who choose to beg rather than work.[147]

Whewell went on to describe his view of the proper aspect that relief of the poor by the state should assume. First, relief for the poor should be conceived as temporary; that is, the poor man should not be barred from resuming work as an independent laborer. Thus he should not be forced to sell tools of his trade, or the furniture of his home, before receiving public support. The laborer should not be forced into workhouses, then, from which he is unlikely ever to emerge alive. On the other hand, the terms of his condition should be somewhat harder than those of independent laborer: wages should be lower (but still sufficient for subsistence). This was so that the state may teach the lessons of foresight and thrift, instead of (by rewarding such behavior) the lessons of improvidence and idleness. What Whewell suggested was—similar to the scheme of radicals such as Robert Owen—a system of public works. The

144. Ibid., pp. 482–83.

145. See G. R. Searle, *Morality and the Market in Victorian Britain*, pp. 186–87.

146. William Whewell, *Six Lectures on Political Economy*, pp. 97ff.

147. William Whewell, *Elements of Morality, Including Polity*, p. 483. Moreover, he argued against the claim that a legal provision for charity causes individuals to be uncharitable. Indeed, "the example of England appears to show that a legal provision for the poor does not extinguish the disposition to spontaneous charity. In this country, all lands and houses are taxed for the relief of the poor. . . . And yet there is, perhaps, no country in which there is more spontaneous charity" (ibid., p. 492). In 1827, while writing the drafts of the sermons that he edited in deference to Jones, Whewell was less optimistic. He criticized the response of many people to the Poor Laws: "We perpetually make our *institutions* a pretext for shutting of not only our pockets but our hearts." He noted that if the Poor Laws were abolished, men would have to be more charitable, "but why *wait* for the abolish for this? It will do just as much good now. More indeed" ("Drafts of Sermons [1827]," WP R. 6.17 f. 23, p. 4).

poor, he argued, should be put to work (at these lower wages) building and re-
pairing roads, draining swamps, and constructing public buildings.[148]

Whewell admitted that such a scheme of public works would be adequate
only when the number of laborers to be so employed was relatively small compared with the total number of laborers. But he recognized that this was not
the case in his time. What was needed, he claimed, was to "facilitate as much
as possible the rise of men from the lower to the higher classes of society."
Since the desire for manufactured goods was increasing, there was need for a
growing class of tradesmen to provide these goods.[149] Whewell argued that if
more men moved to this line of work, then both production and consumption
would increase, and thus the economic well-being of the nation. Moreover, he
believed that induction supported the view that this was a possible solution.
Based on Jones's work, he claimed that increasing rents were caused by technological improvements, not the extension of agriculture to worse soils, as
Ricardo had claimed. Because of these improvements, more people were
being released from the land, since fewer people were needed to grow the same
amount of food. Further, new technologies in other areas created new opportunities for employment, opportunities that would raise the standard of living and change the economic and class status of those who took advantage of
them. Economic growth would thus spread from the agricultural to other sec-
tors, which would ultimately allow for upward economic mobility of many
individuals, as well as increased economic well-being of the nation.

To this end, Whewell suggested the education of members of the lower
classes so that they could rise to the newly growing middle or "intermediate"
classes.[150] Unlike Malthus, who believed that education of the lower classes
would be useful in helping them to realize the necessity for their own "moral
restraint," or Martineau, who thought that this education would not improve
the lot of the poor, but that by helping them to see the necessity of their state
would make it more palatable to them, Whewell proposed a much more progressive solution: education could help lift the poor out of the laboring class
and into the growing middle class.

Whewell also gave a moral justification for rejecting the workhouse re-

148. William Whewell, *Elements of Morality, Including Polity*, pp. 484–85. Robert Owen, in
his *New View of Society* (1812–16), had argued that through education of the poor the *need* for the
Poor Laws would be eliminated, but that meanwhile there should be some system of public works,
especially the creation of new communities, to employ the poor. Some conservative authors, such
as Coleridge, also urged that public employment be created as a means to alleviate the distress of
the poor. Malthus complained about such schemes in his 1826 essay (see J. R. Poynter, *Society and
Pauperism*, p. 254).
149. William Whewell, *Elements of Morality, Including Polity*, p. 488.
150. Ibid., p. 490.

quirement. "If a man, by accepting public relief, is placed in a condition in which there is a permanent bar to his becoming again an independent and thriving labourer, the object of humanity is defeated, and the man is reduced to a kind of servitude." In such a case, "he has not the means of self-guidance and advancement, which are requisite to his moral being."[151] The workhouse requirement was therefore to be rejected because of its negative consequences for man's ability to act as a moral agent. Just as enslaving a black man is immoral, so too is reducing a pauper to lifelong servitude.

As we have seen, Whewell and Jones embraced Malthus's view that political economy must be concerned with moral, political, and social issues, as well as with the bare economic ones. They rejected the opposing Ricardian position. They also took Malthus's side in the methological debate between Malthus and Ricardo over the proper methodology of political economy. Like Malthus, Whewell and Jones strongly argued that political economy is an inductive science; this was consistent with their general "induction project." But they rejected Malthus's population principle, seeing it as a noninductive anticipatory thesis; indeed, they criticized Malthus for having forgotten his own methodological injunctions. Therefore it is not surprising that Whewell and Jones rejected the political consequences drawn from Malthus's principle, especially the New Poor Law of 1834. Whewell's criticisms of this law, and his proposed solution—which stressed the upward class mobility of the impoverished laborer, by the proper kind of education—highlight the difficulties with previous categorizations of Whewell as an archconservative. He was not arguing for maintaining the status quo, in which paupers had to struggle just to remain at the level of mere subsistence. We turn now to an examination of Mill's more complicated relationship with the Ricardian system. We will see that by the end of his career, Mill's view of political economy came to resemble, in some crucial ways, that of Whewell and Jones.

Mill on Political Economy

In the *Autobiography*, Mill recounted that he "went through a complete course in political economy," reading Ricardo and Smith, in 1819 (that is, when he was thirteen years old). His first published writings were letters in the *Traveler* in 1822 defending his father and Ricardo against the criticisms of Robert Torrens on the theory of economic value.[152] He went on to write

151. Ibid., p. 484.
152. See J. Viner, "Bentham and J. S. Mill," p. 154, and Elie Halévy, *The Growth of Philosophical Radicalism*, p. 351.

reviews, essays, and newspaper articles on political economy from the '20s until the late '40s. His *System of Logic*, which contained discussion of the method of political economy, and his *Principles of Political Economy* were reissued in successive editions up until 1872.

By examining these works, it is possible to see the shift in Mill's interests and views from his early Bentham-inflected days to his later period, considered by Mill himself as more "radical" because of his growing desire to reform society by reforming the minds of individuals. Over the course of his career, he moved away from his early uncritical acceptance of Bentham's deductive, geometrical methodology. But Mill was ultimately unable to combine inductive and deductive methodologies, as he had wished to do after his pair of essays on Bentham and Coleridge in 1838 and 1840. Rather, in the *System of Logic*, he continued to endorse the geometrical method for political economy, but only as a temporary measure, until it becomes possible to separate the many individual causes of man's behavior. By the time he began writing the *Principles*, in the mid-1840s, Mill was less concerned with the methodology of political economics as a science; he concentrated instead on how the art of political economy could be utilized to bring about the cultivated society he sought.

In his 1828 review of Whately's *Elements of Logic*, Mill agreed with Whately's view that the method of political economy is deductive and syllogistic.[153] However, by the time he wrote his essay "On the Definition of Political Economy; and on the Method of Investigation Proper to It," Mill's view had changed. This article appeared in the *London and Westminster Review* in 1836, but it is known that he had originally written the article in 1831 and revised it in 1833 (the year he wrote his first article critical of Bentham).[154] In this essay he abandoned his earlier belief in the use of syllogistic method in the human sciences; he later attributed this shift to Macaulay's criticisms of his father, which had appeared in 1829. Nevertheless, Mill maintained his belief in the deductive nature of reasoning in political economy, claiming that the political economist reasons deductively from assumptions about human nature.[155] Calling the method of political economy the "a priori"

---

153. John Stuart Mill, "Review of Whately's Introductory Lectures on Political Economy," *CW* 11:1–35.

154. See John Stuart Mill, "On the Definition of Political Economy; and on the Method of Investigation Proper to It," *CW* 4:309.

155. There has been some confusion over the methodology presented by Mill in the essay "On the Definition of Political Economy," due to the fact that, at one point in the essay, he claimed that the a priori method in political economy "involves a mixed method of induction and ratiocination" (see *CW* 4:324). However, a closer look at the essay shows that it is not the case that Mill presented political economy as being both inductive and deductive, as other commentators claim (see, e.g.,

method, he explained that "by the method a priori we mean (what has commonly been meant) reasoning from an *assumed* hypothesis; which is not a
practice confined to mathematics, but is of the essence of all science which
admits of general reasoning at all."[156] Like geometry, political economy is built
upon arbitrary definitions: "Geometry presupposes an arbitrary definition of
a line, 'that which has length but not breadth.' Just in the same manner does
Political Economy presuppose an arbitrary definition of man, as a being who
invariably does that by which he may obtain the greatest amount of neces-
saries, conveniences, and luxuries, with the smallest quantity of labour and
physical self-denial with which they can be obtained in the existing state of
knowledge."[157] But Mill admitted that such arbitrary assumptions "might be
totally without foundation in fact."[158] He confessed that other laws of motives
counteract man's purely economic behavior. (Recall that in his 1833 essay on
Bentham, he had criticized his former mentor for ignoring that men have
other motives for their actions besides utility.) Just as the mathematician does
not believe that his definition of a line corresponds to any actual line, the
political economist does not imagine that real men have no desire but
wealth.[159] Yet the a priori method of assuming a purely "economic man" must
be used, because it is impossible to proceed by the a posteriori inductive
method. There are too many disturbing causes or, as Mill calls them here,
"influencing circumstances," for the a posteriori method to be successful.[160]

In this essay, Mill demonstrated clearly the debt he owed to Dugald Stewart, who, as we saw earlier, had trained Mill's father in political economy.
Stewart argued that axioms of political economy were established by "universal experience," and could not be disproved by particular facts. He claimed
in his *Elements of the Philosophy of the Human Mind* that "the premises . . .
from which these conclusions [of political economy] are deduced, are neither
hypothetical assumptions, nor metaphysical abstractions. They are practical
axioms of good sense, approved by the experience of men in all ages of the
world; and of which, if we wish for any further [confirmation], *we have only
to retire within our bosoms* or open our eyes on what is passing around us."[161]
Similarly, Mill contended that the a priori method embraces a "wider field of

Oskar Kubitz, *The Development of John Stuart Mill's "System of Logic,"* pp. 137–38, and Samuel
Hollander, *The Economics of John Stuart Mill*, 1:68).

156. John Stuart Mill, "On the Definition of Political Economy; and on the Method of Investigation Proper to It," *CW* 4:325; emphasis added.

157. Ibid., 4:326.

158. Ibid.

159. Ibid., 4:327.

160. Ibid., 4:328.

161. Dugald Stewart, *Elements of the Philosophy of the Human Mind*, 2:451.

experience" than the a posteriori method, because in the former case experience includes "what passes in our own minds."[162] He accepted the position
of his father and Stewart that the experience called upon in political economy
includes that gained by introspection. Indeed, in *System of Logic* Mill later
referred to inductions from facts furnished to us by our senses *"or by our
internal consciousness."*[163]

In "On the Definition of Political Economy," Mill distinguished between
the "science" and the "art" of political economy: "Science takes cognizance
of a *phenomenon,* and endeavours to discover its *law;* art proposes to itself an
end, and looks out for *means* to effect it."[164] He opposed the common view
that the science of political economy teaches a nation how it may increase its
wealth; rather, such considerations are part of the art of political economy.[165]
While in this essay, unlike in the review of Whately, Mill discussed the verification of statements in political economy, it is clear that verification plays no
part in the science of political economy, only in its application.[166] Verification
is an empirical process; we must compare predicted results with what does
happen.[167] But we *cannot* verify our economic hypotheses by reference to empirical facts, because we *know* that what happens will *not* accord with our
prediction. Predictions are made on the basis only of an assumed economic
motive—with which the science of political economy is exclusively concerned—while what happens in actual fact is determined by many other
motives as well.[168] On the other hand, the practical rules constituting the art
of political economy may be verified empirically. If the art of political economy suggests means for effecting a particular end, such as the goal of increasing the wealth of a nation, then it is possible to compare the predicted effect
with what does, in fact, happen once the suggested method is attempted.

Around the time that Mill wrote this essay, he was arguing in favor of the
New Poor Law in a series of newspaper articles published in 1834. He particularly approved of the workhouse provision of the new statute. In these
articles, Mill expressed the Benthamite justification for the workhouse provision. Years before, Bentham himself, in a number of pamphlets and unpub-

162. John Stuart Mill, "On the Definition of Political Economy; and on the Method of Investigation Proper to It," *CW* 4:324–25.
163. John Stuart Mill, *System of Logic, CW* 7:252.
164. John Stuart Mill, "On the Definition of Political Economy; and on the Method of Investigation Proper to It," *CW* 4:312.
165. Ibid.
166. Ibid., *CW* 4; see p. 325.
167. Ibid., 4:331–32.
168. Ibid., 4:322–33. See Abraham Hirsch, "John Stuart Mill on Verification and the Business
of Science," pp. 847–48.

lished papers, had addressed the issue of poor relief. He gave a utilitarian motive for labor: men endure the physical pain of labor only to satisfy their desire for the pleasures of food and shelter. (Indeed, he argued that it was necessary that there would always be poverty, in order that laborers have the sanction of hunger to make them work.)[169] Laborers naturally seek the "most eligible" condition, that is, the one in which they can receive the most pleasure with the least amount of pain. If they could receive the same aid in poor relief without working as they could earn as independent laborers, then clearly (according to Bentham) they would choose not to work. It was thereby necessary to render the condition of the relief-receiving pauper "less eligible" than that of the independent laborer. Bentham suggested that the pauper should wear a badge, which would render the "condition of the man of industry more eligible than that of the man of non-industry: it consequently tends to dispose men to embrace the former condition in preference to the latter."[170]

In his 1797 *Observations on the Poor Law Bill Introduced by the Right Honorable William Pitt*, Bentham argued that Pitt's Bill was a violation of the utilitarian principle. It reduced the incentive of labor by granting relief to the able-bodied idle; Bentham insisted that no man would work unless he had to, since work is painful. It granted outdoor relief, which Bentham claimed was wasteful; only in workhouses under the watchful eye of an administrator could subsistence be carefully measured in order to obtain the greatest amount of labor for the least amount of food. Moreover, Bentham argued against the bill's allowance of relief even for individuals possessing property valued up to thirty pounds. Instead, he suggested that the principle for relief should be, "Come in and give all up, or stay out and starve."[171] It was obvious as well to Bentham that conditions in workhouses needed to be worse than on the outside, or else all paupers would prefer to go in rather than work. Thus, before a Parliamentary Commission in 1811, he proposed large workhouses on his Panopticon model, which included separation of the sexes and the requirement of work, even for children. This plan was rejected, in part due to memories of the failure of the workhouse system in the eighteenth century. But Bentham's proposal was ultimately revived by the committee that recommended the New Poor Law in 1834, which included Bentham's former secretary, Edwin Chadwick.[172]

169. See Karl Polanyi, *The Great Transformation*, p. 121.

170. See Jeremy Bentham, *Badging the Poor*; quoted in J. R. Poynter, *Society and Pauperism*, p. 126.

171. Jeremy Bentham, *Collected Works*, 8:451; see Raymond Cowherd, *Political Economists and the English Poor Laws*, pp. 89–90.

172. See J. R. Poynter, *Society and Pauperism*, pp. 133–42.

Following Bentham, Mill explained that "pauper labour anywhere but in the workhouse is merely a particular kind of idleness. A person who is sure of employment whether his labour be efficient or only nominal, will make no exertion that he can possibly avoid."[173] Although he had criticized Bentham in his 1833 essay for having an overly simplistic view of man's motives for acting, Mill agreed that in the science of legislation, it was appropriate to deduce principles of law from the utility principle.[174] Thus he noted, "Relief must be given only in exchange for labour, and labour at least as irksome and severe as that of the least fortunate among independent labourers."[175] He later rejected outdoor relief for the Irish during the Potato Famine, claiming that it would "bid fair to involve the whole labouring population with the feelings of sturdy beggars."[176]

Like the utilitarian reformers, Mill accepted Malthus's population principle. (Indeed, in the *Logic* he uses arguments against Malthus's principle as examples of the *ignoratio elenchi*, the fallacies of confusion.)[177] This contributed to his view that the New Poor Law was necessary in order to keep the population from expanding beyond the food supply. However, he and the utilitarian radicals departed from Malthus in believing that the negative consequences of the population principle could be avoided. For Malthus, the population principle leads inevitably to the consequence that a large portion of society is doomed to suffer. It is relevant to recall here that his *Essay on the Principle of Population* is an anti-Utopian tract; Malthus argued, against Godwin and Condorcet, that it is impossible for human society ever to reach a stage in which there was no more poverty and misery.[178] For Mill and the utilitarians, on the other hand, the poverty caused by the results of the pop-

<hr/>

173. John Stuart Mill, "Walter on the Poor Law Amendment Bill," *CW* 23:712.

174. Mill praised Bentham for being "the first who attempted regularly to deduce all the secondary and intermediate principles of law, by direct and systematic inference from the one great axiom or principle of general utility" ("Remarks on Bentham," *CW* 10:10).

175. John Stuart Mill, "The Proposed Reform of the Poor Laws," p. 361. Elsewhere, in support of the measure, he noted that "it affords the means by which society may guarantee a subsistence to every one of its members, without producing any of the fatal consequences to their industry and prudence" that had been produced by the outdoor parish relief system ("Lord Brougham's Speech on the Poor Law Amendment Bill," p. 597).

176. John Stuart Mill, "The Condition of Ireland," *CW* 24:972. After the passage of the New Poor Law, the alliance between middle- and working-class radicals broke down, as the latter believed themselves to have been betrayed by the the former, who had supported the "starvation law." This created the impetus for the Chartist movement. See N. Kirk, *Labour and Society in Britain and the USA*, p. 121.

177. See John Stuart Mill, *System of Logic, CW* 8:829.

178. Indeed, for Malthus the population principle functioned as a final cause: the pressure brought to bear on individuals by the limit of resources compared to population was a spur to exertion and virtue, and thus a gift of Providence.

ulation principle was a problem that could be solved.[179] In the *Autobiography*, Mill explained their use of Malthus's population principle: "This great doctrine, originally brought forward as an argument against the indefinite improvability of human affairs, we took up with ardent zeal in the contrary sense, as indicating the sole means of realizing that improvability by securing full employment at high wages to the whole labouring population through a voluntary restriction in the increase of their numbers."[180] Following the lead of Francis Place, Mill advocated for the use of birth control within marriage to bring about this "voluntary restriction" of the numbers of the poor. (In 1823, he was arrested and served two days in jail for distributing a pamphlet by Place endorsing the use of contraceptives by married women.)[181] Malthus, an ordained minister, always saw his task from an explicitly Christian perspective; because of this, he did not consider contraception as a possible means of "moral restraint."[182] This is one of the reasons that he could never be very optimistic about the possibilities for the success of moral restraint. The other is that he did not believe that the poor could ever learn to give up instant gratification for the long-term benefits. Mill departed from Malthus in his confidence in the modifiability of the poor if their standard of living and their education were improved.[183]

Mill's next major consideration of political economy occurred in book 6 of the *System of Logic*, "The Logic of the Moral Sciences." The section on political economy is often described as a complete development of the method Mill already had developed for political economy in the earlier essay "On the Definition of Political Economy."[184] This reading is supported by the facts that in 1844 he republished the essay, basically unchanged, in his collection

179. See T. Sowell, "Malthus and the Utilitarians," p. 211.

180. John Stuart Mill, *Autobiography*, CW 1:107.

181. See J. R. Poynter, *Society and Pauperism*, p. 266. The utilitarian reformers were criticized for this solution by other radicals; thus much of Mill's early journalism was devoted to convincing other radicals that birth control was not "a device of the rich to oppress the poor," but rather a means of empowering laborers to force a higher wage from their wealthy employers. See Donald Winch, *Riches and Poverty*, p. 283, and CW 22:80–100. On the story of Mill's arrest, see Nicholas Capaldi, *John Stuart Mill*, p. 41.

182. See Donald Winch, *Riches and Poverty*, p. 241. In fact, Malthus had two reasons for condemning birth control as a means to moral restraint, as he made clear in his 1817 edition of the *Essay:* "I should always particularly reprobate any artificial and unnatural modes of checking population, both on account of their immorality and their tendency to remove a necessary stimulus to industry. If it were possible for each married couple to limit by a wish the number of their children, there is certainly reason to fear that the indolence of the human race would be very greatly increased" (appendix to *An Essay on the Principle of Population*, 1817 ed., pp. 368–69).

183. See Leslie Stephen, *The English Utilitarians*, vol. 3, *John Stuart Mill*, p. 181, and Stefan Collini et al., *That Noble Science of Politics*, pp. 71–72.

184. See, for example, Oskar Kubitz, *The Development of John Stuart Mill's "System of Logic,"* p. 242.

Essays on Some Unsettled Questions of Political Economy, and that he
quoted extensively from the essay in his discussion of political economy in
the *Logic.* Nevertheless, there are two important differences in the position
Mill presented in the *Logic.* First of all, he rejected the geometrical method
of Bentham, at least as a permanent methodology. Second, he added an a posteriori verification step to his methodology in political economy.

In the *Logic,* Mill noted that there are two incorrect modes of social science: the chemical, or experimental, method, and the abstract, or geometrical, mode.[185] The chemical method corresponds to the Baconian method used
by Coleridge and his school, and was criticized by Mill in his 1840 essay on
Coleridge. It is also, Mill explained in the *Autobiography,* the mistaken
method used by Macaulay when arguing against Mill's father.[186] In the *Logic,*
he scathingly pointed out that "the vulgar notion, that the safe methods on
political subjects are those of the Baconian induction—that the true guide is
not general reasoning, but specific experience—will one day be quoted as
among the most unequivocal marks of a low state of the speculative faculties
in any age in which it is accredited."[187] The chemical method is incorrect
because in the social sciences, experiments are not possible.[188] Indeed, even
chemistry itself is now a deductive (that is, nonexperimental) science; Mill
claimed, "Bacon's conception of scientific inquiry has done its work and . . .
science has now advanced into a higher stage."[189] But, he quickly added, the
geometrical or abstract method—favored by his father and Bentham—is also
faulty. Those who endorse this view are correct in characterizing social science as *deductive,* but they wrongly assimilate deductive science with geometry rather than with astronomy and natural philosophy.[190] The geometrical
model is incorrect because geometry does not allow for taking account of
counteracting causes.[191] And indeed, Mill explained, those who have used
this method have built all upon an overly narrow notion of the causes at work
in political society: Hobbes made all depend on fear; the Benthamites, on
self-interest.[192] But neither of these views takes account of other factors at
work in motivating human actions. "The phenomena of society do not
depend, in essentials, on some *one* agency or law of human nature," Mill

185. John Stuart Mill, *System of Logic, CW* 8:878.
186. See John Stuart Mill, *Autobiography, CW* 1:96.
187. John Stuart Mill, *System of Logic, CW* 7:452.
188. Ibid., 8:882; see also John Stuart Mill, "On the Definition of Political Economy; and on
the Method of Investigation Proper to It," *CW* 4:328.
189. John Stuart Mill, *System of Logic, CW* 8:886.
190. Ibid., 8:887.
191. Ibid., 8:887–88.
192. Ibid., 8:889–90.

insisted.[193] It is necessary to take account of a fuller notion of human motivation.

Mill countered that the proper method for the social sciences is the concrete deductive method.[194] As he had pointed out in book 3, and as we have seen in chapter 2, this method is necessary for any inquiry in which there is a composition of causes, that is, when there are two or more causes interacting and interfering with each other to produce a single effect.[195] Recall that the deductive method proceeds by three steps: induction, deduction, verification. Thus in the *Logic*, in contrast with his discussion in his earlier essay "On the Definition of Political Economy," Mill clearly included a step of verification as part of the science of political economy. Moreover, unlike in the earlier essay, he was attempting to build a bridge between the inductive (chemical) methodology of Coleridge and the deductive (geometrical) methodology of Bentham; this is the project he announced in the 1838 and 1840 essays on Bentham and Coleridge. The success at this project would enable Mill to claim that he had "possessed the philosophy of the age." Unfortunately, however, as we will see, this project ultimately failed, at least in the case of the social sciences.

The inductive step of the deductive method consists in the application of the methods of experimental inquiry (that is, the methods of agreement, difference, residues, concomitant variation, and the joint method of agreement and difference), by which we attempt, in the case of the social sciences, to discover the laws of human action that combine to produce behavior in a particular case. In the next step, that of ratiocination, it is necessary to calculate deductively the effect of the combination of causes discovered in the first step. In the third step, this calculated effect must be verified by being found to accord with the results of direct observation.

Mill noted that in political economy and all social sciences, verification can only be *indirect*.[196] This is because in the social sciences—unlike in the physical sciences—we cannot make predictions; there are too many possible counteracting causes.[197] (Here Mill was in accord with Ricardo, who had similarly claimed that political economy could not be a predictive science be-

193. Ibid., 8:894.

194. Mill explicitly denied that in political economy, the inverse deductive method described by Comte is useful; that is, he rejected the claim that social science consists in "generalizations from history, verified, not originally suggested, by deduction from the laws of human nature" (ibid., 8:897). (Yet he admitted that this inverse or historical method is useful for general sociological studies, and the science of history; see pp. 911ff).

195. Ibid., 7:439.
196. Ibid., 8:909.
197. Ibid., 8:898.

cause of the inevitable intervention of numerous "disturbing causes.") More-over, because men are free to change their character and hence their motivations for acting at any time, the laws of human behavior can only be "laws of tendencies." Thus Mill characterized the social sciences as "sciences of tendencies." Because predictions are not possible, the conclusions of the first two steps of the deductive method in social science must be verified by "retrodictions"; that is, we must see that our deductions "would have enabled us to predict the present and the past."[198] For example, if we wish to verify our deductions in political economy regarding a certain country, "we must be able to explain all the mercantile or industrial facts of a general character, appertaining to the present state of that country: to point out the causes sufficient to account for all of them, and prove, or show good ground for supposing, that these causes have really existed." If there remains anything we could not have predicted, this is a "residual phenomena" requiring further study.[199]

But there is yet another difference between the application of the deductive method in the social sciences and in the physical sciences. In book 3 of the *Logic*, Mill had admitted that "in the cases, unfortunately very numerous and important, in which the causes do not suffer themselves to be separated and observed apart, there is much difficulty in laying down with due certainty the inductive foundation necessary to support the deductive method."[200] In book 6, it becomes clear that the social sciences generally consist in cases in which the numerous interacting causes cannot be separated into the "simple instances" required for the deductive method. In the social sciences, causes are *always* compounded.[201] Men have many motivations for their actions, and generally more than one is at work at any time. Further, it is not possible to "separat[e] and observ[e] apart" these multiple motivations, precisely because, as Mill had earlier claimed, the social sciences are *not* experimental sciences.

What, then, is the political economist to do? In providing an alternative means of proceeding, Mill declared that there is

---

198. Ibid., 8:910.
199. Ibid.
200. Ibid., 7:456.
201. Ibid., 8:896. Mill attempted to escape this conclusion by noting that there is an "appropriate remedy" for the difficulty in separating out individual causes: namely, the verification step, which can save us from faulty deductions. As we saw in chapters 2 and 3, Mill also at times suggested this kind of remedy for empirical hypotheses regarding unobservable entities or causes; but yet, as we also saw there, this contradicts what he claimed about the inability of deductive verification to prove the truth of hypotheses not derived inductively. There would appear to be a similar problem here.

one large class of social phenomena, in which the immediately determining causes are principally those which act through the desire of
wealth; and in which the psychological law mainly concerned is the
familiar one, that a greater gain is preferred to a smaller. . . . . By reasoning from that one law of human nature, and from the particular outward
circumstances . . . which operate upon the human mind through that law,
we may be enabled to explain and predict this portion of the phenomena
of society, so far as they depend on that class of circumstances only.[202]

Mill claimed, in other words, that the political economist must assume that
man has only a single motivation for his economic actions, namely the preference for a greater gain over a smaller one. Thus, as he had asserted in the
earlier essay "On the Definition of Political Economy," Mill held that the
political economist reasons from the "assumed" and "arbitrary" hypothesis
of the purely rational economic man. Thus political economy is a "hypothetical or abstract science," which begins from an assumed hypothesis and
then deduces the consequences that follow from it in particular circumstances.[203] Its conclusions, therefore, are also hypothetical, in that "they are
grounded on some supposition set of circumstances, and declare how some
given cause would operate in those circumstances, supposing that no others
were combined with them."[204]

In the end, then, Mill characterized political economy as the geometrical
science of his father and Bentham, without successfully uniting to it the
inductive element of the Coleridgian method. Ironically, the realization of
man's many motivations for acting and the alterability of his character caused
Mill to retain the deductive methodology of social science for which he had
criticized his father and Bentham in the 1838 and 1840 essays on Bentham
and Coleridge. Because of the complexity of the laws of human behavior, he
was not optimistic that the individual causes could be separated so as to
apply his methods of experimental inquiry. However, he recognized that
because of this, political economy could only be a hypothetical science.
Moreover, Mill suggested that one day it might be possible for political economists to use the joint inductive-deductive method; he claimed that in these
complex cases in which the individual causes cannot be separated, we need
to wait for a "later period of inquiry" before "the consideration of the [other]

202. Ibid., 8:901.
203. Ibid., 8:904. In his 1867 "Inaugural Address to the University of St. Andrews," Mill similarly noted that "all true political science is . . . *à priori*" (*CW* 21:237).
204. John Stuart Mill, *System of Logic*, *CW* 8:900.

causes which act through them, or in concurrence with them." Thus Mill
hinted that the present hypothetical, geometrical nature of political economy was a temporary result of our current ignorance, and could be overcome
in the future. For this reason he concluded his discussion of political economy in the *Logic* by calling it a "preliminary science."[205]

I turn now to Mill's most widely read work on political economy, the
*Principles of Political Economy with Some of Their Applications to Social
Philosophy*, which was published in 1848. It was extremely successful: the
first print run of 1,000 copies sold out in less than a year, followed by another
2,250 by early 1852. A later inexpensive "People's edition" sold an impressive
10,000 copies. Mill was called upon to give evidence to five Parliamentary
Select Committees between 1850 and 1861, and his election to Parliament in
1865 was in part due to his reputation in political economy. It has been said
that after he published the *Principles*, "English economists, for one generation, were men of one book."[206] For this reason, in March of 1868, *Punch* satirized Mill as the "Father of Political Economy" holding, a bit awkwardly, the
infant discipline.

Principles of Political Economy has struck many commentators as dia-
metrically opposed to Mill's other writings on the subject. In fact, there is a
difference: in this work Mill's concern was not with *method* in political economy, but rather (as the subtitle indicates) with *applications* of the laws of
political economy to practical questions. He was interested here in what he
had earlier called the "art," rather than the "science," of political economy.
He admitted that he was not particularly concerned with methodology in
this book; as he explained to his friend Henry Chapman, the *Principles*,
"while embodying all the abstract science in the completest form yet attained," was to be "essentially a book of applications exhibiting the principles
of the science in the concrete."[207] Moreover, in the work he expressed his
growing concern with improving not only the economic and political state of
society, but also the moral and intellectual state—that is, creating the conditions for the type of cultivated society he later argued for in "Utilitarianism" (which he began writing shortly after publishing the *Principles*). As he
wrote to Edward Herford, "At present I expect very little from any plans which
aim at improving even the economical state of the people by purely economic
or political means. We have come, I think, to a period, when progress, even of
a political kind, is coming to a halt, by reason of the low intellectual and

205. See ibid., 8:905.
206. See Neil DeMarchi, "The Success of Mill's *Principles*," p. 119.
207. John Stuart Mill to Henry Chapman, March 9, 1847, *CW* 13:708–9.

FIGURE 6. "Mill, the Father of Political Economy."
Cartoon from *Punch*, March 21, 1868. Author's
collection.

moral state of all classes."[208] In the *Autobiography*, Mill noted the turn that
his views on political economy began to take in the mid-1840s:

> In [my early days] I had seen little further than the old school of political
> economists into the possibilities of fundamental improvements in social
> arrangements. Private property . . . and inheritance, appeared to me as to
> them, the *dernier mot* of legislation: and I looked no further than to mit-
> igating the inequalities consequent on these institutions. . . . The notion
> that it was possible to go further than this in removing the injustice . . .
> involved in the fact that some are born to riches and the vast majority to
> poverty, I then reckoned chimerical.[209]

208. John Stuart Mill to Edward Herford, January 22, 1850, *CW* 14:45.
209. John Stuart Mill, *Autobiography*, *CW* 1:239.

He claimed that by the mid-1840s, he and Harriet Taylor had become optimistic about the possibility of reforming society in a more radical way. He described himself and Taylor at this time as being not democrats but rather socialists, with the caveat that "we repudiated with the greatest energy that tyranny of society over the individual which most Socialist systems are supposed to involve." The way to make a socialist system succeed, Mill explained, was to cause the laboring classes as well as their employers to work for the greatest good of the whole society, rather than their own narrowly focused interest. This result could be achieved by "education, habit, and the cultivation of the sentiments." That human nature is alterable by education and this kind of self-cultivation led the Mills to the belief that the current economic, political, and social system was "merely provisional."[210]

Thus Mill's primary concern in the *Principles* was not with the laws governing the production of wealth, but with practical questions regarding what a society should do with its wealth, how wealth should be distributed among the different classes. An insight he developed from his reading of Auguste Comte was that laws of distribution are not immutable laws in the sense that the laws of production are; rather, they are determined partly by "the laws and customs of societies."[211] That is, a society may decide which system of distribution it prefers. In order to make this decision, a society must consider what type of culture, with what kind of citizens, it wants to create. For this reason Mill noted in his preface to the *Principles* that he was taking Adam Smith as his model. The work of Smith is important to follow because he recognized that, as Mill put it, "there are perhaps no practical questions . . . which admit of being decided on economic premises alone," and therefore "Political Economy is inseparably intertwined with many other branches of social philosophy."[212] Mill explained that in order to write a book concerning practical questions in political economy, it was necessary for him to engage in a broader task encompassing more than just political economy narrowly defined. He was, then, taking on a project more like that of Malthus (and Jones) than that of his father and Ricardo.[213] Mill admitted this shift in his interest, claiming

---

210. Ibid., 1:241. Mill criticized the Benthamites for not seeing the mutability of social arrangements as early as his "Miss Martineau's Summary of Political Economy" in 1834 (see *CW* 4:225).

211. See John Stuart Mill, *Principles of Political Economy, CW* 2:21, 199; and Mark Blaug, *Ricardian Economics*, pp. 165–66.

212. John Stuart Mill, *Principles of Political Economy, CW* 2:xci.

213. John Neville Keynes points out that one contrast between Mill's *Logic* and the *Principles* is that in the latter, he gives an "ethical," as opposed to "abstract," account of economic problems (*The Scope and Method of Political Economy*, pp. 19–20).

to find the "abstract investigations" of political economy to be of little import relative to the "great practical questions" of the day.[214]

Indeed, in the *Principles*, Mill concentrated much of his effort on discussing the type of moral, social, and intellectual transformation of individuals that was necessary in order for society to be improved. For example, he noted that "in England, it is not the desire for wealth that needs to be taught, but the use of wealth, and appreciation of the objects of desire which wealth cannot purchase. . . . Every real improvement in the character of the English . . . must necessarily moderate the ardour of their devotion to the pursuit of wealth."[215] In contradistinction to earlier writers on political economy, he did not fear the arrival of the "stationary state," in which there is no further increase in capital and wealth. This condition was actually desirable, as it would allow for a greater amount of "human improvement." Once a stationary state had occurred, the energies of mankind would no longer be devoted solely to the "struggle for riches." Instead, these energies could focus on a more equitable distribution of wealth as well as an improved culture. The desirable outcome Mill envisioned was one in which there would be two classes: "a well-paid and affluent body of labourers" and "a much larger body of persons than at present, not only exempt from the coarser toils, but with sufficient leisure, both physical and mental, from mechanical details, to cultivate freely the graces of life."[216] He explained that in the stationary state, in which "minds ceased to be engrossed by the art of getting on," there would be much room "for all kinds of mental culture, and moral and social progress."[217] While Ricardo and the Benthamite radicals believed that civilization would cease to advance once a stationary state occurred, Mill held that a stationary state would be conducive to the advance of civilization, for it would allow the type of cultivation of minds that real progress required.[218]

In his chapter "On the Probable Future of the Labouring Classes," Mill considered the moral, political, social, and intellectual consequences of treating the laborer as an independent, educated person in his own right. On these grounds he discussed the proper relation that ought to exist between the laboring classes and their employers, concluding that it should not be one of dependence, but rather independence, or "self-dependence." The poorer classes, in at least the more modern parts of the nation, have had enough edu-

214. John Stuart Mill to Karl D. Heinrich, March 20, 1852, *CW* 14:87.
215. John Stuart Mill, *Principles of Political Economy, CW* 2:105.
216. Ibid., 3:755.
217. Ibid., 3:756.
218. On this point see Jonathan Riley, "Mill's Political Economy," p. 314.

cation to think for themselves.[219] They can therefore no longer be led by their employers the way children are—and must be—led by their parents. Unlike the "backward societies" that Mill exempted from the requirements of liberty in his *On Liberty*, "the poor have come out of leading-strings, and cannot any longer be governed or treated like children."[220] Yet though they are ready for the "virtues of independence," they still must attain them; and so the laboring poor require education and "cultivation" of the kind Mill later discussed in "Utilitarianism."[221] Further, in this chapter he noted the importance of the social independence of women, claiming that the inequality of women "must ere long be recognized as the greatest hindrance to moral, social, and even intellectual improvement."[222] He linked the independence of women with the diminution of the "evil of over-population," but this is in addition to the moral, social, and intellectual consequences noted first.[223]

In his discussions, Mill stressed the provisional nature of the current institutions, and indicated the need to take into account the possibility for greater moral and intellectual cultivation in the future. For example, in his chapter "Of Property," he claimed that in discussing the respective merits of capitalism, communism, and other forms of socialism, it is necessary to consider "our present imperfect state of moral and intellectual cultivation."[224] Arguments against the impracticality of communism based on our current motives and habits are not adequate arguments against communism as a system that might be successfully applied in the future. Mill noted that a common argument against communism was that it could not work, because men are motivated only by their personal interest, and communism dampens or destroys this motivation. But he pointed out, "Mankind are capable of a far greater amount of public spirit than the present age is accustomed to suppose possible." This public spirit—another way of calling the desire for the general good—could be cultivated by education and public opinion. Thus Mill noted that "history bears witness to the success with which large bodies of human beings may be trained to feel the public interest their own."[225] The present

219. John Stuart Mill, *Principles of Political Economy*, CW 3:762.
220. Ibid., 3:763.
221. See his discussion in ibid., 3:763–64.
222. Ibid., 3:765.
223. Ibid., 3:766.
224. Ibid., 2:213.
225. Ibid., 2:205. In this passage, Mill highlighted the power of public opinion to train the motives of individuals: he called public opinion "the most universal, and one of the strongest, of personal motives," noting "the force of this motive in deterring from any act or omission positively reproved by the community" as well as "the power also of emulation, in exciting to the most strenuous exertions for the sake of the approbation and admiration of others" (ibid).

unfitness of the laboring classes for communism or other forms of socialism did not render these systems inapplicable for all time.[226] It is likely that in a cultivated state of moral and intellectual ability, both the private property system and communism could work well; thus the decision between them must ultimately rest on which system is "consistent with the greatest amount of human liberty and spontaneity," because "after the means of subsistence are assured, the next in strength of the personal wants of human beings is liberty."[227] Moreover, the question is not whether communism will provide more "personal and mental freedom" than people in the lowest ranks of society have at the present time, but whether it will provide more than such people could have in an *improved* capitalist society.[228]

Mill's attitude toward socialism became more positive between the first and third editions of the *Principles*.[229] This had mainly to do with his increased confidence in the ability of socialist systems to allow for the proper kind of development of individuals, the "complete renovation of the human mind" that he sought. Indeed, he came to believe that this regeneration could be achieved more fully in a socialist rather than a capitalist society. Inklings of this can be seen in opinions he expressed on the "laissez-faire" principle throughout his career.

In the concluding chapter of the *Principles*, Mill argued that the "let alone" doctrine should be the general rule, because "a people among whom there is no habit of spontaneous action for a collective interest . . . have their faculties only half-developed."[230] If the state decides everything, then individuals cannot develop themselves. Yet there are important exceptions. In the essay on Coleridge, Mill had claimed that in general, Coleridge was an "errant driveller" when he wrote on political economy; nevertheless, he had praised Coleridge for rejecting the laissez-faire doctrine in favor of a view of the importance of the role of the state in, among other things, "the development of those faculties which are essential to [man's] humanity, that is to his

---

226. See also Mill's preface to the third edition of the *Principles* of 1852, *CW* 2:xcii.

227. Ibid., 2:208.

228. Ibid., 2:209.

229. In his "Chapters on Socialism," which he was working on at the time of his death, and which was published in 1879, Mill noted that the question of socialism versus capitalism could not be decided a priori, but required experiment: there must be time "for the question to work itself out on an experimental scale, by actual trial ("Chapters on Socialism," *CW* 5:736). He ultimately rejected state socialism, in which the state owns the means of production, but approved of those forms of socialism that depend on voluntary association in small communities. Marx was little known in England up to the time of Mill's death, and Mill does not seem to have read any of the writings of either Marx or Engels (see John M. Robson, *The Improvement of Mankind*, pp. 275–76).

230. John Stuart Mill, *Principles of Political Economy, CW* 3:943.

rational and moral being."[231] In the *Principles*, Mill clarified his own view of the important role the state must play in creating the cultivated mind. "The uncultivated," he explained, "cannot be competent judges of cultivation. Those who most need to be made wiser and better, usually desire it the least, and if they desired it, would be incapable of finding their way to it by their own lights."[232] Government, therefore, can take an active role in promoting the proper kind of cultivation required for the renovation of the human mind. It seems that Mill came to believe that socialist systems were better equipped than were purely capitalist ones in allowing the kind of active governmental role in the cultivation process that he hoped would lead to a reformed society. His earlier concerns about the loss of individualism in socialist societies were, by this time, trumped by his interest in individuals becoming cultivated in the proper way. In his late essay "Centralization" (1862), Mill noted that the amount of government intervention allowable depends on the level of cultivation a society has attained; it varies "not only with the wants of every country and age, and the capabilities of every people, but with the requirements of every kind of work to be done."[233] At certain stages in the development of the cultivated society, socialism would be the preferable system.[234] Mill came to believe that his own time was one that would benefit from such a system.

We have seen that in his *Principles of Political Economy*, Mill was concerned with the type of reform of society—through the moral and intellectual cultivation of its citizens—that he endorsed by the 1840s. However, neither his interest in the distribution of wealth, nor his consequent attention to moral, social, and political issues as well as economic ones, indicates that he had abandoned the deductive methodology in the science of political economy. He continued to believe that this science was deductive. Indeed, while there was no deductive way to choose the desired system of wealth distribution, the *consequences* of any chosen distribution system followed as

231. See John Stuart Mill, "Coleridge," *CW* 10:155–56; the quotation is from Coleridge's *Second Lay Sermon*. Mill also praised Coleridge for his insistence on "the idea of a *trust* inherent in landed property," i.e., the rejection of the Lockean notion of property as absolute proprietorship of land apart from any produce of labor (p. 157).

232. John Stuart Mill, *Principles of Political Economy, CW* 3:947.

233. John Stuart Mill, "Centralization," *CW* 19:581.

234. Similarly, in "A Few Words on Non-Intervention," Mill claimed that strict *laissez-faire* policies are also inappropriate in international affairs, because nations, like individuals, "have duties . . . towards the weal of the human race." The question of intervention is "a really moral question." A civilized nation has the moral duty to help other nations become more civilized ("A Few Words on Non-Intervention," *CW* 21; quotations from pp. 116 and 118). See also Stefan Collini, introduction, *CW* 21:xxix.

certainly as the effects of physical laws.[235] These consequences were based, he believed, on the axiomatic principles of Ricardo and the laws of production deduced from them.

Thus Mill's art of political economy rests on a foundation of deductive science. As he wrote to John Austin in February 1848, "I doubt if there will be a single opinion (on pure political economy) in [my book] which may not be exhibited as a corollary from [Ricardo's] doctrines."[236] This is why Mill never altered the sections of the *Logic* in which he claimed that political economy is an a priori, hypothetical science, even though there were five more editions after the first appearance of the *Principles*, and even though he made numerous other types of changes to those editions (especially to the one appearing in 1851, in which he responded to Whewell's recently published criticisms of his view of physical science). In his *Autobiography*, Mill continued to refer to political economy as an "abstract science," noting that his *Principles* had the success it did because it was a book on both the abstract science and its applications.[237] So the difference between the *System of Logic* and the *Principles of Political Economy* is that in the later work, Mill emphasized, as he had not in the earlier one, his belief that this hypothetical science could be the foundation for an art of political science that would help bring about the reforms of society he sought.[238]

Responses to Mill

Whewell and Jones, not surprisingly, were highly critical of the deductive view of political economy expressed in the *Logic*. After reading it, Jones wrote to Whewell about Mill's view of social science:

235. See John Stuart Mill, *Principles of Political Economy*, CW 2:21, 199–200 and CW 3:754; and Neil DeMarchi, "The Success of Mill's *Principles*," p. 137.

236. John Stuart Mill to John Austin, February 22, 1848, CW 13:731.

237. John Stuart Mill, *Autobiography*, CW 1:243–45.

238. Mark Blaug claims that Mill was able to remain "a faithful advocate of Ricardian economics" by embracing various "immunizing strategems," including the liberal use of *ceteris paribus* clauses empty of specific content. Moreover, in each successive edition of the *Principles* Mill extended the short-run period required for the effects of technical progress in agriculture to work themselves out beyond Ricardo's period of twenty-five years, so that Ricardo's law would not be disconfirmed by the evidence. See Blaug, *The Methodology of Economics*, p. 76, and Neil DeMarchi, "The Empirical Content and Longevity of Ricardian Economics," p. 66. Even in defending Mill against Blaug's criticism, DeMarchi admits, "It cannot be said that Mill always attempted to test his theory against the facts. . . . Mill was sometimes willing to live with a gap between his deductive theory and the facts" (ibid., p. 64).

Practically he prefers I presume the smallest possible quantity of induction and the greatest possible of ratiocination—It would serve him right
to take some of the social science in the probable progress of which he dis-
cards induction and shew where ratiocination led in other days his Papa
and himself—How moderate an induction would have been their salva-
tion and how little when reasoning had led them by the nose into a slough
they were in any plight to save themselves by a verification of facts.[239]

Jones seems to be referring to the support of the utilitarian radicals for the
workhouse provision of the New Poor Law, which by the time of this comment was already coming under attack for the terrible conditions in which
the inhabitants of workhouses suffered. The utilitarians, Jones asserted, had
been led to support this provision because of their "miserable logic." The next
year, when Mill's *Essays on Some Unsettled Questions of Political Economy*
appeared, Jones further complained, "Young Mill has been publishing a paper
to prove that a priori reasoning is not only good in Pol. Econ. [*sic*] but the *only*
reasoning applicable to it—God help him and those this belief leads to trust
in him[,] his Papa and his school."[240]

However, when Mill published his *Principles of Political Economy* in
1848, neither Whewell nor Jones criticized it as harshly as they had the earlier works. Whewell wrote to Jones, "Of course you will soon look at Mill's
book on *Political Economy*. It is full of interesting discussions on all the great
social and economic questions of the day, and there are arguments and views
extremely well put throughout."[241] In his 1862 *Lectures on Political Economy*, Whewell quoted copiously from Mill's book, praising its analysis of several key points. Although Mill continued to support a deductive, rather than
inductive, methodology for political economy, his *Principles* contained much
with which Whewell and Jones could agree.

First of all, Mill had accepted Jones's categorization of the different forms
that rent takes throughout the globe.[242] (Whewell still complained that Mill

239. Richard Jones to William Whewell, April 15, 1843, WP Add.Ms.c.52 f. 80.

240. Richard Jones to William Whewell, May 10, 1844, WP Add.Ms.c.52 f. 92.

241. William Whewell to Richard Jones, April 30, 1848, in Isaac Todhunter, *William Whewell,
D.D.*, 2:345–46. This is not to suggest that Whewell agreed with everything in the book; he criticized Mill's Ricardian positions, such as the conclusion that "the proportional produce of land
necessarily decreases with the extension of agriculture," complaining, "how can we say, with any
sound sense or use, that the produce of land increases universally in a diminishing ratio, when we
have to allow that there is a principle, which we call 'the progress of civilization,' skill and the like,
which may prevent this diminishing ratio for centuries, and during the whole life of a nation?"
(p. 346)

242. Neil DeMarchi describes this as one example of Mill's "conciliatory approach" in the
Principles: "Mill expounded Ricardo's law of distribution, while accepting Richard Jones's obser-

did not give Jones enough credit for this scheme.)[243] Moreover, his *Principles* included some historical detail of the kind that Jones had urged political economists to consider.[244] For example, in his discussion of peasant proprietorship, Mill used evidence of peasant properties in Switzerland, Norway, Germany, Belgium, the Channel Islands, and France.[245] He elsewhere adduced empirical evidence for the Malthusian population principle, pointing specifically to the cases of France after the Revolution and England during 1715–65.[246] In a discussion of Ricardo on wages, Mill also indicated the importance of taking note of historical circumstances. "The conclusion Mr. Ricardo draws from [his deductions], namely, that wages in the long run rise and fall with the permanent price of food, is, like almost all his conclusions, true hypothetically, that is, granting the suppositions from which he sets out. But in the application to practice, it is necessary to consider the minimum of which he speaks [that is, the minimum wages necessary for subsistence and population replacement] . . . is itself liable to vary." He claimed that the history of English peasantry shows that this minimum has varied over time.[247]

Further, although Mill reiterated his support for the 1834 Poor Law, much of what he wrote in the *Principles* about poor relief is consistent with Whewell's position in the *Elements of Morality, Including Polity* (which had been published three years before the first edition of the *Principles*). In the *Principles*, Mill, like Whewell earlier, focused on the issue of educating the poor, not the workhouse requirement. Indeed, he hardly mentions the workhouse at all. This may be because in the mid-1840s, numerous abuses of the workhouse system had been made public. For example, the Andover workhouse scandal of 1845—in which starving inmates were found fighting over the putrid gristle and marrow of animal bones they were engaged in crushing—had led to the formation of a New Poor Law Board in 1847. Within five

vations on different forms of peasant rent as useful facts, showing the limiting influence of custom on the universality—though not, within their own sphere, the correctness—of economic principles based upon the supposition of competition" ("The Success of Mill's *Principles*," p. 141; see also John Stuart Mill, *Principles of Political Economy*, CW 2:239–44, 247–48).

243. See William Whewell to Richard Jones, April 30, 1848, in Isaac Todhunter, *William Whewell, D.D.*, 2:345–46.

244. Henry Sidgwick, in his 1883 *Principles of Political Economy* (pp. 37–38n), pointed to some inductive passages in Mill's *Principles*. John Neville Keynes also noted that Mill used an inductive method, for instance in his discussion of peasant proprietorship (Keynes, *The Scope and Method of Political Economy*, pp. 17–18).

245. John Stuart Mill, *Principles of Political Economy*, CW 2:254–77.

246. Ibid., 2:342. Although, as Mark Blaug has noted, Mill ignored the fact that population had increased more slowly in Britain from 1815 to 1848 than it had from 1793 to 1815 (see Blaug, *The Methodology of Economics*, p. 73).

247. John Stuart Mill, *Principles of Political Economy*, CW 2:341.

years local guardians were again providing outdoor relief at their discretion.[248] Mill alluded to the workhouse requirement in the chapter "Popular Remedies for Low Wages," claiming that the guarantee of support was not injurious to the "minds and habits of the people," as long as "the relief, though ample in respect to necessities, was accompanied with conditions which they disliked, consisting of some restraints on their freedom, and the privation of some indulgences."[249] But again, he did not directly mention the workhouse requirement, concentrating instead on education as the means to lifting the poor out of their unhappy state. In dealing with the poor, Mill maintained, "unless either by their general improvement in intellectual or moral culture, or at least by raising their habitual standard of comfortable living, they can be taught to make a better use of favourable circumstances, nothing permanent can be done for them; the most promising schemes end only in having a more numerous, but not a happier, people."[250] So, it follows that "no remedies for low wages have the smallest chance of being efficacious, which do not operate on and through the minds and habits of the people."[251] Education has the best chance of changing minds and habits.[252]

Again, as we have seen before, Mill stressed the need for moral education in order to create cultivated minds, who would associate their own personal interests with the good of the whole. Public opinion could play an important role in this as well. Since, on Mill's view, the chief explanation of low wages was offered by the Malthusian population principle, public opinion should be used to discourage people from producing children they could not support. Thus he noted, "Little improvement can be expected in morality until the producing large families is regarded with the same feeling as drunkenness or any other physical excess."[253] Granting equal civil rights to women could help as well, because "it is seldom by the choice of the wife that families are too numerous."[254]

On the issue of charity, Mill also made arguments in the *Principles* similar to those made earlier by Whewell. Mill noted that the state must be responsible for guaranteeing subsistence to the destitute, even the able-bodied.[255] He adduced several reasons for this position, which marked a change from his earlier views. One is that the state provides subsistence for the criminal in

248. See R. A. Soloway, *Prelates and People*, p. 188.
249. John Stuart Mill, *Principles of Political Economy*, CW 2:360.
250. Ibid., 2:159.
251. Ibid., 2:366.
252. Ibid., 2:372.
253. Ibid., 2:368n.
254. Ibid.
255. Ibid., 3:962.

prison, so that "not to do the same for the poor who have not offended is to give a premium on crime." Moreover, the state cannot take on the role of "inquisitor" in determining who are among the "deserving" poor and who are not. Private charity can make these distinctions, however, and thus bestow more on the deserving poor, raising them even above the level of mere subsistence. Mill insisted that relief to the poor must be given in such a manner as to render the condition of the recipient less desirable than that of the independent laborer; however, in the *Principles* he gave a reason for this that does not stress the Benthamite principle that men will work as little as necessary for the same gain. Rather, as Whewell had done, he emphasized the *educational* value of this doctrine. If men can expect to be supported as well by not working as by working, they will not learn "energy and self-dependence."[256]

But I think the main reason Whewell and Jones were not terribly critical of Mill's final work on political economy is that Mill's focus on the art of political economy, and his recognition of political economy as a branch of social philosophy, brought him closer to the "ethical school" of political economy than he had been before. Thus it must have seemed to Whewell and Jones that they had won the war on some important fronts: though Mill still adhered to the deductive methodology, he accepted the broader scope of political economy that Whewell and Jones endorsed. Like Whewell, Jones, Malthus, and Smith, but unlike Ricardo, Mill agreed with the importance of admitting moral, social, and political considerations into his discussion of political economy. By noting that the system of wealth distribution a state prefers depends on the type of society it wishes to create, he allowed that the political economist should speak about the moral and social conditions that might be brought about by different economic systems. By the end of his career, Mill had come to believe, with Whewell and Jones, that the political economist must be concerned with reforming society at large.

256. Ibid., 3:961.

The Debate's Legacy

> Great men and writers outlive the ideas and most of the monuments of
> their time, and descend to posterity disjoined from the element in which
> they lived, and by which their thoughts ought to be interpreted. This is
> especially the case with great reformers.
> —*John Stuart Mill, "Grote's Plato," 1866*

Even Whewell's death in 1866 at the age of seventy-two (from injuries
sustained after being thrown from his horse) did not end his debate with
Mill: in 1865 Mill brought out an essay on Auguste Comte, which Whewell
criticized in his last paper, published soon after his death. By this time Mill
was sitting in Parliament and making the rounds as a much-sought-after
speaker on political topics, particularly woman's suffrage. When he did not
win a second term in the 1868 elections, he retired to Avignon, France, where
he spent most of the time in relative seclusion near Harriet Taylor's grave until
his death in 1873. Mill and his adversary had lived long lives, and had spent a
good portion of them pursuing the project of reforming philosophy.

Both men suffered a surprising fate (though one that Mill seemed to an-
ticipate): after having been lauded for so long as authorities in their multiple
fields of expertise, each was considered somewhat obsolete by the time he
died. Many of Whewell's obituary writers, while praising his "omniscience,"
indicated their skepticism that his reforms of philosophy would live on. They
frequently suggested that his greatest legacy might be the inspiration to follow his example by reforming philosophy with "zeal and vigor." His student
and early biographer Isaac Todhunter, for example, wrote of Whewell that "his
monument will be his memory: his universal knowledge, his readiness to
undergo any labour for the advancement of science, his punctual and zealous
discharge of all that he thus undertook, and his generous sympathy with the

FIGURE 7. Portrait of Mill by John and Charles Watkins, 1865. By permission of
National Portrait Gallery, London.

pursuits and successes of his friends."[1] Another writer suggested that Whewell's legacy would be similar to Francis Bacon's, noting, "Bacon has produced an effect upon scientific thought which no one would care to measure
by the amount of actual reading which his works receive."[2] As with Bacon's

1. Isaac Todhunter, *William Whewell, D.D.*, 1:416.
2. Harvey Carlisle, "William Whewell," p. 144.

FIGURE 8. Photograph of Whewell toward the end of his life by J. Rylands. By permission of the Master and Fellows of Trinity College, Cambridge.

books, it was not expected that Whewell's would actually be read after his death. Mill too was praised in his obituaries for his "zeal" more than his particular philosophical views. In one postmortem "sketch of his life," it was claimed that "he worked with the self-sacrificing zeal of an apostle . . . for the good of society," even though "he may have blundered and stumbled in his pursuit of truth."[3] The economist Alfred Marshall similarly noted—although in a private context—that "Mill was not a constructive genius of the first order . . . generally the most important benefits he has conferred on the science [of political economy] are due rather to his character than his intellect."[4] In the period immediately following his death, it was Mill's character, more than his philosophical system, that was most often commended.[5]

Why were Mill and Whewell eulogized in this rather ambivalent fashion? To a certain extent, each had—as Mill claimed is common for "great reformers"—outlived important parts of his philosophical system. For example, when Whewell died in 1866, Mill was one of the preeminent political economist of the day, and the criticisms of Ricardianism made decades earlier by Whewell and Jones had faded into the background. However, by the time of Mill's death in 1873, the economic system of Ricardo and Mill had come under attack on several fronts, not only for its methodology but also for many of its most central theoretical points. Mill's political economy was no longer the reigning system. Again, at the time of Whewell's death, his own epistemological and moral positions—which were seen by many as the intuitionist approach Mill had described them as being—seemed out of step with the dominant empiricist and utilitarian outlook of the time. (His philosophy was later enshrined in the *Dictionary of National Biography* as an "old-fashioned form of intuitionism.")[6] Mill appeared to have given intuitionism a death-blow the year before with his book on William Hamilton. Ironically, by the time Mill died, only seven years later, his own empiricist position was no longer so secure. The British tide was beginning to turn toward German ideal-

3. E. H. Fox Bourne et al., *John Stuart Mill*, pp. 25 and 26. John Morley's review of the *Autobiography* for the *Fortnightly Review* also focused on Mill's character rather than his intellectual achievements (see Morley, "Mr. Mill's Autobiography").

4. Alfred Marshall to W. S. Jevons, February 4, 1874, in R. D. C. Bleck, ed., *Papers and Correspondence of William Stanley Jevons* (London: Macmillan, 1977), 4:100; quoted in Jeff Lipkes, *Politics, Religion, and Classical Political Economy in Britain*, pp. 25–26.

5. Nonetheless, Mill's character was not sacrosanct. Abraham Hayward, for example, wrote a "sneeringly hostile" obituary for *The Times*, which was criticized from the pulpit by Stopford Brooke, a liberal cleric; in the controversy that ensued between them, Hayward issued a widely circulated letter in which he condemned Mill for his youthful activity in support of birth control as well as for his relationship with Harriet Taylor during her marriage to John Taylor. For details of this controversy, see Stefan Collini, "From Sectarian Radical to National Possession," p. 381.

6. See Leslie Stephen, "Whewell, William," p. 1371.

ism. This veering toward idealism was exhibited explicitly the following year, when T. H. Green published his extensive critique of empiricism in the introduction to the edition of Hume he edited with T. H. Grose—a criticism that strongly influenced the next generation of philosophers in Britain, as well as John Dewey and William James in the United States.[7]

In the twentieth century, both Mill and (to a lesser extent) Whewell were "rediscovered" as major figures of the nineteenth century. However, when their writings have been studied, they have often been—as Mill also believed is typical for the works of great reformers—"disjoined from the element in which they lived, and by which their thoughts ought to be interpreted." Over the course of this examination, I have shown that because of this, common ways of understanding the ideas of Mill and Whewell are flawed. By rejoining their works to their "element," as Mill put it, we can better comprehend not only the works themselves, but also the relevance these works may have for theorists working on similar problems today. This point is important, because one of the reasons for studying past figures such as Whewell and Mill is the value this kind of investigation can have for our current thinking about issues that remain contested. If we deny that there is any such worth, then we are consigning thinkers of the past to the museum of history, to be studied merely as quaint productions of a distant time. It has been an implicit theme of this book that neither Mill nor Whewell should be studied in this way. Both philosophers expressed views that can be useful to those of us interested in reforming society and science today.

For example, interpretations generally given to Mill's writings on politics and morality have not always been informed by readings of them embedded in their proper context. The typical view of Mill is that he was a proponent of individualistic liberalism, with its emphasis on "negative liberty," or the freedom from interference when one's actions do not harm others; and hence that he drew a strong distinction between self- and other-regarding actions. The overwhelming consensus among commentators holds that in *On Liberty*, Mill rejected any interference at all in self-regarding conduct. As we have seen, however, even in *On Liberty* Mill did not draw a clear distinction between actions that are self-regarding and those that regard others. Most actions, it turns out, "concern" or "interest" society, on his view. Even the typical examples used to illustrate Mill's alleged view of purely self-regarding action—such as the bachelor who gets drunk nightly, but still keeps his job and so is no burden on society—are condemned by Mill, and are put in the realm of actions that may be censured and punished by public opinion, dis-

7. See James T. Kloppenberg, *Uncertain Victory*, p. 50.

approbation, and ostracism. Those who argue that on Mill's view, "conduct must be harmful to others before it can be the subject of legal penalties or coercive social pressures"[8] ignore his claim that "a person may suffer very severe penalties at the hands of others for faults which directly concern only himself."[9]

Writing to his friend John Sterling in 1831, Mill described the position of "liberalism" in a manner consistent with the modern view, which focuses on negative liberty:

> Liberalism . . . is for making every man his own guide and sovereign master, and letting him think for himself and do exactly as he judges best for himself . . . but forbidding him to give way to authority; and still less allowing [others] to constrain him more than the existence and toler-able security of every man's person and property renders indispensably necessary.

But Mill then gave his opinion of this type of liberalism: "It is difficult to conceive a more thorough ignorance of man's nature, and of what is necessary for his happiness or what degree of happiness and virtue he is capable of attaining than this system implies."[10] According to Mill, man's happiness does *not* depend upon being left alone to "think for himself" and "do exactly as he judges best for himself." True happiness, on Mill's view, requires proper cultivation to appreciate the higher pleasures and the desire to act toward the good of the whole. This cultivation requires moral education and even soci-etal constraint. Thus he explicitly *rejected* the view that individuals should be constrained as little as was consistent with the personal safety and private property of others.

We have seen that between the early 1830s and the 1859 publication of *On Liberty*, Mill became more, rather than less, concerned with the develop-ment of individuals necessary in order to create a truly cultivated society. The need to direct this development always mediated his desire to allow diversity and individuality. In *On Liberty* there is no sudden break from this concern. I have shown that in this work, Mill located the benefits of diversity in a particular historical context, namely nineteenth-century England, when custom was founded on mediocrity rather than on genius. Diversity was use-ful in this time primarily as a means of presenting a multiplicity of opposing

---

8. The quotation is from C. L. Ten, "Mill's Place in Liberalism," p. 193, though it is represen-tative of many who write on Mill.

9. John Stuart Mill, *On Liberty, CW* 18:278.

10. John Stuart Mill to John Sterling, October 20–22, 1831, *CW* 12:84.

positions, some subset of which would together comprise part of the truth. Thus Mill stressed the need for "geniuses," or the intellectual class, to provide guidance and leadership. But in this work he also clearly expressed his view that over time, an improving society would exhibit decreasing diversity of opinion and lifestyle, partly because of the societal constraints—in the exercise of public opinion—that would and should be used to cause people to abandon the "brutish" pleasures in favor of the higher ones, and to act toward the public good rather than for their own "miserable individuality." In his later *Auguste Comte and Positivism*, Mill explained that "the domain of moral duty, in an improving society, is always widening. When what was once uncommon virtue becomes common virtue, it comes to be numbered among obligations."[11] While a certain kind of diversity of lifestyle may continue to be an essential part of human happiness, he suggested that the scope of this liberty becomes narrower over time as a society progresses; one gets the sense that he allowed that people should always have the liberty to choose between reading Wordsworth and Shelley, but that they should *not* be free to choose between being a drunkard and being sober, or between pushpin and poetry. Choices incompatible with a cultivated society, even if not forbidden by law, may be—and should be, at a certain point—condemned by public opinion. But such condemnation will not always be necessary. Mill believed that the type of moral education he proposed would naturally bring about a convergence of lifestyle and opinion. As more and more people develop, through moral education, the taste for the higher pleasures, as well as the motive of acting toward the public good, they will see that their happiness depends on abjuring the brutish pleasures in favor of the higher ones, and on abandoning their "miserable individuality."

Thus it is clear that Mill endorsed a notion of liberty at odds with that which is generally attributed to him. Noticing this has both historical and theoretical significance. Many liberal authors—notably Isaiah Berlin, Friedrich von Hayek, and John Rawls—associate liberalism with "negative liberty." They stress the freedom from constraint as the defining characteristic of liberalism. They do so, in part, for historical reasons: it is claimed that the ancestral originators of modern liberalism, and its most important proponents—Hobbes, Locke, Hume, and Mill—all endorsed the view that liberty was primarily the notion of freedom from oppression or constraint. However, as my examination has shown, Mill had a more complex view of liberty, one that did not focus on negative liberty. He was concerned not only with the freedom from illegitimate restraint, but also, and more important, the "free-

11. John Stuart Mill, *Auguste Comte and Positivism*, CW 10:338.

dom" to live in a certain kind of society, one in which people preferred the
higher pleasures to the lower, in which all were qualified and allowed to be
part of the political process, and in which all worked toward the good of the
whole. Achieving this freedom requires, on Mill's view, a certain kind of education and even, for a time, a measure of societal constraint.[12]

Thus my portrayal of Mill calls into question his position as the ancestor of modern views of liberty that emphasize merely negative liberty. This
has theoretical significance as well as historical importance. Mill's analysis
can show political thinkers of today problems arising from the negative liberty position: tyranny of the majority over the minority, societal stasis rather
than development and cultivation, limitations on originality and freedom
of thought. But at the same time, his works, and the "strange confusions"
within them, highlight the difficulties in trying to construct a conception of
liberty based on a more complex notion of the relation between the individual and society. In the end, both Mill's moral and political philosophy seem
destined to lead to the tyranny of the majority by the minority, the minority
being the educated elites, those who do not suffer from "miserable individuality" and who appreciate the "higher" pleasures. Thus it is no surprise, as
we have seen, that Mill's politics as well as his moral philosophy ended up far
closer to those of Whewell's than Mill imagined.[13]

12. Joseph Hamburger claims that, on Mill's conception of society, "surveillance, intrusion,
and restriction" will always be necessary, and for this reason removes Mill entirely from the family of liberal thinkers (see *John Stuart Mill on Liberty and Control*). Hamburger is led to this conclusion by his interpretation of Mill as believing in the inherent selfishness of human nature. However, as we have seen, Mill held that we are endowed with an innate "natural sympathy" for our
fellow humans, and explicitly denied the "universal selfishness" of individuals. It was, rather, the
"present standard of morality, and . . . the existing social institutions" that formed selfish characters (*Considerations on Representative Government*, CW 19:405). Mill's incredible optimism
about the improvability of human character by the proper education, and the correct social institutions, led him to the belief that, at some point in history, individuals with liberty will inevitably
make unselfish choices, and be motivated to act for the common good. Like Plato, Mill believed
that once every citizen was educated in the good, there would be a natural conformity to unanim-
ity, without the need for societal constraint.

13. Today, many political theorists are searching for a conception of liberty based on a more
complex and realistic view of the relation between the individual and society. For example, communitarian critics of Rawls's political theory, such as Michael Sandel, William Galston, and
Charles Taylor, and recent proponents of "republicanism," such as Quentin Skinner and Philip Pettit, locate the liberty of the individual in his or her relation to society, but do so in different ways.
While the communitarians emphasize the influence of the society on the individual, by helping
individuals develop morally so that they will be more capable of living a good life, the republicans
focus on the influence of the individual on society, by his or her participation in the political process. As we have seen, Mill showed that both types of positions can be seen as "half-truths" that
need to be combined. Yet he himself was unable to do so successfully. See Michael J. Sandel, *Liberalism and the Limits of Justice*, William Galston, *Justice and the Good*, and Charles Taylor, the
papers in *Philosophical Papers of Charles Taylor*, vol. 2. For a discussion of some differences

Our other reformer has, like Mill, suffered the fate of having his ideas disjoined from their element. By examining Whewell's writings on science in their context, we have seen that contrary to most other commentators in the twentieth century, his view of scientific method was indeed—as he always claimed—an inductive one. By attending to his writings in all areas, as well as to his correspondence with Jones and Herschel, we can see clearly that Whewell never abandoned his lifelong interest in renovating Baconian inductivism, even while grafting to it his own antithetical epistemology. Indeed, most of his other work was inspired, if not motivated by, this interest in renovating Bacon. For instance, we have seen that Whewell's work in political economy was triggered by his antipathy to the deductivism of the Ricardian system. His moral philosophy was intended to provide an inductive system of ethics, as opposed to what he saw as the deductive (and thus ultimately arbitrary) system of the utilitarians. Even his later *Of the Plurality of Worlds* was provoked to some degree by his annoyance at the noninductive "assumptions" and "speculations" of those who claimed that there was intelligent life on other planetary bodies.[14]

Correctly interpreting Whewell's view of scientific method is not only advantageous in the interest of historical accuracy; it also has theoretical value. One legacy of the influence of Mill's *System of Logic* has been the view that *induction* refers to a narrow logical operation involving only enumerative or eliminative forms of reasoning. Mill, almost single-handedly, convinced the Anglo-American philosophical world that induction could not involve any contribution from the mind, or any type of creativity, but rather was an extremely narrow logic. Moreover, his view of induction precluded the possibility that the operation could reach unobservable or theoretical entities or properties. (Indeed, this is why commentators reading Whewell through their Mill-colored glasses have seen him as a noninductivist.) The influence of the Millian view of induction has had unfortunate effects in the development of philosophy of science. In the twentieth century, this influence led to the delineation of a false dichotomy: between a narrow Millian inductivism and hypothetico-deductivism. It was argued by some proponents of hypothetico-deductivism that scientific discovery could not be a matter of merely calculating enumerations of observed instances, perhaps together with some eliminative process; there was, rather, an element of creativity involved, so that discovery could not proceed by a logical "rulebook." Further, it was often

---

between individualist and communitarian liberalisms, see Taylor, "Cross Purposes." See also Quentin Skinner, *Liberty before Liberalism*, and Philip Pettit, *Republicanism*.

    14. For more on this point see my "'Gegen alle vernunftbegabten Bewohner anderer Welten'."

noted that theoretical entities were an important part of modern science, and since Mill's methods could not reach theoretical entities, these methods could not be the proper path to discovery. In this way it was concluded that scientific discovery was not, and could not be, inductive. Instead, it must consist in a noninferential process, one that does not involve any "logic."

I call this a "false dichotomy," because these proponents of the hypothetico-deductive method supposed that the only alternative to a nonrational process of scientific discovery was one similar to Mill's inductive logic. But as we have seen, Whewell had earlier—and more consistently with previous understandings of the term—proposed a type of inductive methodology that was not subject to the criticisms leveled at Mill's view. Whewell's discoverers' induction is a richer form of reasoning, one that involves not only enumeration and elimination of observed instances, but also causal and analogical reasoning, the application of which generally involves a creative element, specifically the selection of an appropriate conception. Because of this broader conception of induction, Whewell's view allows the inductive inference to theoretical entities and properties. Thus the choice is not between a narrow Millian form of reasoning in scientific discovery and no reasoning at all. Moreover, it does seem that important instances of discovery in the history of science (only some of which have been discussed in this book) have conformed to Whewell's methodology. Whewell's discoverers' induction therefore seems to be a good starting point for developing a modern philosophy of scientific discovery.[15]

∽∾

Both Mill and Whewell believed that controversy, debate, and free discussion were necessary in order for truth to emerge. This is the central theme of the famous second chapter of Mill's *On Liberty*, where he argued that, at least in

15. We have seen, however, that Whewell's epistemology is grounded upon a theological foundation, which might seem to undercut my claim that his view is relevant today. However, I believe that the theological foundation can be replaced with a naturalistic one, without significantly altering Whewell's conclusions. Specifically, we can substitute an evolutionary epistemology for his theologically based one. Though Whewell himself would not have made this move, it is noteworthy that an evolutionary approach to epistemology captures well his notion of the progressive clarification and unfolding of Ideas and conceptions in the course of our intercourse with the natural world over the history of science. It also fits well with Whewell's emphasis on the moral and intellectual evolution of man, which became an integral part of his philosophical system by the time he wrote *Of the Plurality of Worlds* and *On the Philosophy of Discovery* in the 1850s. For an early proposal on how an evolutionary epistemology can be substituted for the theological element of Whewell's philosophy, using recent studies in cognitive psychology, see the concluding chapter of my "The Method of Induction."

the current state of British society, any interference in this process—for example, by silencing minority voices—was an infringement not only of individual liberty but also of society's ability to discover truth.[16] He noted, "The peculiar evil of silencing the expression of an opinion is that it is robbing the human race. . . . If the opinion is right, they are deprived of the opportunity of exchanging error for truth; if wrong, they lose, what is almost as great a benefit, the clearer perception and livelier impression of truth, produced by its collision with error."[17] Mill explicitly included science in his claim that there should be "absolute freedom of opinion and sentiment on all subjects, practical or speculative, scientific, moral, or theological."[18]

Though Mill seems not to have realized it, his antagonist Whewell had earlier expressed a very similar view. As we have seen, a central tenet of Whewell's inductive methodology concerns the need for conceptions to be explicated before they can be applied to the facts in order to discover empirical laws. Discussion, debate, and controversy are necessary parts of this process. "In such controversies," Whewell explained in *The Philosophy of the Inductive Sciences*, "the conceptions in question are turned in all directions, examined on all sides; the strengths and weaknesses of the maxims which men apply to them are fully tested; the light of the brightest minds is diffused to others."[19] Even correctly explicated conceptions benefit when they are opposed by alternative opinions. As Whewell somewhat picturesquely described it, "The tendency of all such controversy is to diffuse truth and to dispel errour. Truth is consistent, and can bear the tug of war; Errour is incoherent, and falls to pieces in the struggle. True conceptions can endure the sun, and become clearer as a fuller light is obtained; confused and inconsistent notions vanish like visionary spectres at the break of a brighter day."[20] Though he made this point explicitly in the context of his discussion of the physical sciences, we should recall that Whewell's antithetical epistemology, which requires the explication of conceptions, applies to all areas of knowledge, including morality and the human sciences. Thus his claim about the impor-

16. Mill had also made this point in one of his essays on "The Spirit of the Age," where he noted, "When all opinions are questioned, it is in time found out what are those which will not bear a close examination. . . . It is by discussion, also, that true opinions are discovered and diffused" (see *CW* 22:233). Later, in his *Representative Government*, he wrote, "In all human affairs, conflicting influences are required, to keep one another alive and efficient even for their own proper uses" (*CW* 19:439). In his review of Grote's *Aristotle*, Mill similarly stressed the importance of debate (in the context of Aristotle's discussion of the Socratic Elenchus in the *Topics*) as a means of discovering truth (see *CW* 11:505–10).

17. John Stuart Mill, *On Liberty*, *CW* 18:229.

18. Ibid., *CW* 18:225.

19. See William Whewell, *The Philosophy of the Inductive Sciences*, 2:9–10.

20. Ibid., 2:7–8.

tance—indeed, the necessity—of the free conflict between opposing opinions pertains to all realms of thought.

Whewell and Mill believed that the opposition of conflicting opinions could lead to truth. It is not surprising, then, that both of them supposed that by debating each other on the important issues of the day, truth would emerge. The revelation of truth would, they expected, lead to the reform of science and society. In these final pages I have suggested that, even today, we can use their insights in attempting to achieve these reforms.

BIBLIOGRAPHY

Archives

John Herschel Papers, Royal Society of London [JHP]
Whewell Papers, Trinity College, Cambridge [WP]

Published Sources

Achinstein, Peter. *Particles and Waves.* New York: Oxford University Press, 1991.

———. "Inference to the Best Explanation: Or, Who Won the Mill-Whewell Debate?" *Studies in History and Philosophy of Science* 23 (1992): 349–64.

Adamson, R. "Bacon." *Encyclopaedia Britannica,* 9th ed. (1875), 3:186–87.

Airy, George Bidell. "Report on the Progress of Astronomy during the Present Century." *Report on the First and Second Meetings of the British Association for the Advancement of Science,* p. 189. London: John Murray, 1833.

Albee, Ernest. *A History of English Utilitarianism.* London: George Allen & Sons, 1901.

Anschutz, R. P. *The Philosophy of J. S. Mill.* Oxford: Oxford University Press, 1953.

Ashcraft, Richard. "Class Conflict and Constitutionalism in J. S. Mill's Thought." In Rosenblum, *Liberalism and the Moral Life,* pp. 105–26.

Babbage, Charles. *On the Economy of Machinery and Manufactures.* London: C. Knight, 1832.

Bacon, Francis. *The Works of Francis Bacon.* Collected and edited by J. Spedding, R. L. Ellis, and D. D. Heath. 14 vols. London: Longman & Co., 1857–61; new edition, 1877–89.

Bagehot, Walter. *Biographical Studies.* London: Longmans, Green, 1895.

Bailey, Samuel. *Discourse on Various Subjects: Read before Literary and Philosophical Societies.* 1829. Reprint, London: Longman, Brown, Green, and Longmans, 1852.

Bain, Alexander. "On the Constitution of Matter." *Westminster Review* (1841), pp. 69–86.

———. *James Mill.* London: Longmans, Green, 1882.

———. *John Stuart Mill: A Criticism, with Personal Reflections*. London: Longmans, Green, 1882.

———. *Autobiography*. London: Longmans, Green and Co., 1904.

Becher, Harvey W. "Woodhouse, Babbage, Peacock and Modern Algebra." *Historia Mathematica* 7 (1980): 389–400.

———. "William Whewell and Cambridge Mathematics." *Historical Studies in the Physical Sciences* 11 (1981): 1–48.

———. "Voluntary Science in Nineteenth-Century Cambridge University to the 1850s." *British Journal for the History of Science* 19 (1986): 57–87.

———. "Whewell's Odyssey: From Mathematics to Moral Philosophy." In Fisch and Schaffer, *William Whewell*, pp. 1–29.

Belsham, Thomas. *Elements of the Philosophy of the Mind, and of Moral Philosophy, to Which Is Prefixed a Compendium of Logic*. London: J. Johnson, 1801.

Benson, Donald R. "Facts and Constructs: Victorian Humanists and Scientific Theorists on Scientific Knowledge." In Paradis and Postlewait, *Victorian Science and Victorian Values*, pp. 299–318.

Bentham, Jeremy. *Works of Jeremy Bentham*. Edited by John Bowring. 11 vols. Edinburgh: Tait, 1838–43.

Berg, Maxine. *The Machinery Question and the Making of Political Economy*. Cambridge: Cambridge University Press, 1980.

Berkeley, George. *A Treatise concerning the Principles of Human Knowledge*. Edited by Jonathan Dancy. Oxford: Oxford University Press, 1998.

Berlin, Isaiah. "Two Concepts of Liberty." In *Four Essays on Liberty*, pp. 118–72. Oxford: Oxford University Press, 1992.

Birks, T. R. *Modern Utilitarianism; or, The Systems of Paley, Bentham, and Mill Examined and Compared*. London: Macmillan, 1874.

Blanché, Robert. *Le rationalisme de Whewell*. Paris: F. Alcan, 1935.

Blaug, Mark. *Ricardian Economics: A Historical Study*. New Haven, CT: Yale University Press, 1958.

———. *The Methodology of Economics; or, How Economists Explain*. 2nd ed. Cambridge: Cambridge University Press, 1988.

Boyd, Richard. "What Realism Implies and What It Does Not." *Dialectica* 43 (1989): 5–29.

Brent, Richard. *Liberal Anglican Politics: Whiggery, Religion and Reform 1830–1841*. Oxford: Clarendon, 1987.

Brewster, David. "On the History of the Inductive Sciences." *Edinburgh Review* 66 (1837): 110–51.

———. "Whewell's Philosophy of the Inductive Sciences." *Edinburgh Review* 74 (1842): 139–61.

Broad, C. D. *Kant, an Introduction*. Cambridge: Cambridge University Press, 1978.

Brock, W. H., and R. M. Macleod. "The Scientists' Declaration: Reflections on Science and Belief in the Wake of *Essays and Reviews*, 1864–5." *British Journal for the History of Science* 9 (1976): 39–66.

Brooke, John H. "Natural Theology and the Plurality of Worlds: Observations on the Brewster-Whewell Debate." *Annals of Science* 34 (1977): 221–86.

———. "Richard Owen, William Whewell, and *The Vestiges.*" *British Journal for the History of Science* 10 (1977): 132–45.

———. "Indications of a Creator: Whewell as Apologist and Priest." In Fisch and Schaffer, *William Whewell*, pp. 149–73.

———. *Science and Religion: Some Historical Perspectives.* Cambridge: Cambridge University Press, 1991.

Brown, James Robert, and J. Mittelstrass, eds. *An Intimate Relation.* Dordrecht: Kluwer, 1989.

Browne, Janet. *Charles Darwin: The Power of Place. Volume II of a Biography.* New York: Alfred A. Knopf, 2002.

Brush, Stephen G. "Prediction and Theory-Evaluation: The Case of Light Bending." *Science* 246 (1989): 1124–29.

Buchdahl, Gerd. "Deductivist versus Inductivist Approaches in the Philosophy of Science as Illustrated by Some Controversies between Whewell and Mill." In Fisch and Schaffer, *William Whewell*, pp. 311–44.

Buchwald, Jed Z. *The Rise of the Wave Theory of Light.* Chicago: University of Chicago Press, 1989.

———. "Kinds and the Wave Theory of Light." *Studies in History and Philosophy of Science* 23 (1992): 39–74.

Burnyeat, Myles F. "What Was the 'Common Arrangement'? An Inquiry into John Stuart Mill's Boyhood Reading of Plato." *Utilitas* 13 (2001): 1–32.

Burrow, John W. *A Liberal Descent: Victorian Historians and the English Past.* Cambridge: Cambridge University Press, 1981.

———. *Whigs and Liberals: Continuity and Change in English Political Thought.* Oxford: Clarendon, 1988.

Butts, Robert E. "Necessary Truth in Whewell's Theory of Science." *American Philosophical Quarterly* 2 (1965): 161–81.

———. "On Walsh's Reading of Whewell's View of Necessity." *Philosophy of Science* 32 (1965): 175–81.

———. "Professor Marcucci on Whewell's Idealism." *Philosophy of Science* 34 (1967): 175–83.

———. "Reply to David Wilson: Was Whewell Interested in True Causes?" *Philosophy of Science* 40 (1973): 125–28.

———. "Whewell's Logic of Induction." In Giere and Westfall, *Foundations of Scientific Method*, pp. 53–85.

———. "Consilience of Inductions and the Problem of Conceptual Change in Science." In *Logic, Laws and Life,* edited by Robert G. Colodny, pp. 71–88. Pittsburgh: University of Pittsburgh Press, 1977.

———. "Pragmatism in Theories of Induction in the Victorian Era: Herschel, Whewell, Mach and Mill." In *Pragmatik: Handbuch Pragmatischen Denkens,* edited by Herbert Stachowiak, pp. 40–58. Hamburg: F. Meiner, 1987.

———. *Historical Pragmatics: Philosophical Essays*. Dordrecht: Kluwer, 1993.

———. "Induction as Unification: Kant, Whewell, and Recent Developments." In *Kant and Contemporary Epistemology*, edited by Paolo Parrini, pp. 273–89. Dordrecht: Kluwer, 1994.

Campanelli, G. "Whewell, William." In *The New Palgrave: A Dictionary of Economics*, edited by John Eatwell, Murray Milgate, and Peter Newman, 4:900–901. London: Macmillan Press, 1987.

Campos Boralevi, Lea. *Bentham and the Oppressed*. Berlin: Walter de Gruyter, 1984.

Cannon, Susan F. "The Whewell-Darwin Controversy." *Journal of the Geological Society of London* 132 (1976): 377–84.

———. *Science in Culture: The Early Victorian Period*. New York: Science History Publications, 1978.

Cannon, Walter F. [Later, Susan F.] "The Problem of Miracles in the 1830s." *Victorian Studies* 4 (1960): 5–32.

———. "The Impact of Uniformitarianism: Two Letters from John Herschel to Charles Lyell, 1836–37." *Proceedings of the American Philosophical Society* 105 (June 1961): 301–14.

———. "Science and Broad Churchmen: An Early Victorian Intellectual Network." *Journal of British Studies* 4 (1964): 65–88.

———. "William Whewell: Contributions to Science and Learning." *Notes and Records of the Royal Society* 19 (1964): 176–91.

Cantor, Geoffrey. "The Reception of the Wave Theory of Light in Britain: A Case Study of the Role of Methodology in Scientific Debate." *Historical Studies in the Physical Sciences* 4 (1975): 109–32.

———. *Optics after Newton: Theories of Light in Britain and Ireland, 1704–1840*. Manchester: Manchester University Press, 1983.

———. "Between Rationalism and Romanticism: Whewell's Historiography of the Inductive Sciences." In Fisch and Schaffer, *William Whewell*, pp. 67–86.

Capaldi, Nicholas. *John Stuart Mill: A Biography*. Cambridge: Cambridge University Press, 2004.

Carlisle, Harvey. "William Whewell." *Macmillan's Magazine* 45 (1882): 138–44.

Carlyle, Thomas. "Signs of the Times." *Edinburgh Review* 49 (1829): 439–59.

———. "Boswell's *Life of Johnson*." *Fraser's Magazine* 5 (May 1832): 379–413.

Carpenter, William Benjamin. "Darwin on the Origin of Species." *National Review* 10 (1860): 188–214. Reprinted in Hull, *Darwin and His Critics*, pp. 88–114.

Cartwright, David E. *Tides: A Scientific History*. Cambridge: Cambridge University Press, 1999.

Chalmers, Thomas. *On the Power, Wisdom, and Goodness of God as Manifested in the Adaptation of External Nature to the Moral and Intellectual Constitution of Man*. Bridgewater Treatise I. 2 vols. London: William Pickering, 1833.

———. "Morell's Modern Philosophy." *North British Review* 6 (1847): 307.

Chambers, Robert. *Vestiges of the Natural History of Creation.* London: J. Churchill, 1844.

Checkland, S. G. "The Advent of Academic Economics in England." *Manchester School of Economic and Social Studies* 19 (1951): 43–79.

Clark, J. W. "Thirlwall, (Newall) Connop." Revised by H. C. G. Matthew. *Oxford Dictionary of National Biography.* Oxford: Oxford University Press, 2004. http://www.oxforddnb.com/view/article/27185.

Clark, W. G. "William Whewell." *Macmillan's Magazine* 13 (1866): 545–52.

Coburn, Kathleen. "Coleridge: A Bridge between Science and Philosophy." In *Coleridge's Variety: Bicentenary Studies,* edited by John Beer, pp. 81–100. Pittsburgh: University of Pittsburgh Press, 1975.

Cochrane, J. A. "The First Mathematical Ricardian Model." *History of Political Economy* 2 (1970): 419–31.

Cohen, I. Bernard, and Richard S. Westfall. *Newton: Texts, Backgrounds, Commentaries.* New York: W. W. Norton & Company, 1995.

Coleridge, Samuel Taylor. *Preliminary Treatise on Method: General Introduction to the "Encyclopedia Metropolitana."* 3rd ed. London: J. J. Griffen, 1849.

———. *The Collected Works of Samuel Taylor Coleridge.* Edited by Katherine Coburn and Burt Winer. 16 vols. London: Routledge and Kegan Paul; Princeton, NJ: Princeton University Press, 1965–80.

Collini, Stefan. Introduction to John Stuart Mill, *The Collected Works of John Stuart Mill,* vol. 21, pp. vii–lvi.

———. "The Ordinary Experience of Civilized Life: Sidgwick's Politics and the Method of Reflective Analysis." In *Essays on Henry Sidgwick,* edited by Bart Schultz, pp. 333–68. Cambridge: Cambridge University Press, 1992.

———. "From Sectarian Radical to National Possession: John Stuart Mill in English Culture, 1873–1945." in G. W. Smith, *John Stuart Mill's Social and Political Thought,* 4:380–405.

Collini, Stefan, Donald Winch, and John Burrow. *That Noble Science of Politics: A Study in Nineteenth-Century Intellectual History.* Cambridge: Cambridge University Press, 1983.

Comte, Auguste. *The Positive Philosophy of Auguste Comte.* Freely translated and condensed by Harriet Martineau. New York: C. Blanshard, 1855.

Corsi, Pietro. "The Heritage of Dugald Stewart: Oxford Philosophy and the Method of Political Economy, 1809–1832." *Nuncius* 2 (1987): 89–144.

———. *Science and Religion: Baden Powell and the Anglican Debate, 1800–1860.* Cambridge: Cambridge University Press, 1988.

Cowherd, Raymond G. *Political Economists and the English Poor Laws.* Athens: Ohio University Press, 1977.

Cremaschi, Sergio. "Ricardo's Philosophy: Between Jeremy Bentham and Thomas Belsham." Paper presented at the International Philosophy Conference, University of Rijeka, Croatia, May 2001.

Cremaschi, Sergio, and Marcelo Dascal. "Malthus and Ricardo on Economic Methodology." *History of Political Economy* 28 (1996): 475–511.

Cunningham, Andrew, and Nicholas Jardine, eds. *Romanticism and the Sciences.* Cambridge: Cambridge University Press, 1990.

Curtis, Ronald. "Darwin as an Epistemologist." *Annals of Science* 44 (1987): 379–407.

Curtis, Simon. "The Philosopher's Flowers: John Stuart Mill as Botanist." *Encounter* 70 (1988): 26–33.

Darwin, Charles. *On the Origin of Species.* 1859. A Facsimile of the first edition with an introduction by Ernst Mayr. Cambridge, MA: Harvard University Press, 1964.

———. *On the Origin of Species.* 2nd ed. 1860. Reprint, Oxford: Oxford University Press, 1996.

———. *Autobiography.* Edited by N. Barlow. London: Collins, 1958.

———. *The Correspondence of Charles Darwin.* Edited by Frederick H. Burkhardt, Sydney Smith, Duncan M. Porter, Janet Browne, Marsha Richmond, Joy Harvey, Sheila Ann Dean et al. 14 vols. Cambridge: Cambridge University Press, 1983–2004.

———. *The Works of Charles Darwin.* Edited by Paul H. Barrett and R. B. Freeman. 29 vols. London: William Pickering, 1986–89.

Darwin, Francis, ed. *The Life and Letters of Charles Darwin.* In 2 vols. New York: D. Appleton and Co., 1891.

Deas, H. "Crystallography and Crystallographers in England in the Early 19th Century: A Preliminary Survey." *Centaurus* 6 (1959): 129–48.

DeMarchi, Neil B. "The Empirical Content and Longevity of Ricardian Economics." *Economica* 37 (1970): 257–76. Reprinted in Wood, *David Ricardo: Critical Assessments*, 3:53–71.

———. "The Success of Mill's *Principles.*" *History of Political Economy* 6 (1974): 119–57.

DeMarchi, N. B., and R. P. Sturges. "Malthus and Ricardo's Inductivist Critics: Four Letters to William Whewell." *Economia* 40 (1973): 379–93.

DeMorgan, Augustus. "Review of the *Philosophy of the Inductive Sciences.*" *Athenaeum,* no. 672 (September 12, 1840): 707–9.

———. *Formal Logic; or, The Calculus of Inference.* London: Taylor & Welton, 1847.

———. "The Works of Francis Bacon." *Athenaeum,* nos. 1611 and 1612 (September 11 and 18, 1858): 332–33, 367–68.

———. "Review of Whewell's *Novum Organum* [sic] *Renovatum.*" *Athenaeum,* no. 1628 (1859): 42–44.

Desmond, Adrian. "Richard Owen's Reaction to Transmutation in the 1830s." *British Journal for the History of Science* 18 (1985): 25–50.

Desmond, Adrian, and James Moore. *Darwin.* New York: Warner Books, 1992.

Devlin, Patrick. "Mill on Liberty in Morals." In *The Enforcement of Morals.* London: Oxford University Press, 1965.

Distand, N. Merrill. "Hare, J. C." *Oxford Dictionary of National Biography.* Oxford: Oxford University Press, 2004. http://www.oxforddnb.com/view/article12304.

Donner, Wendy. *The Liberal Self: John Stuart Mill's Moral and Political Philosophy.* Ithaca, NY: Cornell University Press, 1991.

Ducasse, Curt J. "John Stuart Mill's *System of Logic.*" In *Theories of Scientific Method: The Renaissance through the Nineteenth Century,* edited by Edward H. Madden, pp. 218–32. Seattle: University of Washington Press, 1960.

———. "Francis Bacon's Philosophy of Science." In *Theories of Scientific Method,* edited by R. M. Blake, C. J. Ducasse, and E. H. Madden, pp. 50–74. Seattle: University of Washington Press, 1969.

Duncan, Graeme. *Marx and Mill: Two Views of Social Conflict and Social Harmony.* Cambridge: Cambridge University Press, 1973.

———. "John Stuart Mill and Democracy." In G. W. Smith, *John Stuart Mill's Social and Political Thought,* 3:69–87.

Edgeworth, F. Y. "Jones, Richard." In *Palgrave's Dictionary of Political Economy,* edited by Henry Higgs, 2:490–91. 3 vols. First ed., 1894–99; rev. ed., 1925–26. Reprint of rev. ed., New York: A. M. Kelley, 1963.

Eisenach, Eldon J. *Mill and the Moral Character of Liberalism.* University Park: Pennsylvania State University Press, 1998.

Ellgård, Alvar. *Darwin and the General Reader: The Reception of Darwin's Theory of Evolution in the British Periodical Press, 1859–1872.* With a new foreword by David L. Hull. Chicago: University of Chicago Press, 1990.

Ellis, Robert L. "General Preface to the Philosophical Works." In Francis Bacon, *The Works of Francis Bacon,* 1:21–67. London: Longman & Co., 1857.

Espinasse, Francis. *Lancaster Worthies.* L. P., 2nd ser. London: Simpkin, Marshall and Co., 1877.

Faraday, Michael. *Experimental Researches in Electricity.* 1855. 3 vols. bound as 2. New York: Dover Publications, 1965.

———. *Experimental Researches in Chemistry and Physics.* London: Richard Taylor and William Francis, 1859.

Fine, Arthur. "The Natural Ontological Attitude." In *Scientific Realism,* edited by Jarrett Leplin, pp. 83–107. Berkeley and Los Angeles: University of California Press, 1984.

Fisch, Max H., Christian J. W. Kloesel, Edward C. Moore, Don D. Roberts, Lynn A. Ziegler, Nathan Houser, Marc Simon, Ursula Niklas, André De Tienne, Cornelis de Waal, Jonathan K. Eller, and Albert Lewis, eds. *Writings of Charles S. Peirce: A Chronological Edition.* 6 vols. Bloomington: Indiana University Press, 1982–2000.

Fisch, Menachem. "Necessary and Contingent Truth in William Whewell's Antithetical Theory of Knowledge." *Studies in History and Philosophy of Science* 16 (1985): 275–314.

———. "Whewell's Consilience of Inductions: An Evaluation." *Philosophy of Science* 52 (1985): 239–55.

———. "A Physicist's Philosopher: James Clerk Maxwell on Mathematical Physics." *Journal of Statistical Physics* 51 (1988): 309–19.

———. "A Philosopher's Coming of Age: A Study in Erotetic Intellectual History." In Fisch and Schaffer, *William Whewell*, pp. 31–66.

———. *William Whewell, Philosopher of Science.* Oxford: Oxford University Press, 1991.

———. "'The Emergency Which Has Arrived': The Problematic History of Nineteenth-Century British Algebra—A Programmatic Outline." *British Journal for the History of Science* 27 (1994): 247–76.

Fisch, Menachem, and Simon Schaffer, eds. *William Whewell: A Composite Portrait.* Oxford: Oxford University Press, 1991.

Forbes, Duncan. *The Liberal Anglican Idea of History.* Cambridge: Cambridge University Press, 1952.

Forbes, J. D. "On the Refraction and Polarization of Heat." *Transactions of the Royal Society of Edinburgh* 13 (1835): 131–68.

Forster, Malcolm R. "Unification, Explanation, and the Composition of Causes in Newtonian Mechanics." *Studies in History and Philosophy of Science* 19 (1988): 55–101.

Fox, Caroline. *Memories of Old Friends.* 2nd ed. 2 vols. London: Smith, Elder, 1882.

Fox Bourne, E. H., W. T. Thornton, Herbert Spencer, Henry Trimen, W. Minto, J. H. Levy, W. A. Hunter, J. E. Cairnes, Henry Fawcett, Millicent Fawcett, and Frederic Hansen. *John Stuart Mill: His Life and Works.* New York: Henry Holt, 1873.

Fresnel, Augustin. "Memoir on the Diffraction of Light." 1818. Reprinted in *The Wave Theory of Light*, edited by Henry Crew, pp. 81–144. New York: Cincinnati American, 1900.

Friedman, Richard. "A New Exploration of Mill's Essay *On Liberty.*" *Political Studies* 14 (October 1966): 281–304.

Galston, William. *Justice and the Good.* Chicago: University of Chicago Press, 1980.

Garland, Martha M. *Cambridge before Darwin: The Ideal of a Liberal Education, 1800–1860.* Cambridge: Cambridge University Press, 1980.

Gash, Norman. *Reaction and Reconstruction in English Politics 1832–1852.* Oxford: Clarendon, 1965.

———. *Sir Robert Peel: The Life of Sir Robert Peel after 1830.* London: Longman, 1972.

Gaukroger, Stephen. *Francis Bacon and the Transformation of Early-Modern Philosophy.* Cambridge: Cambridge University Press, 2001.

Ghiselin, Michael. *The Triumph of the Darwinian Method.* Berkeley and Los Angeles: University of California Press, 1969.

Giere, Ronald N., and Richard S. Westfall, eds. *Foundations of Scientific Method: The Nineteenth Century.* Bloomington: Indiana University Press, 1973.

Gillespie, Charles C. *Genesis and Geology: A Study in the Relation of Scientific Thought, Natural Theology, and Social Opinion in Great Britain, 1790–1850.* Cambridge, MA: Harvard University Press, 1951.

Goldman, Lawrence. "The Origins of British 'Social Science': Political Economy, Natural Science and Statistics, 1830–5." *The Historical Journal* 26 (1983): 587–616.

——. *Science, Reform and Politics in Victorian Britain: The Social Science Association, 1857–1886.* Cambridge: Cambridge University Press, 2002.

Grampp, W. D. "Malthus and His Contemporaries." *History of Political Economy* 6 (1974): 278–304.

Gray, Asa. "Charles Robert Darwin." *Nature* 10 (June 4, 1874): 79–81.

Gray, John. *Mill on Liberty: A Defense.* London: Routledge, 1983.

Gray, John, and G. W. Smith, eds. *John Stuart Mill on Liberty: In Focus.* London: Routledge, 1991.

Green, Joseph Henry. *Vital Dynamics: The Hunterian Oration before the Royal College of Surgeons in London, 14 February 1840.* London: William Pickering, 1840.

Green, Michele. "The Religion of Sympathy: J. S. Mill." *The European Legacy* 1 (1996): 1705–15.

Gregory, C. A. "Jones, Richard." In *The New Palgrave: A Dictionary of Economics,* edited by John Eatwell, Murray Milgate, and Peter Newman, 2:1035–36. London: Macmillan Press, 1987.

Grote, John. *Exploratio Philosophica.* Cambridge: The University Press, 1865.

Gruber, Howard E., and Paul H. Barrett, eds. *Darwin on Man: A Psychological Study of Scientific Creativity* by Howard E. Gruber, together with Darwin's early and unpublished notebooks, transcribed and annotated by Paul H. Barrett. New York: E. P. Dutton, 1974.

Haac, Oscar. *The Correspondence of John Stuart Mill and Auguste Comte.* New Brunswick, NJ: Transaction, 1995.

Hacking, Ian. "A Tradition of Natural Kinds." *Philosophical Studies* 61 (1991): 109–26.

——. "Working in a New World: The Taxonomic Solution." In Horwich, *World Changes,* pp. 275–310.

Halévy, Elie. *The Growth of Philosophical Radicalism.* 1928. Translated by Mary Morris. London: Faber and Faber, 1972.

Hall, A. Rupert. *The Cambridge Philosophical Society: A History, 1819–1969.* Cambridge: Cambridge Philosophical Society, 1969.

Hamburger, Joseph. *Intellectuals in Politics: John Stuart Mill and the Philosophical Radicals.* New Haven, CT: Yale University Press, 1965.

——. *John Stuart Mill on Liberty and Control.* Princeton, NJ: Princeton University Press, 1999.

Hamilton, William. "Review of [Whewell's] *Thoughts on the Study of Mathematics as Part of a Liberal Education.*" *Edinburgh Review* 62 (1836): 218–52.

——. *Lectures on Metaphysics and Logic.* Edited by John Veitch and Henry L. Mansel. 4 vols. Edinburgh: Blackwood, 1859–60.

Hanson, N. R. *Patterns of Discovery: An Inquiry into the Conceptual Foundations of Science.* Cambridge: Cambridge University Press, 1958.

Harman, Gilbert. "The Inference to the Best Explanation." *Philosophical Review* 74 (1965): 88–95.

Harman, P. M. *Energy, Force and Matter: The Conceptual Development of Nineteenth-Century Physics.* Cambridge: Cambridge University Press, 1982.

————, ed. *Wranglers and Physicists: Studies on Cambridge Physics in the Nineteenth Century.* Manchester: Manchester University Press, 1985.

————. *The Natural Philosophy of James Clerk Maxwell.* Cambridge: Cambridge University Press, 1998.

Harper, William. "Consilience and Natural Kind Reasoning." In Brown and Mittelstrass, *An Intimate Relation*, pp. 115–52.

Harrison, Brian. *The Transformation of British Politics, 1860–1995.* Oxford: Oxford University Press, 1996.

Hartley, David. *Observations on Man, his frame, his duty, and his expectations.* London: J. Johnson, 1749.

Harvey-Phillips, M. B. "Malthus' Theodicy: The Intellectual Background of His Contribution to Political Economy." *History of Political Economy* 16 (1984): 591–608.

Hayek, F. A. von. *The Constitution of Liberty.* Chicago: University of Chicago Press, 1960.

Hazlitt, William. *A Reply to the Essay on Population.* London, 1807.

Hempel, Carl G. *Philosophy of Natural Science.* Englewood Cliffs, NJ: Prentice-Hall, 1966.

Henderson, James P. "Induction, Deduction and the Role of Mathematics: The Whewell Group vs. The Ricardian Economists." *Research in the History of Economic Thought and Methodology* 7 (1990): 1–36.

————. *Early Mathematical Economics: William Whewell and the British Case.* Lanham: Rowman and Littlefield, 1996.

Herschel, John F. W. *Preliminary Discourse on the Study of Natural Philosophy.* London: Longman, Rees, Orme, Brown and Green and John Taylor, 1830.

————. "On the Absorption of Light by Coloured Media, Viewed in Connexion with the Undulatory Theory." *Report of the Third Meeting of the British Association for the Advancement of Science, Held at Cambridge in 1833*, pp. 373–74. London: John Murray, 1834.

————. "Whewell on Inductive Sciences." Review of the *History* and the *Philosophy. Quarterly Review* 68 (1841): 177–238.

————. "Address of the President." *Report of the Fifteenth Meeting of the British Association for the Advancement of Science, Held at Cambridge in June 1845*, pp. xxvii–xliv. London: John Murray, 1846.

————. "The Reverend William Whewell, DD." *Proceedings of the Royal Society of London* 65 (1867–68): li–lxi.

————. *Calendar of the Correspondence of Sir John Herschel.* Edited by Michael J. Crowe. Cambridge: Cambridge University Press, 1998.

Hesse, Mary B. "Consilience of Inductions." In *The Problem of Inductive Logic*, edited by Imre Lakatos, pp. 232–47. Amsterdam: North Holland Publication Co., 1968.

————. "Francis Bacon's Philosophy of Science." In *Essential Articles for the Study of Francis Bacon*, edited by B. Vickers, pp. 114–39. Hamden, CT: Archon Books, 1968.

———. "Whewell's Consilience of Inductions and Predictions [Reply to Laudan]." *Monist* 55 (1971): 520–24.

Heyck, T. W. *The Transformation of Intellectual Life in Victorian England*. Chicago: Lyceum Books, 1982.

Hilton, Boyd. *The Age of Atonement: The Influence of Evangelicalism on Social and Economic Thought, 1785–1865*. Oxford: Clarendon, 1988.

Himmelfarb, Gertrude. *On Liberty and Liberalism: The Case of John Stuart Mill*. New York: Knopf, 1974.

Hirsch, Abraham. "John Stuart Mill on Verification and the Business of Science." *History of Political Economy* 24 (1992): 843–66.

Hodge, M. J. S. "Darwin, Whewell and Natural Selection." *Biology and Philosophy* 6 (1991): 457–60.

———. "The History of the Earth, Life, and Man: Whewell and Paletiological Science." In Fisch and Schaffer, *William Whewell*, pp. 255–88.

———. "Darwin's Argument in the *Origin*." *Philosophy of Science* 59 (1992): 461–64.

Hollander, Samuel. "William Whewell and John Stuart Mill on the Methodology of Political Economy." *Studies in the History and Philosophy of Science* 14 (1983): 127–68.

———. *The Economics of John Stuart Mill*. 2 vols. Oxford: Blackwell, 1984.

Hollis, Patricia. "Anti-Slavery and British Working Class Radicalism in the Years of Reform." In *Anti-Slavery, Religion and Reform*, edited by Christine Bolt and Seymour Dressler. Folkestone, UK: William Dawson and Sons, 1980.

Horton, M. "In Defense of Francis Bacon." *Studies in History and Philosophy of Science* 4 (1973): 241–78.

Horwich, Paul, ed. *World Changes: Thomas Kuhn and the Nature of Science*. Cambridge, MA.: MIT Press, 1993.

Houghton, Walter E. *The Victorian Frame of Mind, 1830–70*. New Haven, CT: Yale University Press, 1957.

Hudson, W. D. *A Century of Moral Philosophy*. Guildford, UK: Lutterworth Press, 1980.

Hull, David, ed. *Darwin and His Critics*. Cambridge, MA: Harvard University Press, 1973.

Hume, David. *An Enquiry concerning the Principles of Morals*. London: A. Miller, 1751.

Hutton, R. H. "Mill and Whewell on the Logic of Induction." *The Prospective Review* 6 (1850): 77–111.

Huzel, J. P. "The Demographic Impact of the Old Poor Law: More Reflections on Malthus." In Wood, *Thomas Robert Malthus*, 4:270–85.

———. "Malthus, the Poor Law, and Population in Early 19th Century England." In Wood, *Thomas Robert Malthus*, 4:138–65.

Jacobs, Struan. "John Stuart Mill on Induction and Hypotheses." *Journal of the History of Ideas* 29 (1991): 69–83.

———. *Science and British Liberalism: Locke, Bentham, Mill and Popper*. Aldershot, UK: Avebury, 1991.

James, Frank A. J. L. *The Correspondence of Michael Faraday*. 4 vols. London: The Institution of Electrical Engineers, 1996.

Jardine, Lisa. *Francis Bacon: Discovery and the Art of Discourse*. Cambridge: Cambridge University Press, 1974.

———. "*Experientia Literata* or *Novum Organum*? The Dilemma of Bacon's Scientific Method." In *Francis Bacon's Legacy of Texts*, edited by W. Sessions, pp. 47–67. New York: AMS Press, 1990.

Jones, Richard. *An Essay on the Distribution of Wealth, and on the Sources of Taxation, Part I: Rent*. London: Murray, 1831.

———. *Literary Remains, Consisting of Lectures and Tracts on Political Economy of the Late Rev. Richard Jones*. Edited with a preface by William Whewell. London: Murray, 1859.

Kahan, Alan S. *Aristocratic Liberalism: The Social and Political Thought of Jacob Burkhardt, John Stuart Mill, and Alexis de Tocqueville*. New York: Oxford University Press, 1992.

Kant, Immanuel. *Critique of Pure Reason*. Edited by Norman Kemp Smith. London, 1929.

———. *Prolegomena to Any Future Metaphysics*. Translated by P. Carus and revised by J. Ellington. Indianapolis: Hackett, 1977.

Kavaloski, Vincent Carl. "The Vera Causa Principle: A Historico-Philosophical Study of a Metatheoretical Concept from Newton through Darwin." Ph.D. diss., University of Chicago, 1974.

Kepler, Johannes. *New Astronomy*. 1609. Translated by William H. Donahue. Cambridge: Cambridge University Press, 1992.

Kern, Paul N. B. "Universal Suffrage without Democracy: Thomas Hare and John Stuart Mill." In G. W. Smith, *John Stuart Mill's Social and Political Thought*, 3:165–78.

Keynes, John M. *A Treatise on Probability*. London: Macmillan, 1921.

Keynes, John Neville. *The Scope and Method of Political Economy*. 4th ed. London: Macmillan Press, 1917.

Kinzer, Bruce L. "J. S. Mill and the Secret Ballot." In G. W. Smith, *John Stuart Mill's Social and Political Thought*, 3:179–96.

Kinzer, Bruce L., A. T. Robson, and J. M. Robson. *A Moralist In and Out of Parliament: John Stuart Mill at Westminster 1865–1868*. Toronto: University of Toronto Press, 1992.

Kirk, N. *Labour and Society in Britain and the USA: Capitalism, Custom and Protest*. Hants, UK: Scolar Press, 1994.

Kitcher, Philip. "Arithmetic for the Millian." *Philosophical Studies* 10 (1980): 215–36.

Kleiner, Scott. "A New Look at Kepler and Abductive Argument." *Studies in History and Philosophy of Science* 14 (1983): 279–313.

Kloppenberg, James T. *Uncertain Victory: Social Democracy and Progressivism in*

European and American Thought, 1870–1920. New York: Oxford University Press, 1986.

Koestler, Arthur. *The Watershed: A Biography of Johannes Kepler.* Garden City, NY: Academic Press of America, 1960.

Kornblith, Hilary. *Inductive Inference and Its Natural Ground: An Essay in Naturalistic Epistemology.* Cambridge, MA: MIT Press, 1993.

Kozhamthadam, Job. *The Discovery of Kepler's Laws: The Interaction of Science, Philosophy, and Religion.* Notre Dame, IN: Notre Dame University Press, 1994.

Kubitz, Oskar A. *The Development of John Stuart Mill's "System of Logic."* Illinois Studies in the Social Sciences 18, nos. 1–2 (Urbana: University of Illinois, 1932): 1–310.

Kuhn, Thomas. "Afterwords." In Horwich, *World Changes*, pp. 311–42.

Laine, M., ed. *A Cultivated Mind: Essays on J. S. Mill Presented to J. M. Robson.* Toronto: University of Toronto Press, 1991.

Laudan, Larry. "Theories of Scientific Method from Plato to Mach." *History of Science* 7 (1969): 1–63.

———. "William Whewell on the Consilience of Inductions." *Monist* 55 (1971): 368–91.

———. "Why Was the Logic of Discovery Abandoned?" In *Scientific Discovery: Case Studies*, edited by Thomas Nickles, pp. 173–83. Dordrecht: Reidel, 1980.

Leary, John E. Jr. *Francis Bacon and the Politics of Science.* Ames, IA: Four State University Press, 1994.

Lees, Lynn Hollen. *The Solidarities of Strangers: The English Poor Laws and the People, 1700–1948.* Cambridge: Cambridge University Press, 1998.

Leplin, Jarrett. *A Novel Defense of Scientific Realism.* New York: Oxford University Press, 1997.

Letwin, Shirley R. *The Pursuit of Certainty.* Cambridge: Cambridge University Press, 1965.

Levere, T. R. *Poetry Realized in Nature: Samuel Taylor Coleridge and Early 19th Century Science.* New York: Cambridge University Press, 1981.

———. "Samuel Taylor Coleridge on Nature and Reason: With a Response from William Whewell." *The European Legacy* 1 (1996): 1683–93.

Levy, David. *How the Dismal Science Got Its Name: Classical Economics and the Ur Text of Radical Politics.* Ann Arbor: University of Michigan Press, 2001.

Lipkes, Jeff. *Politics, Religion, and Classical Political Economy in Britain.* London: Macmillan, 1999.

Lipton, Peter. *Inference to the Best Explanation.* London: Routledge, 1991.

Lloyd, Elisabeth A. "Feyerabend, Mill, and Pluralism." In *The Worst Enemy of Science?* edited by John Preston, Gonzalo Munévar, and David Lamb, pp. 155–24. New York: Oxford University Press, 2000.

Lloyd, Humphrey. "Report on the Progress and Present State of Physical Optics." *Report of the Fourth Meeting of the British Association for the Advancement of Science, Held in Edinburgh in 1834*, pp. 295–413. London: John Murray, 1835.

Lloyd, Trevor. "John Stuart Mill and the East India Company." In Laine, *A Cultivated Mind*, pp. 44–79.

Locke, John. *An Essay concerning Human Understanding*. Edited by P. H. Nidditch. Oxford: Oxford University Press, Clarendon Press, 1979.

Losee, John. "Whewell and Mill on the Relation between Philosophy of Science and History of Science." *Studies in History and Philosophy of Science* 14 (1983): 113–26.

Lugg, Andrew. "The Process of Discovery." *Philosophy of Science* 52 (1985): 207–20.

———. "History, Discovery and Induction: Whewell on Kepler on the Orbit of Mars." In Brown and Mittelstrass, *An Intimate Relation*, pp. 283–98.

Lyell, Charles. "State of the Universities." *Quarterly Review* 36 (1827): 216–68.

———. *Travels in North America, with Geological Observations on the United States, Canada and Nova Scotia*. 2 vols. London: J. Murray, 1845.

Lyell, Katherine M., ed. *Life, Letters and Journals of Sir C. Lyell*. 2 vols. London: John Murray, 1881.

Macaulay, T. B. "Lord Bacon." *Edinburgh Review* 65 (1837): 1–104.

———. "Utilitarian Theory of Government." In *Macaulay's Miscellaneous Writings and Speeches*, pp. 205–25. New York: Longmans, Green and Co., 1889.

———. "Westminster Reviewer's Defense of Mill." In *Macaulay's Miscellaneous Writings and Speeches*, pp. 184–204. New York: Longmans, Green and Co., 1889.

Macfarlane, Alexander. *Lectures on Ten British Physicists of the Nineteenth Century*. New York: John Wiley, 1919.

MacIntyre, Alasdair. *After Virtue*. London: Duckworth, 1981.

Macleod, Roy M. "Whigs and Savants: Reflections on the Reform Movement in the Royal Society, 1838–48." In *Metropolis and Province*, edited by J. Inkster and J. Morrell, pp. 55–90. London: Hutchinson, 1983.

Malthus, T. R. *An Essay on the Principle of Population; or, A View of Its Past and Present Effects on Human Happiness; with an Inquiry into Our Prospects respecting the Future Removal or Mitigation of the Evils Which It Occasions*. Selected and introduced by Donald Winch, using the text of the 1803 edition as prepared by Patricia James for the Royal Economics Society, 1990. Cambridge: Cambridge University Press, 1992.

———. *The Grounds of an Opinion on the Policy of Restricting the Importation of Foreign Corn*. 1815. Reprinted in *The Pamphlets of T. R. Malthus*, pp. 137–73. New York: Augustus M. Kelley, 1970.

———. *The Works of Thomas Robert Malthus*. Edited by E. A. Wrigley and D. Souden. 8 vols. London: Pickering, 1986.

———. *Principles of Political Economy*. Variorum Edition. Edited by John Pullen. Cambridge: Cambridge University Press for the Royal Economic Society, 1989.

Mandler, Peter. "Tories and Paupers: Christian Political Economy and the Making of the New Poor Law." *The Historical Journal* 33 (1990): 81–103.

Manier, Edward. *The Young Darwin and His Cultural Circle*. Dordrecht: Reidel, 1978.

Mansel, Henry Longueville. *Prolegomena Logica: An Inquiry into the Psychological Character of Logical Processes.* 2nd ed. Boston: Gould & Lincoln, 1860.

Marcucci, Silvestro. *L' "Idealismo" Scientifico di William Whewell.* Pisa: Pubblicazioni dell'Istituto di Filosofia, 1963.

———. "William Whewell: Kantianism or Platonism?" *Physis* 12 (1970): 69–72.

Marshall, James Garth. *Minorities and Majorities: Their Relative Rights; A Letter to The Lord John Russell on Parliamentary Reform.* London, 1853.

Martin, Julian. *Francis Bacon, the State, and the Reform of Natural Philosophy.* Cambridge: Cambridge University Press, 1992.

Martineau, Harriet. *Cousin Marshall.* London: Charles Fox, 1832.

Masson, David. *Recent British Philosophy.* New York: Appleton, 1866.

Maxwell, J. C. *Treatise on Electricity and Magnetism.* 3rd ed. 2 vols. Slightly altered reprint of the 1891 ed. New York: Dover, 1954.

———. *Scientific Letters and Papers of James Clerk Maxwell.* Edited by Peter M. Harman. Cambridge: Cambridge University Press, 1990.

McKerrow, Raymie E. "Richard Whately and the Revival of Logic in Nineteenth-Century England." *Rhetorica* 5 (1987): 163–85.

McMullin, Ernan. "Conceptions of Science in the Scientific Revolution." In *Reappraisals of the Scientific Revolution,* edited by D.C. Lindberg and R. S. Westman, pp. 27–92. New York: Cambridge University Press, 1990.

———. *The Inference That Makes Science.* Milwaukee: Marquette University Press, 1992.

———. "The Impact of Newton's *Principia* on the Philosophy of Science." *Philosophy of Science* 68 (2001): 279–310.

McNally, David H. "Science and the Divine Order: Law, Idea and Method in William Whewell's Philosophy of Science." Ph.D. diss., University of Toronto, 1982.

Mehta, Uday Singh. *Liberalism and Empire: A Study in 19th Century British Liberal Thought.* Chicago: University of Chicago Press, 1999.

Metcalfe, John E. "Whewell's Developmental Psychologism: A Victorian Account of Scientific Progress." *Studies in History and Philosophy of Science* 22 (1991): 117–39.

Merz, J. T. *A History of European Thought in the Nineteenth Century.* 4 vols. Edinburgh: Blackwood, 1896–1914.

Metz, Rudolf. *One Hundred Years of British Philosophy.* London: George Allen & Unwin Ltd, 1938.

Milgate, Murray, and Shannon C. Stimson. *Ricardian Politics.* Princeton, NJ: Princeton University Press, 1991.

Mill, James. *Analysis of the Phenomena of the Human Mind.* 2 vols. London: Baldwin and Cradock, 1829.

———. *Analysis of the Phenomena of the Human Mind. New Edition, with Notes Illustrative and Critical by Alexander Bain, Andrew Findlater, and George Grote, Edited, with Additional Notes, by John Stuart Mill.* 2 vols. London: Longmans et al., 1869. Unless otherwise noted, this is the edition cited in the text.

Mill, John Stuart. "Lord Brougham's Speech on the Poor Law Amendment Bill." *Monthly Repository* 7 (1833): 597.

———. "Notes on the Newspapers." *Monthly Repository* 8 (1834): 592.

———. "The Proposed Reform of the Poor Laws." *Monthly Repository* 8 (1834): 361.

———. *The Collected Works of John Stuart Mill.* John M. Robson, general editor. 33 vols. Toronto: University of Toronto Press; London: Routledge and Kegan Paul, 1963–91. [CW]

Miller, William L. "Richard Jones: A Case Study in Methodology." *History of Political Economy* 3 (1972): 198–207.

Milton, J. R. "Induction before Hume." *British Journal for the Philosophy of Science* 38 (1987): 49–74.

Morley, John. "Mr. Mill's Autobiography." *Fortnightly Review* 21 (January 1, 1874): 1–20.

Morrell, Jack. "The Judge and Purifier of All." *History of Science* 30 (1992): 97–113.

Morrell, Jack, and Arnold Thackray. *Gentlemen of Science: Early Years of the British Association for the Advancement of Science.* Oxford: Clarendon Press, 1981.

Morrison, Margaret. "Unification, Realism and Inference." *British Journal for the Philosophy of Science* 41 (1990): 305–32.

———. "Whewell on the Ultimate Problem of Philosophy." *Studies in History and Philosophy of Science* 28 (1997): 417–37.

Nesbitt, G. L. *Benthamite Reviewing.* New York: Columbia University Press, 1934.

Newton, Isaac. *Mathematical Principles of Natural Philosophy.* 1687. Translated by Andrew Motte, 1729, with an introduction by I. Bernard Cohen. In 2 vols. London: Dawson, 1968.

———. *Opticks.* Based on the 4th ed. (London, 1730). With a preface by I. Bernard Cohen. New York: Dover Publications, 1979.

———. *The Correspondence of Isaac Newton.* Edited by H. W. Turnbull. 2 vols. Cambridge: Cambridge University Press, 1957–60.

Nicholson, Peter. "The Reception and Early Reputation of Mill's Political Thought." In Skorupski, *The Cambridge Companion to Mill,* pp. 464–96.

Nickles, Thomas. "Beyond Divorce: Current Status of the Discovery Debate." *Philosophy of Science* 52 (1985): 177–206.

———. "From Natural Philosophy to Metaphilosophy of Science." In *Kelvin's Baltimore Lectures and Modern Theoretical Physics,* edited by Robert Kargon and Peter Achinstein, pp. 507–41. Cambridge, MA: MIT Press, 1987.

———. "Enlightenment versus Romantic Models of Creativity in Science—and Beyond." *Creativity Research Journal* 7 (1994): 277–314.

Niiniluoto, Ilkka. "Notes on Popper as a Follower of Whewell and Peirce." *Ajatus* 37 (1977): 272–327.

———. "Hintikka and Whewell on Aristotelian Induction." *Grazer Philosophische Studien* 49 (1994/95): 49–61.

Nockles, Peter B. "Rose, Hugh James." *Oxford Dictionary of National Biography.* Oxford: Oxford University Press, 2004. http://www.oxforddnb.com/view/article/24094.

Nussbaum, Martha. *Cultivating Humanity.* Cambridge, MA: Harvard University Press, 1997.

O'Hanlon, H. F. *A Criticism of John Stuart Mill's Pure Idealism; and an Attempt to Shew That, If Logically Carried Out, It Is Pure Nihilism.* Oxford: Parker, 1866.

Oldroyd, D. R. "How Did Darwin Arrive at His Theory? The Secondary Literature to 1982." *History of Science* 22 (1984): 325–74.

Olson, Richard. *Scottish Philosophy and British Physics, 1750–1880.* Princeton, NJ: Princeton University Press, 1975.

Orsini, G. N. G. *Coleridge and German Idealism.* Carbondale: Southern Illinois University Press, 1969.

Ospovat, Dov. *The Development of Darwin's Theory.* Cambridge: Cambridge University Press, 1981.

Owen, Richard. *Lectures on the Comparative Anatomy and Physiology of the Invertebrate Animals.* London: Longmans, Green, 1843.

———. *On the Nature of Limbs.* London, John van Voorst, 1849.

———. "Review of *Origin of Species* and Other Works." *Edinburgh Review, or Critical Journal* 111 (1860): 487–532. Reprinted in Hull, *Darwin and His Critics,* 175–213.

Owen, Robert. *New View of Society.* London: Longman, 1817.

Packe, Michael St. John. *The Life of John Stuart Mill.* New York: The Macmillan Co., 1954.

Paley, William. *Moral and Political Philosophy.* London: T. and J. Allman, 1826.

———. *Natural Theology; or, Evidences of the Existence and Attributes of the Deity, Collected from the Appearances of Nature.* Cambridge: Hilliard and Brown, 1830.

Paradis, James, and Thomas Postlewait, eds. *Victorian Science and Victorian Values: Literary Perspectives.* New Brunswick, NJ: Rutgers University Press, 1985.

Park, Katherine, Lorraine J. Daston, and Peter L. Galison. "Bacon, Galileo and Descartes on Imagination and Analogy." *Isis* 75 (1984): 287–326.

Pattison, Mark. "J. S. Mill on Hamilton." *The Reader,* May 20, 1865.

Peirce, Charles S. *Collected Papers of Charles Sanders Peirce.* Edited by Charles Hartshorne and Paul Weiss. Vol. 1. Cambridge, MA: Harvard University Press, 1960.

———. "Lecture on the Theories of Whewell, Mill, and Comte." In *Writings of Charles S. Peirce: A Chronological Edition,* edited by Max H. Fisch, 1:205–23. Bloomington: Indiana University Press, 1982.

———. "Whewell." In Fisch, *Writings of Charles S. Peirce,* 2:337–45. Bloomington: Indiana University Press, 1984.

Perez-Ramos, Antonio. *Francis Bacon's Idea of Science and the Maker's Knowledge Tradition.* Oxford: Clarendon Press, 1988.

———. "Francis Bacon and the Disputations of the Learned (Review Article)." *British Journal for the Philosophy of Science* 42 (1991): 577–88.

Pettit, Philip. *Republicanism: A Theory of Freedom and Government.* Oxford: Oxford University Press, 1997.

Pictet, François Jules. "Sur l'Origine de l'espèce." *Bibliothèque Universelle. Revue Suisse et Etrangère. Archives des Sciences Physique et Naturelles,* n.s., 7 (1860): 233–55. Reprinted and translated in Hull, *Darwin and His Critics,* pp. 142–52.

Pocock, J. G. A. *Politics, Language and Time. Essays in Political Thought and History.* Chicago: University of Chicago Press, 1989.

Polanyi, Karl. *The Great Transformation.* New York: Rinehart, 1957.

Popper, Karl R. *The Logic of Scientific Discovery.* New York: Basic Books, 1959.

———. *Conjectures and Refutations: The Growth of Scientific Knowledge.* London: Routledge and Kegan Paul, 1963.

Porter, Theodore M. *The Rise of Statistical Thinking, 1820–1900.* Princeton, NJ: Princeton University Press, 1986.

Poynter, J. R. *Society and Pauperism: English Ideas on Poor Relief, 1795–1834.* London: Routledge and Kegan Paul, 1969.

Preyer, Robert O. "The Romantic Tide Reaches Trinity: Notes on the Transmission and Diffusion of New Approaches to Traditional Studies at Cambridge, 1820–40." In Paradis and Postlewait, *Victorian Science and Victorian Values,* pp. 39–68.

Priestley, Joseph. *Hartley's Theory of the Human Mind: or, The Principle of Association of Ideas.* London, 1775.

Pullen, J. M. "Malthus' Theological Ideas and Their Influence on His Principle of Population." *History of Political Economy* 13 (1981): 39–54.

Quinton, Anthony. *Francis Bacon.* Oxford: Oxford University Press, 1980.

Randall, J. H. "John Stuart Mill and the Working-Out of Empiricism." *Journal of the History of Ideas* 26 (1965): 59–88.

Rashid, Salim. "Dugald Stewart, 'Baconian' Methodology, and Political Economy." *Journal of the History of Ideas* 46 (1985): 245–57.

———. "Malthus' *Principles* and British Economic Thought, 1820–1835." In Wood, *Thomas Robert Malthus,* 2:217–37.

Rauch, Alan. *Useful Knowledge: The Victorians, Morality, and the March of the Intellect.* Durham, NC: Duke University Press, 2001.

Rawls, John. *A Theory of Justice.* Cambridge, MA: Harvard University Press, 1971.

———. *Collected Papers.* Edited by Samuel Freeman. Cambridge, MA: Harvard University Press, 1999.

Recker, Doren. "Causal Efficacy: The Structure of Darwin's Argument Strategy in the *Origin of Species.*" *Philosophy of Science* 54 (1987): 147–75.

Redman, Deborah. *The Rise of Political Economy as a Science: Methodology and the Classical Economists.* Cambridge, MA: MIT Press, 1997.

Reid, Thomas. *Essays on the Intellectual Powers of Man.* Edinburgh: John Bell, 1785.

Ricardo, David. *The Works and Correspondence of David Ricardo.* Edited by Piero Sraffa. 11 vols. Cambridge: Cambridge University Press, 1951–73.

Richards, Joan L. *Mathematical Visions: The Pursuit of Geometry in Victorian England.* Boston: Academic Press, 1988.

Richards, Robert J. *Darwin and the Emergence of Evolutionary Theories of Mind and Behavior.* Chicago: University of Chicago Press, 1987.

———. "Kant and Blumenbach on the *Bildungstrieb:* A Historical Misunderstanding." *Studies in History and Philosophy of Science Part C* 31 (2000): 11–32.

———. *The Romantic Conception of Life: Science and Philosophy in the Age of Goethe.* Chicago: University of Chicago Press, 2002.

Riley, Jonathan. *Liberal Utilitarianism: Social Choice Theory and J. S. Mill's Philosophy.* Cambridge: Cambridge University Press, 1988.

———. "Mill's Political Economy: Ricardian Science and Liberal Utilitarian Art." In Skorupski, *The Cambridge Companion to Mill,* pp. 293–337.

Robson, John M. Introduction to vol. 31 of John Stuart Mill, *The Collected Works of John Stuart Mill,* pp. vii–l.

———. Textual introduction to Mill's *Examination.* In John Stuart Mill, *The Collected Works of John Stuart Mill,* 9:lxix–cii.

———. *The Improvement of Mankind: The Social and Political Thought of John Stuart Mill.* London: Routledge and Kegan Paul, 1968.

Robson, Robert. "William Whewell, FRS: Academic Life." *Notes and Records of the Royal Society of London* 19 (1964): 168–76.

———. "Trinity College in the Age of Peel." In *Ideas and Institutions of Victorian Britain,* pp. 312–35. London: G. Bell and Sons, 1967.

Roebuck, John. "Democracy in America." In *Pamphlets for the People,* edited by John Roebuck. 2 vols. London: Charles Ely, 1835.

Rosenblum, Nancy, ed. *Liberalism and the Moral Life.* Cambridge, MA: Harvard University Press, 1989.

Ross, Sidney. "Faraday Consults the Scholars: The Origin of the Terms of Electrochemistry." *Notes and Records of the Royal Society* 16 (1961): 187–220.

———. "'Scientist': The Story of a Word." *Annals of Science* 18 (1962): 65–85.

Rossi, Paolo. "Ants, Spiders, Epistemologists." In *Francis Bacon: Terminologia e Fortuna nel XVII Secolo,* edited by M. Fattori, pp. 245–60. Rome: Edizioni dell'Ateneo, 1984.

———. "Bacon's Idea of Science." In *The Cambridge Companion to Bacon,* edited by M. Peltonen, pp. 25–46. Cambridge: Cambridge University Press, 1996.

Rothblatt, Sheldon. *Tradition and Change in English Liberal Education.* London: Faber and Faber, 1976.

———. *The Revolution of the Dons: Cambridge and Society in Victorian England.* 2nd ed. Cambridge: Cambridge University Press, 1981.

Rupke, Nicolaas. *Richard Owen: Victorian Naturalist.* New Haven, CT: Yale University Press, 1994.

Ruse, Michael. "Darwin's Debt to Philosophy: An Examination of the Influence of the Philosophical Ideas of John F. W. Herschel and William Whewell on the Development of Charles Darwin's Theory of Evolution." *Studies in History and Philosophy of Science* 6 (1975): 159–81.

————. "Charles Lyell and the Philosophers of Science." *British Journal for the History of Science* 9 (1976): 121–31.

————. "The Scientific Methodology of William Whewell." *Centaurus* 20 (1976): 227–57.

————. "William Whewell and the Argument from Design." *Monist* 60 (1977): 244–68.

————. *The Darwinian Revolution: Science Red in Tooth and Claw*. Chicago: University of Chicago Press, 1979.

————. "Biological Species: Natural Kinds, Individuals, or What?" *British Journal of the Philosophy of Science* 38 (1987): 225–42.

————. *The Darwinian Paradigm: Essays on Its History, Philosophy, and Religious Implications*. London: Routledge, 1989.

————. "William Whewell: Omniscientist." In Fisch and Schaffer, *William Whewell*, pp. 87–116.

Ryan, Alan. Introduction to Mill's *Examination*. In John Stuart Mill, *The Collected Works of John Stuart Mill*, 9:vii–lxvii.

————. *John Stuart Mill*. New York: Pantheon, 1970.

————. *J. S. Mill*. London: Routledge, 1974.

————. *The Philosophy of John Stuart Mill*. Atlantic Highlands, NJ: Humanities Press International, 1990.

————. "Two Concepts of Politics and Democracy: James and John Stuart Mill." In G. W. Smith, *John Stuart Mill's Social and Political Thought*, 3:138–61.

Sandel, Michael J. *Liberalism and the Limits of Justice*. Cambridge: Cambridge University Press, 1982.

Santurri, E. N. "Theodicy and Social Policy in Malthus's Thought." In Wood, *Thomas Robert Malthus*, 1:402–18.

Sargent, Rose-Mary. *The Diffident Naturalist: Robert Boyle and the Philosophy of Experiment*. Chicago: University of Chicago Press, 1995.

Scarre, Geoffrey. *Logic and Reality in the Philosophy of John Stuart Mill*. Dordrecht: Reidel, 1989.

————. *Utilitarianism*. London: Routledge, 1996.

Schabas, Margaret. *A World Ruled by Number: William Stanley Jevons and the Rise of Mathematical Economics*. Princeton, NJ: Princeton University Press, 1990.

Schaffer, Simon. "Genius in Natural Philosophy." In Cunningham and Jardine, *Romanticism and the Sciences*, pp. 82–98.

————. "The History and Geography of the Intellectual World: Whewell's Politics of Language." In Fisch and Schaffer, *William Whewell*, pp. 201–31.

Schagrin, Morton L. "Whewell's Theory of Scientific Language." *Studies in the History and Philosophy of Science* 4 (1973): 231–40.

Schipper, F. "William Whewell's Conception of Scientific Revolutions." *Studies in History and Philosophy of Science* 19 (1988): 43–53.

Schneewind, Jerome B. "Whewell's Ethics." In *Studies in Moral Philosophy*, edited by Nicholas Rescher, pp. 108–41. American Philosophical Quarterly Monograph Series, no. 1. Oxford: Basil Blackwell, 1968.

————. *Sidgwick's Ethics and Victorian Moral Philosophy.* Oxford: Oxford University Press, 1977.

Schumpeter, Joseph A. *Economic Doctrine and Method: An Historical Sketch.* London: George Allen & Unwin Ltd, 1954.

Schweber, S. S. "The Origin of the *Origin* Revisited." *Journal of the History of Biology* 10 (1977): 229–316.

————. "Darwin and the Political Economists." *Journal of the History of Biology* 13 (1980): 195–289.

————, ed. *Aspects of the Life and Thought of Sir John Frederick Herschel.* New York: Arno Press, 1981.

————. "Scientists as Intellectuals: The Early Victorians." In Paradis and Postlewait, *Victorian Science and Victorian Values,* pp. 1–37.

Searle, G. R. *Morality and the Market in Victorian Britain.* Oxford: Clarendon Press, 1998.

Sedgwick, Adam. *A Discourse on the Studies of the University of Cambridge.* 3rd ed. Cambridge: J. & J. J. Deighton, 1834.

————. "Review of *Vestiges of the Natural History of Creation.*" *Edinburgh Review* 82 (1845): 1–85.

————. "Objections to Mr. Darwin's Theory of the Origin of Species." *Spectator,* March 24, 1860, pp. 285–86; reprinted, with corrections, April 7, 1860, pp. 334–35. Reprinted in Hull, *Darwin and His Critics,* pp. 159–66.

Semmel, Bernard. *John Stuart Mill and the Pursuit of Virtue.* New Haven, CT: Yale University Press, 1983.

Senior, Nassau. *Four Introductory Lectures on Political Economy.* 1852. Reprinted in Nassau Senior, *Selected Writings on Economics.* New York: A. M. Kelley, 1966.

Shairp, John Campbell, P. J. Tate, and A. Adams Reilly, eds. *Life and Letters of James David Forbes, FRS.* London: Macmillan, 1873.

Shapiro, Barbara. "Sir Francis Bacon and the Mid-17th Century Movement for Law Reform." *American Journal of Legal History* 24 (1980): 331–62.

Sidgwick, Henry. "Philosophy at Cambridge." *Mind* 1 (1876): 235–46.

————. *Principles of Political Economy.* London: Macmillan & Co., 1883.

————. *The Methods of Ethics.* Foreword by John Rawls. Indianapolis: Hackett Publishing Co., 1981.

Skinner, Quentin. "Meaning and Understanding in the History of Ideas." *History and Theory* 8 (1969): 3–53.

————. *The Foundations of Modern Political Thought.* 2 vols. Cambridge: Cambridge University Press, 1978.

————. *Liberty before Liberalism.* Cambridge: Cambridge University Press, 1997.

Skorupski, John. *John Stuart Mill.* London: Routledge, 1989.

————. *English Language Philosophy: 1750–1945.* Oxford: Oxford University Press, 1993.

————, ed. *The Cambridge Companion to Mill.* Cambridge: Cambridge University Press, 1998.

Sloan, Phillip R. "Preforming the Categories: Eighteenth-Century Generation The-
ory and the Biological Roots of Kant's A Priori." *Journal of the History of Philos-
ophy* 40 (2002): 229–53.
———. "Whewell's *Philosophy of Discovery* and the Archetype of the Vertebrate
Skeleton: The Role of German Philosophy of Science in Richard Owen's Biology."
Annals of Science 60 (2003): 39–61.
Small, Robert. *An Account of the Astronomical Discoveries of Kepler.* 1804. Reprint,
Madison: University of Wisconsin Press, 1963.
Smith, G. W. "J. S. Mill on Edger and Réville: An Episode in the Development of
Mill's Conception of Freedom." In Wood, *John Stuart Mill,* 4:550–66.
———. "The Logic of J. S. Mill on Freedom." In Wood, *John Stuart Mill,* 4:534–49.
———, ed. *John Stuart Mill's Social and Political Thought.* 4 vols. London: Rout-
ledge, 1998.
Smith, Robert. "The Cambridge Network in Action: The Discovery of Neptune." *Isis*
80 (1989): 395–422.
Snyder, Laura J. "It's *All* Necessarily So: William Whewell on Scientific Truth." *Stud-
ies in History and Philosophy of Science* 25 (1994): 785–807.
———. "The Method of Induction." Ph.D. diss., Johns Hopkins University, 1996.
———. "Discoverers' Induction." *Philosophy of Science* 64 (1997): 580–604.
———. "The Mill-Whewell Debate: Much Ado about Induction." *Perspectives on
Science* 5 (1997): 159–98.
———. "Is Evidence Historical?" In *Philosophy of Science: The Central Issues,*
edited by Martin Curd and J. A. Cover, pp. 460–80. New York: W. W. Norton &
Company, 1998.
———. "Renovating the *Novum Organum:* Bacon, Whewell, and Induction." *Studies
in History and Philosophy of Science* 30A (1999): 531–57.
———. "Whewell, William." *Stanford University Online Encyclopedia of Philosophy,* December 2000; enlarged and updated edition, January 2004. http://plato
.stanford.edu.
———. "Whewell and the Scientists: Science and Philosophy of Science in 19th Century Britain." In *History of Philosophy of Science: New Trends and Perspectives,*
edited by M. Heidelberger and F. Stadler, pp. 81–94. Dordrecht: Kluwer, 2002.
———. "'Gegen alle vernunftbegabten Bewohner anderer Welten': William Whewell
und die Debatte um die Vielzahl der Welten." In *Science & Fiction II: Leben auf
anderen Sternen,* edited by Thomas Weber, pp. 89–112. Munich: Fischer Verlag,
2004.
———. "Consilience, Confirmation, and Realism." In *Scientific Evidence: Philosophical Theories and Applications,* edited by Peter Achinstein, pp. 129–48.
Baltimore: Johns Hopkins University Press, 2005.
Soloway, R. A. *Prelates and People: Ecclesiastical and Social Thought in England
1783–1852.* London: Routledge & Kegan Paul; Toronto: University of Toronto
Press, 1969.

Sowell, Tom. "Malthus and the Utilitarians." In Wood, *Thomas Robert Malthus*, 1:210–16.

Spencer, Herbert. "Mill *versus* Hamilton—The Test of Truth." *Fortnightly Review* 1 (July 15, 1865): 531–50.

Stair Douglas, Janet Mary. *The Life and Selections from the Correspondence of William Whewell, D.D.* London: C. Kegan and Paul, 1882.

Staley, Kent W. "Logic, Liberty, and Anarchy: Mill and Feyerabend on Scientific Method." *The Social Sciences Journal* 35 (1999): 603–14.

Stephen, James Fitzjames. "Mr. Mill on Political Liberty." *Saturday Review*, February 12, 1859.

———. "Money and Money's Worth." *Cornhill Magazine* 9 (1864): 97–109.

Stephen, Leslie. "Whewell, William." In *Dictionary of National Biography*, L. Stephen, general editor, 20:1365–74. Oxford: Oxford University Press, 1889.

———. *The Life of James Fitzjames Stephen.* London: Smith, Elder, 1895.

———. *The English Utilitarians.* 3 vols.: 1: *Jeremy Bentham*; 2: *James Mill*; 3: *John Stuart Mill*. London: Duckworth, 1900.

Stephenson, Bruce. *Kepler's Physical Astronomy.* Princeton, NJ: Princeton University Press, 1987.

Stewart, Dugald. *Elements of the Philosophy of the Human Mind.* In 3 vols. London: A. Strahan and T. Cadell; Edinburgh: W. Creech, 1792–1827.

Stillinger, Jack. "John Mill's Education: Fact, Fiction and Myth." In Laine, *A Cultivated Mind*, pp. 19–43.

Stoll, Marion Rush. *Whewell's Philosophy of Induction.* Lancaster, PA.: Lancaster Press, 1929.

Strong, E. W. "William Whewell and John Stuart Mill: Their Controversy over Scientific Knowledge." *Journal of the History of Ideas* 16 (1955): 209–31.

Strong, John Vincent. "Studies in the Logic of Theory Assessment in Early Victorian Britain, 1838–1860." Ph.D. diss., University of Pittsburgh, 1978.

Taylor, Charles. *Philosophical Papers of Charles Taylor.* 2 vols. Cambridge: Cambridge University Press, 1985.

———. "Cross Purposes: The Liberal-Communitarian Debate." In Rosenblum, *Liberalism and the Moral Life*, pp. 159–82.

Ten, C. L. "Mill's Place in Liberalism." *Political Science Reviewer* 24 (1995): 179–204.

———. *Mill's Moral, Political, and Legal Philosophy.* Aldershot, UK: Ashgate, 1999.

Tennyson, Hallam. *Alfred, Lord Tennyson: A Memoir.* New York: Macmillan, 1897.

Thagard, Paul. "Darwin and Whewell." *Studies in the History and Philosophy of Science* 8 (1977): 353–56.

Tholfsen, Trygve R. *Working Class Radicalism in Mid-Victorian England.* New York: Columbia University Press, 1977.

Thompson, Dennis F. *John Stuart Mill and Representative Government.* Princeton, NJ: Princeton University Press, 1979.

Thompson, E. P. *The Making of the Working Class in England*. London: Victor Gol-
   lancz, 1963.
Thompson, T. Perronet. *The True Theory of Rent in Opposition to Mr. Ricardo and
   Others*. London: Hatchard and Son; C. & J. Rivington, 1826.
Todhunter, Isaac. *William Whewell, D.D. An Account of His Writings, with Selec-
   tions from His Literary and Scientific Correspondence*. 2 vols. London: Macmil-
   lan, 1876.
Tricker, R. A. R. *The Contributions of Faraday and Maxwell to Electrical Science*.
   Oxford: Pergamon, 1966.
Tully, James, ed. *Meaning and Context: Quentin Skinner and His Critics*. Princeton,
   NJ: Princeton University Press, 1988.
Turk, Christopher. *Coleridge and Mill*. Aldershot, UK: Avebury, 1988.
Turner, Frank M. *Between Science and Religion: The Reaction to Scientific Natural-
   ism in Late Victorian England*. New Haven, CT: Yale University Press, 1974.
———. *The Greek Heritage in Victorian Britain*. New Haven, CT: Yale University
   Press, 1981.
———. *Contesting Cultural Authority: Essays in Victorian Intellectual Life*. New
   York: Cambridge University Press, 1993.
Turton, Thomas. *Thoughts on the Admission of Persons without Regard to Their
   Religious Opinions to Certain Degrees in the Universities of England*. Cam-
   bridge and London, 1834.
Tyndall, John. *Faraday as a Discoverer*. 1870. Reprint, New York: D. Appleton & Co.,
   1873.
Urbach, Peter. *Francis Bacon's Philosophy of Science*. La Salle, IL: Open Court Press,
   1987.
Urbinati, Nadia. *Mill on Democracy: From Athenian Polis to Representative Gov-
   ernment*. Chicago: University of Chicago Press, 2002.
Valone, David. "The Dark and Tangled Recesses of Knowledge: Theology and Moral
   Sciences at Cambridge, 1812–1837." Ph.D. diss., University of Chicago, 1994.
Van Fraassen, Bas C. *The Scientific Image*. Oxford: Oxford University Press, 1980.
———. "Empiricism in Philosophy of Science." In *Images of Science: Essays in Real-
   ism and Empiricism*, edited by Paul M. Churchland and Clifford A. Hooker, 245–
   308. Chicago: University of Chicago Press, 1985.
Venn, John. *The Logic of Chance*. 3rd ed. 1888. Reprint, New York: Macmillan, 1962.
———. *The Principles of Empirical or Inductive Logic*. 2nd ed. 1889. Reprint, Lon-
   don: Macmillan, 1907.
Viner, J. "Bentham and J. S. Mill: The Utilitarian Background." In G. W. Smith, *John
   Stuart Mill's Social and Political Thought*, 1:145–64.
Waldron, Jeremy. "Mill and Moral Distress." In *Liberal Rights: Collected Papers,
   1981–1991*. Cambridge: Cambridge University Press, 1993.
Walsh, H. T. "Whewell on Necessity." *Philosophy of Science* 29 (1962): 139–45.
Warwick, Andrew. *Masters of Theory: Cambridge and Rise of Mathematical Physics*.
   Chicago: University of Chicago Press, 2003.

Webster, Charles. *The Great Instauration: Science, Medicine and Reform, 1626–1660.* London: Duckworth, 1975.

Wellek, René. *Immanuel Kant in England: 1793–1838.* Princeton, NJ: Princeton University Press, 1931.

Wettersten, John. "Discussion: William Whewell; Problems of Induction vs. Problems of Rationality." *British Journal for the Philosophy of Science* 45 (1994): 716–42.

Wettersten, John, and Joseph Agassi. "Whewell's Problematic Heritage." In Fisch and Schaffer, *William Whewell,* pp. 345–69.

Whately, Richard. *Elements of Logic, with Additions.* 2nd ed. London, 1827.

———. *Introductory Lectures on Political Economy.* 2nd ed. London: B. Fellowes, 1832.

———. *Paley's "Moral and Political Philosophy."* With Annotations. London: J. W. Parker & Son, 1859.

Whewell, William. *An Elementary Treatise on Mechanics.* Cambridge: J. Deighton and Sons, 1819.

———. "On the Double Crystals of Flour Spa," paper presented November 26, 1821. *Transactions of the Cambridge Philosophical Society* 1, pt. II (1822): 331–42.

———. "On the Angle Made by Two Planes, or Two Straight Lines, Referred to Three Oblique Coordinates," paper presented November 24, 1823. *Transactions of the Cambridge Philosophical Society* 2, pt. I (1827): 197–202.

———. "A General Method of Calculating the Angles Made by Any Plane of Crystals, and the Laws according to Which They Are Formed," paper presented November 25, 1824. *Philosophical Transactions of the Royal Society of London* 115, pt. I (1825): 87–130.

———. *Account of Experiments Made at Dolcoath Mine, in Cornwall, in 1826 and 1828.* Cambridge: J. Smith, 1828.

———. *Commemoration Sermon Preached in the Chapel of Trinity College, Cambridge, December 17, 1828.* Cambridge: J. Smith, 1828.

———. *An Essay on Mineralogical Classification and Nomenclature; with Tables of the Orders and Species of Minerals.* Cambridge: J. Smith, 1828.

———. "Mathematical Exposition of Some Doctrines of Political Economy," paper presented March 2 and 4, 1829. *Transactions of the Cambridge Philosophical Society* 3, pt. I (1830): 191–230.

———. "Jones—*On the Distribution of Wealth and the Sources of Taxation.*" *British Critic, Quarterly Theological Review* 10 (1831): 41–61.

———. "Lyell's *Principles of Geology,* Volume 1." *British Critic, Quarterly Theological Review* 9 (1831): 180–206.

———. "Mathematical Exposition of Some of the Leading Doctrines in Mr. Ricardo's *Principles of Political Economy and Taxation,*" paper presented April 18 and May 2, 1831. *Transactions of the Cambridge Philosophical Society* 4, pt. I (1833): 155–198.

———. "Modern Science—Inductive Philosophy." Review of *Preliminary Discourse*

on the Study of Natural Philosophy, by J. Herschel (1830). Quarterly Review 45
(1831): 374–407.

———. "Science of the English Universities." Review of the Transactions of the
Cambridge Philosophical Society, vol. 3, pts. 1, 2, and 3. British Critic 9 (1831):
71–90.

———. "Lyell's Principles of Geology, Volume 2." Quarterly Review 93 (1832):
103–32.

———. Astronomy and General Physics, Considered with Reference to Natural
Theology. Bridgewater Treatise III. London: William Pickering, 1833.

———. "Essay towards a First Approximation to a Map of Cotidal Lines," paper presented May 2, 1833. Philosophical Transactions of the Royal Society of London
123, pt. I (1833): 147–236.

———. Memoranda and Directions for Tide Observations. Cambridge: Hodson and
Brown, 1833.

———. "On the Uses of Definitions." Philological Museum 2 (1833): 263–72.

———. Additional Remarks on Some Parts of Mr. Thirlwall's Two Letters on the
Admission of Dissenters to Academical Degrees. Cambridge: J. & J. J. Deighton,
1834.

———. "Address." Report of the Third Meeting of the British Association for the
Advancement of Science, Held at Cambridge in 1833, pp. xi–xxvi. London: John
Murray, 1834.

———. "Mrs Somerville on the Connexion of the Sciences." Quarterly Review 51
(1834): 54–68.

———. "On the Empirical Laws of the Tides in the Port of London; with Reflexions
on the Theory," paper presented January 9, 1834. Philosophical Transactions of
the Royal Society of London 124, pt. I (1834): 15–45.

———. "On the Nature of the Truth of the Laws of Motion," paper presented February 17, 1834. Transactions of the Cambridge Philosophical Society 5, pt. II
(1835): 149–72. Reprinted in Whewell, The Philosophy of the Inductive Sciences,
pp. 573–94.

———. Remarks on Some Parts of Mr. Thirlwall's Letter on the Admission of
Dissenters to Academical Degrees. Cambridge: J. Smith and J. & J. J. Deighton,
1834.

———. "Researches on the Tides—Fourth Series. On the Empirical Laws of the
Tides in the Port of Liverpool," paper presented November 19, 1835. Philosophical Transactions of the Royal Society of London 126, pt. I (1836): 1–15.

———. "Suggestions Regarding Sir John Herschel's Remarks on the Theory of the
Absorption of Light by Coloured Media." Report of the Fourth Meeting of the
British Association for the Advancement of Science, held in Edinburgh in 1834,
pp. 550–52. London: John Murray, 1835.

———. Thoughts on the Study of Mathematics as Part of a Liberal Education. Cambridge: J. and J. J. Deighton, 1835. Reprinted in Whewell, On the Principles of
English University Education.

———. Preface to *Dissertation on the Progress of Ethical Philosophy, Chiefly during the Seventeenth and Eighteenth Centuries,* by James Mackintosh, pp. 1–46. Edinburgh: Adam and Charles Black, 1836.

———. "Researches on the Tides—6th Series. On the Results of an Extensive System of Tide Observations Made on the Coasts of Europe and America in June 1835," paper presented June 16, 1836. *Philosophical Transactions of the Royal Society of London* 126, pt. II (1836): 289–341.

———. *Thoughts on the Study of Mathematics as Part of a Liberal Education,* 2nd Edition, to Which Is Added a Letter to the Editor of the *"Edinburgh Review"* Occasioned by the Review of the First Edition. Cambridge: J. & J. J. Deighton; London: John W. Parker, 1836.

———. *The History of the Inductive Sciences, from the Earliest to the Present Time.* 1st ed. 3 vols. London: John W. Parker, 1837.

———. *On the Foundations of Morals: Four Sermons Preached before the University of Cambridge, November 1837.* Cambridge: The University Press, 1837.

———. *On the Principles of English University Education.* London: John W. Parker, 1837.

———. *Mechanical Euclid.* 3rd ed., corrected, with "Remarks on Mathematical Reasoning and on the Logic of Induction." Cambridge: J. & J. J. Deighton; London: John W. Parker, 1838.

———. "Presidential Address." Delivered February 16, 1838. *Proceedings of the Geological Society of London* 3 (1839): 624–49.

———. *Two Introductory Lectures to Two Courses of Lectures on Moral Philosophy.* Delivered in 1839 and 1841. Cambridge: John W. Parker, 1841.

———. "Remarks on the Review of the *Philosophy of the Inductive Sciences.*" Reply to DeMorgan. *Athenaeum,* no. 672 (1840): 3–8.

———. "Demonstration That All Matter Is Heavy," paper presented February 22, 1841. *Transactions of the Cambridge Philosophical Society* 7, pt. II (1841): 197–208. Reprinted in Whewell, *On the Philosophy of Discovery,* pp. 523–31.

———. *Architectural Notes on German Churches, with Notes Written during an Architectural Tour in Picardy and Normandy.* 3rd ed. Cambridge: J. & J. J. Deighton, 1842.

———. "On the Fundamental Antithesis of Philosophy," paper presented February 5, 1844. *Transactions of the Cambridge Philosophical Society* 8, pt. II (1844): 170–81.

———. "Remarks on a Review of the *Philosophy of the Inductive Sciences.*" Letter to John Herschel, April 11, 1844. Reprinted in Whewell, *On the Philosophy of Discovery,* pp. 482–91.

———. *Elements of Morality, Including Polity.* 2 vols. London: John W. Parker, 1845. References are to the 4th ed., in 1 vol. (London: John W. Parker, 1864).

———. *Indications of the Creator.* London: John W. Parker, 1845.

———. *Of a Liberal Education in General, and with Particular Reference to the Leading Studies of the University of Cambridge.* London, John W. Parker, 1845.

———. *Lectures on Systematic Morality, Delivered in Lent Term, 1846.* London: John W. Parker, 1846.

———. *The Philosophy of the Inductive Sciences, Founded upon Their History.* 2nd ed. 2 vols. London: John W. Parker, 1847. References are to this edition, unless otherwise noted.

———. *Verse Translations from the German, Including Burger's "Lenore," Schiller's "Song of the Bell," and Other Poems.* London: Murray, 1847.

———. Preface to *Butler's Three Sermons on Human Nature and Dissertation on Virtue,* edited by William Whewell, pp. i–lx. Cambridge: J. & J. J. Deighton; London: John W. Parker, 1848.

———. "Second Memoir on the Fundamental Antithesis of Philosophy," paper presented November 13, 1848. *Transactions of the Cambridge Philosophical Society* 8, pt. V (1849): 614–20.

———. *Of Induction, with Especial Reference to Mr. J. Stuart Mill's "System of Logic."* London: John W. Parker, 1849.

———. "On Hegel's Criticism of Newton's *Principia,*" paper presented May 21, 1849. *Transactions of the Cambridge Philosophical Society* 8, pt. V (1849): 696–701.

———. Preface to *Butler's Six Sermons on Moral Subjects,* edited by William Whewell, pp. i–xxviii. Cambridge: John Deighton; London: John W. Parker, 1849.

———. "Criticism of Aristotle's Account of Induction." Paper presented before the Cambridge Philosophical Society, February 11, 1850. Reprinted in Whewell, *On the Philosophy of Discovery,* pp. 449–61.

———. "Mathematical Exposition of Some Doctrines of Political Economy, Second Memoir," paper presented April 15, 1850. *Transactions of the Cambridge Philosophical Society* 9, pt. I (1856): 128–49.

———. "Mathematical Exposition of Certain Doctrines of Political Economy, Third Memoir," paper presented November 11, 1850. *Transactions of the Cambridge Philosophical Society* 9, pt. II (1856): 1–7.

———. *Of a Liberal Education in General, and with Especial Reference to the University of Cambridge, Part II.* London: John W. Parker, 1850.

———. *Inaugural Lecture: The General Bearing of the Great Exhibition of the Progress of Art and Science.* London, 1851.

———. "Of the Transformation of Hypotheses in the History of Science," paper presented May 19, 1851. *Transactions of the Cambridge Philosophical Society* 9, pt. II (1856): 139–46.

———. *Lectures on the History of Moral Philosophy in England.* London: J. W. Parker and Son, 1852.

———. *Of a Liberal Education in General, and with Especial Reference to the University of Cambridge, Part III.* London, John W. Parker, 1852.

———, trans. *Hugonis Grotii, De Jure Belli et Pacis, libri tres, Accompanied by an Abridged Translation, with the Notes of the Author, Barbeyrac, and others.* 3 vols. Cambridge: The University Press, 1853.

———. *Of the Plurality of Worlds: An Essay. Also, a Dialogue on the Same Subject.* 4th ed. London: J. W. Parker, 1855.

———. "Of the Intellectual Powers According to Plato." Paper presented to the Cambridge Philosophical Society November 10, 1856. Reprinted in Whewell, *On the Philosophy of Discovery,* pp. 440–48.

———. "Of the Platonic Theory of Ideas." Paper presented November 10, 1856. *Transactions of the Cambridge Philosophical Society* 10, pt. I (1864): 94–104. Reprinted in Whewell, *On the Philosophy of Discovery,* pp. 403–16.

———. *History of the Inductive Sciences, from the Earliest to the Present Time.* 3rd ed., with additions. 3 vols. London: J. W. Parker, 1857. References are to this edition, unless otherwise noted.

———. "Spedding's Complete Edition of the Works of Bacon." *Edinburgh Review* 106 (1857): 287–322.

———. *The History of Scientific Ideas.* 2 vols. London: John W. Parker, 1858.

———. *Novum Organon Renovatum.* London: John W. Parker, 1858.

———. Prefatory Notice to *Literary Remains Consisting of Lectures and Tracts on Political Economy, by the Late Rev. Richard Jones,* edited by William Whewell, pp, ix–xl. London: John W. Murray, 1859.

———. *On the Philosophy of Discovery: Chapters Historical and Critical.* London: John W. Parker, 1860.

———, trans. and ed. *The Platonic Dialogues for English Readers.* 2nd ed. 3 vols. Cambridge: Macmillan, 1860–61.

———. *Lectures on the History of Moral Philosophy, 2nd edition, with Additional Lectures on the History of Moral Philosophy.* Cambridge: Deighton, Bell and Company, 1862.

———. *Six Lectures on Political Economy.* Cambridge: The University Press, 1862.

———. "Comte and Positivism." *Macmillan's Magazine* 13 (1866): 353–62.

———. "On the Scientific History of Education." In *The Culture Demanded by Modern Life,* edited by E. L. Youmans, pp. 227–51. New York: D. Appleton and Co., 1867.

Williams, Geraint. "John Stuart Mill and Political Violence." In G. W. Smith, *John Stuart Mill's Social and Political Thought,* 3:237–47.

Williams, L. Pearce. *Michael Faraday.* New York: Basic Books, 1964.

———. *The Origins of Field Theory.* New York: Random House, 1966.

Williams, Perry. "Passing on the Torch: Whewell's Philosophy and the Principles of English University Education." In Fisch and Schaffer, *William Whewell,* pp. 117–47.

Wilson, Curtis. "Kepler's Derivation of the Elliptical Path." *Isis* 59 (1968): 5–25.

Wilson, David B. "Butts on Whewell's View of True Causes." *Philosophy of Science* 40 (1973): 121–24.

———. "Herschel and Whewell's Versions of Newtonianism." *Journal of the History of Ideas* 35 (1974): 79–97.

———. "Convergence: Metaphysical Pleasure versus Physical Constraint." In Fisch and Schaffer, *William Whewell*, pp. 233–54.

Winch, Donald. *Riches and Poverty: An Intellectual History of Political Economy in Britain, 1750–1834.* Cambridge: Cambridge University Press, 1996.

Winstanley, D. A. *Early Victorian Cambridge.* Cambridge: Cambridge University Press, 1940.

———. *Later Victorian Cambridge.* Cambridge: Cambridge University Press, 1947.

Wood, John Cunningham, ed. *David Ricardo: Critical Assessments.* 4 vols. London: Croom Helm, 1985.

———, ed. *Thomas Robert Malthus: Critical Assessments.* 4 vols. London: Croom Helm, 1986.

———, ed. *John Stuart Mill: Critical Assessments.* 4 vols. London: Croom Helm, 1987.

Yeo, Richard. "William Whewell, Natural Theology and the Philosophy of Science in Mid-Nineteenth-Century Britain." *Annals of Science* 36 (1979): 493–512.

———. "An Idol of the Marketplace: Baconianism in Nineteenth Century Britain." *History of Science* 23 (1985): 5–31.

———. "Scientific Method and the Rhetoric of Science in Britain, 1830–1917." In *The Politics and Rhetoric of Scientific Method: Historical Studies,* edited by J. A. Schuster and R. R. Yeo, pp. 259–97. Boston: Reidel, 1986.

———. "William Whewell on the History of Science." *Metascience* 5 (1987): 25–40.

———. "Reviewing Herschel's *Discourse.*" *Studies in History and Philosophy of Science* 20 (1989): 541–52.

———. "William Whewell's Philosophy of Knowledge and Its Reception." In Fisch and Schaffer, *William Whewell*, pp. 175–200.

———. *Defining Science: William Whewell, Natural Knowledge, and Public Debate in Early Victorian Britain.* Cambridge: Cambridge University Press, 1993.

———. "Whewell, William." *Oxford Dictionary of National Biography.* Oxford: Oxford University Press, 2004. http://www.oxforddnb.com/view/article/29200.

Young, R. M. *Darwin's Metaphor: Nature's Place in Victorian Culture.* Cambridge: Cambridge University Press, 1985.

INDEX